Precedents in Zero-Energy Design:

Architecture and Passive Design in the 2007 Solar Decathlon

NEW YORK AND LONDON

by Michael Zaretsky

First published 2010 by Routledge
270 Madison Avenue, New York, NY10016

Simultaneously published in the UK by Routledge
2 Park Square, Milton Park, Abingdon, Oxon, OX14 4RN

Routledge is an imprint of the Taylor & Francis Group, an informa business

Designed and typeset by Michael Zaretsky
Printed and bound in the United States of America on acid-free paper by Edwards Brothers, Inc.

British Library Cataloguing in Publication Data
A catalogue record for this book is available from the British Library

Library of Congress Cataloging-in-Publication Data
A catalog record for this book has been applied for

ISBN10 0-415-77874-3 (hbk)
ISBN10 0-415-77875-1 (pbk)
ISBN10 0-203-86587-1 (ebk)

ISBN13 978-0-415-77874-9 (hbk)
ISBN13 978-0-415-77875-6 (pbk)
ISBN13 978-0-203-86587-3 (ebk)

To Adrian,
who made this possible

Table of Contents

Table of Contents

Acknowledgements

Inspiration

Roger Clark and Michael Pause initially published *Precedents In Architecture* in 1985 (now in its 3rd edition) and when I discovered it in architecture school, I was transformed. This book introduced me to the power that the diagram and case study analysis could play in comparing and understanding architecture. Their book is a graphic analysis of significant examples of architecture throughout history and it has influenced students, faculty and practitioners of architecture and design since its inception.

While *Precedents In Architecture* is a purely graphic analysis of form with no focus on issues of energy, performance and sustainability, I am attempting to further this analytic approach in this book.

When I visited the 2007 Solar Decathlon competition in Washington, DC, I recognized that the labor and knowledge evident on the National Mall had the potential to inspire designers well beyond the three weeks of the event. I taught a case study seminar at the University of Cincinnati in Winter 2008 to ten dedicated graduate students of architecture from the UC School of Architecture and Interior Design. I led these students through an in-depth study of the architecture, passive design systems and active mechanical systems layout of the houses of the 2007 Solar Decathlon.

Erin Connelly, Zhu Dan, Mark Dorsey, Dave Fleming, Dave O'Connell, Jordan Parrott, Carl Sterner, Kelley Romoser, Gregory Tallos and Robbie Zerhusen spent ten weeks drawing, researching and analyzing the houses as part of an investigation into the relationship between architecture and passive design in the Solar Decathlon Houses. The initial research of these students was the seed for research in this book.

All of the drawings in this book were drawn by these students.

Appreciation

This book began with the initial investigations done by the graduate architecture students in the seminar mentioned above with additional research in 2009 by Erin Connelly, Jordan Parrott and Kelley Romoser.

A special thanks goes to graduate research assistant Dave Fleming without whom this book would never have gone to press. He created every chart in the book. He not only worked diligently, well beyond expected hours; but he also challenged me to make this as clear and comprehensible as possible.

My interest in the relationship between architecture and passive design began while studying at the University of Oregon. Howard Davis, Bill Gilland, Alison Kwok, Earl Moursund and John Reynolds were enormous influences on my career.

My interactions with several members of the Society of Building Science Educators (SBSE) has had a huge influence upon my research and knowledge. The annual retreats and listserve are tremendous resources that have been formative in my career. Fellow SBSE member, UC colleague and mentor David Lee Smith has patiently answered hundreds of questions over the years.

John Quale, Rob Peña and Sandy Stannard, all previous authors of Solar Decathlon publications, offered their suggestions and encouragement throughout this process. John Quale looked through several iterations and offered great insights.

Members of nearly every 2007 Solar Decathlon team answered questions and provided resources upon request. Several members of the University of Cincinnati team were especially influential including Anton Harfmann, Luke Field and Carl Sterner.

A special thank you also goes to Francesca Ford and Routledge Press for entrusting me with the opportunity to develop this project.

Most importantly, thank you to Adrian Parr and all of my family who has inspired me and created the space for this to happen.

Analysis and Interpretation

The research and analysis here represents particular perspectives on the architecture and passive design strategies of the 20 houses of the 2007 Solar Decathlon competition. The data behind this research comes from documents provided by the U.S. Department of Energy (DOE), the National Renewable Energy Labs (NREL) and the Energy Efficiency and Renewable Energy (EERE) Associations in addition to the documentation provided by the 20 teams.

Each team published their construction drawings as well as publishing a website that included their documentation on their design process, their goals and their results. I was in contact with members of many teams with additional questions and all were extremely generous with their time.

I have made every attempt to interpret this information in an unbiased and accurate manner. All of the drawings and much of the analysis is based on interpretation.

The hope is that anyone using this book will be inspired to look more deeply at the relationship between architecture and passive design and to design buildings that inspire the next generation of designers of zero-energy buildings.

Foreword by John D. Quale

Every two years, thousands of people descend on the National Mall in Washington, DC to view the prototypical homes entered in the U.S. Department of Energy's (DOE) Solar Decathlon competition. They come to see the innovative homes up close and the many technologies on display. The homes represent thousands of hours of blood, sweat and tears by hard-working university teams from around the country and the world. The event is powerful evidence that we should expect a lot of great things from the next generation of architects and engineers.

Yet the Solar Decathlon represents both the best and the worst of American culture: our collective optimism regarding our ability to solve our nation's environmental and energy problems – and our complete faith in the use of technology as the best way to accomplish this. Too often Americans choose the quick technological fix over the slightly more difficult design or policy fix. The Decathlon organizer's heavy emphasis on photovoltaics puts the focus of the event very much on the side of technology – at the expense of low or no-tech design solutions. In fact, the event perpetuates the falsehood that photovoltaics are the primary solution to energy-inefficient housing. In my opinion, expensive and complex technologies like photovoltaics, while an important strategy, should be the last choice – after designers have exhausted all the other commonsense strategies. We could reduce the energy use of residential buildings in this country by at least half by simply investing in design – everything from super-insulated building envelopes to the careful calibration of a home's response to its unique site and climatic conditions. In some climates, a really good designer might even eliminate the need for renewable energy systems altogether.

This is why Michael Zaretsky's *Precedents in Zero-Energy Design* is such an important book. It thoughtfully documents the homes in the 2007 Solar Decathlon from the point of view of design – including the passive features of each home. It accomplishes this mostly through the graphic language of diagrams – carefully prepared to emphasize the unique characteristics of each design. The result is a document that will help readers recognize that design comes before technology – and renewable energy systems alone can't solve the problems we face.

* * *

During public lectures, I sometimes refer to myself as a recovering Solar Decathlete. It's an easy laugh line – funny for any former Decathletes in the audience, as well as those that have heard about the intense commitment the competition demands of its participants. Although it has been seven years since I helped lead the University of Virginia in the first Solar Decathlon competition in 2002, I still feel the emotional pull of the event with each reoccurrence in '05, '07 and now '09. I imagine it's a little like those retired athletes that work as TV sports commentators for an event they know from experience. It's easy to be a little jealous of the current decathletes, but it is also hard to imagine returning as a participant.

Frankly, the sports metaphor embedded in the Solar Decathlon has always bothered me a little. When the idea was explained to myself and others at a gathering of the first group of teams in 2000, competition founder, Richard King, described how he struggled to come up with ten legitimate events once he settled on the decathlon concept. As I read the early draft of the competition rules and regulations for that first version of the competition, it became clear that the metaphor was stretched a bit too far – pumping up aspects of the design of an 'off-the-grid' house – turning activities such as the production of hot water or the running of a refrigerator into stand-alone events worth 100 points. I found it particularly disturbing to give the same number of points for the architectural design as those given in the vague 'home business' category – judged mostly on having a working computer and printer. (This event has since been eliminated.) I quickly became an advocate for the importance of design within the event, and while other advisors and I were able to convince the Department of Energy and National Renewable Energy Laboratory (NREL) organizers to increase the number of points for architectural design to 200, I never felt like the organizers truly understood or cared about architectural design. In fact, the points for architecture were reduced back to 100 for the 2009 event, and my suggestion that there should be an equal set of 500 points for both the design categories and the energy and water use categories was never taken seriously.

One of the hardest things to believe about the Solar Decathlon is that an event meant to promote highly energy-efficient housing has no specific rewards for teams that use passive design. While it is obvious that the use of natural ventilation or passive heat gain could help the teams reduce their energy load, and therefore get closer to being zero energy, the reality is that teams hoping to win the event are essentially not encouraged to design with the climate, but instead design against it. The homes are continuously monitored, and their performance is judged using very narrow parameters for comfort and energy use established by the DOE and NREL. Successful Decathlon teams cannot take the climate for granted – instead they must close themselves off from the elements even when it is relatively comfortable outside. The slightest change in the dew point can spell disaster for teams with the open windows, and so teams find themselves in one of the few truly comfortable seasons of the year in Washington, DC (early Autumn) sitting inside hermetically sealed 800-square foot boxes trying to game their mechanical systems to achieve operating room climate conditions.

There is an advantage to this strategy: it gives the DOE and NREL somewhat objective criteria for judging the competition. Yet this is the problem with the athletic event metaphor when it comes to designing a home. The design of a home – its human comfort, its energy performance, its engineering sophistication, its efficient use of space, its aesthetic qualities – should be assessed by expert judges using established criteria. Using the Olympic analogy, I would argue that home design is more like diving than swimming – or more like ice skating than speed skating. The successful home must be judged by humans, not by electronic sensors and checklists alone. Not that sensors don't have their place – but like any form of technology, they are only as good as the thought that has gone into the analysis of the data they provide. The Decathlon utilizes judges for a couple of architecture and engineering categories – but ultimately, the data from the monitoring systems and the other 'objective' categories decide the winning teams. Don't get me wrong – I'm a big supporter of active monitoring of buildings – and feel strongly that design benefits from rigorous and thorough post-occupancy evaluation. My concern is less with the method of evaluation and more with the criteria.

Surprisingly, each year many Solar Decathlon teams integrate passive design strategies, even when the guidelines don't actively encourage it. At the original event in 2002, several teams pursued passive strategies – and attempted to widen their occupant's perception of an acceptable temperature and humidity comfort zone, rather than artificially narrow it. The University of Texas at Austin's design from 2002 is an excellent example of a successful passive building that performed poorly in the event because of such 'radical' thinking. My own team's design – nicknamed the Trojan Goat – was organized around a passive heat gain / sun shade space – but we made sure to thermally separate it from the rest of the building during the monitoring phase.

* * *

A few weeks before President Obama's inauguration, I was asked by a freelance writer for a major science magazine to comment on what renewable technology Obama should install at the White House as a symbolic gesture. The writer thought the new administration should do something in response to President Reagan's removal of the solar thermal panels President Carter had installed during his term. As I struggled with the premise of the request, I drafted my top ten list of environmentally responsive technologies, but ultimately suggested that before any of the technologies on my list be used, the White House staff should first re-weatherize the building, and install new blow-in foam to completely fill all insulation cavities. I admit this idea is nowhere near as exciting as the latest generation evacuated solar hot water panels or wind turbines, but high performance insulation will pay for itself long before anything else on my list. And in our current financial crisis, this is clearly the most prudent approach.

Although the article was never published, I was pleased to hear a month or so later that one of the Obama administration's highest priorities was to do complete energy retrofits on federal buildings and in affordable housing across the nation. Perhaps there is reason to hope that when Americans are pushed to the limit, common sense will drive our decisions. Renewable energy technologies are going to play an important role in our response to our current challenges, but solid design choices, while not as exciting as the latest 'eco-bling,' must come before anything else. Americans must begin to look again at the long-term impact of our design decisions.

John D. Quale
Assistant Professor of Architecture and
ecoMOD Project Director
University of Virginia
April 2009

Precedents in Zero-Energy Design:

Introduction

Solar Village on the National Mall, October 2007

Photo from Solar Decathlon 2007 website
https://www.eere-pmc.energy.gov/Solar_Decathlon_07/
Credit: Kaye Evans-Lutterodt/Solar Decathlon

Introduction

From October 3–22, 2007, the third *Solar Decathlon competition* was held on the National Mall in Washington, DC. The Solar Decathlon is an international competition in which university teams from around the world design and construct a one-bedroom, zero-energy house that is built off-site, transported to the National Mall and reconstructed on-site in one week. The houses are judged in ten different contests over a two-week period in early October. In 2007, as in previous years, millions of dollars and thousands of hours went into the design and construction of the 20 houses, and each team developed a unique response to the exact same architectural and engineering challenges.

There is an inherent dichotomy that nearly everyone competing in the event is aware of: although the intent of the competition is to showcase sustainable design, the contests clearly favor technological solutions over the non-mechanized, natural energy flows of *passive design* strategies. As anyone interested in the field of sustainable or *environmental design* is aware, we are facing an environmental crisis as a result of fossil fuel production and increasing greenhouse gas emissions. In the United States, buildings are responsible for 72% of electricity consumption, 39% of energy use and 38% of all carbon dioxide (CO_2) emission[1]. The design and building industries have the potential to reduce worldwide emissions exponentially by reducing our dependence on fossil fuels.

Design and engineering students collaboratively produced 20 examples of zero-energy design houses that offer lessons to everyone. Within the field of design, *zero-energy design* refers to buildings that have annual net-zero energy consumption and emissions. The judges measured the energy use of each team extensively, but there was little assessment of the architectural design and "passive design" strategies employed in each project.

The complexity of decision-making in the planning of a zero-energy house requires tremendous integration of design and engineering. The *diagrams*, comparisons and analyses in this book provide opportunities to see the relationships that emerged between architectural strategies, passive-energy design strategies and mechanical-systems integration strategies – and to assess how these strategies impacted on the rankings in the Decathlon.

This competition requires that each team provide its own power, thereby eliminating *non-renewable energy* consumption. To achieve the energy required to power these houses, designers can turn to expensive, resource-intensive technological solutions or *passive* solutions that significantly reduce energy needs without the costs of *active solar* technologies.

To achieve energy savings in design, we must look beyond pure technology and embrace the opportunities available in every design project from the passive design strategies of *shading*, natural *ventilation*, *thermal storage* and appropriate *insulation*.

Though there were no competition scores for passive design, every team engaged passive strategies to some degree. It was a desire for more analysis of the architecture and passive strategies of the 2007 Solar Decathlon houses that prompted me to lead a seminar in the School of Architecture and Interior

Design at the University of Cincinnati and eventually to pursue this research more deeply, resulting in this book.

According to the 2007 Solar Decathlon competition Rules and Regulations,[2]

> *The Solar Decathlon is an intercollegiate and interdisciplinary design and construction competition that takes up a persistent and age-old question: How do we integrate architecture and technology with a dwelling? In other words, what makes a good house?*

It is my belief that a *"good house"* requires more than the integration of architecture and technology. While I value the positive impacts of the Solar Decathlon competition, any event that encourages environmental consciousness in the design and construction industries needs to openly engage the principles of architecture and passive design as fully as it engages the active technologies.

The rules state that there are three guiding principles behind the Solar Decathlon competition:

> *1. The teams must supply the energy requirements necessary to live and work using only the sunlight shining on their entry – the global solar radiation incident on the house, specifically during the contests.*
>
> *2. The houses will exemplify good design principles that will increase the public's awareness of the aesthetic and energy benefits of solar and energy efficiency design strategies and technologies, which in turn will increase the use of these design principles and technologies.*
>
> *3. The work of the teams, organizers, and sponsors will stimulate accelerated research and development of renewable energy, particularly in the area of building applications.*[3]

The relationship between the need to provide one's own energy and the challenge of designing in a manner that will "*exemplify good design principles"* is an unusual and important combination. The competition attempts to balance design and performance, though anyone involved recognizes that the results are more significantly impacted by engineering than architecture and design.

The Audience

The intent of the competition is to inspire educators, students, industry and homeowners to push the possibilities of design that integrates solar energy as its source of power. There are numerous stories of students whose commitment to environmentally responsible design has been forever changed by their participation in this event. The popularity of the Decathlon is growing exponentially with each year, with an estimated doubling of attendance from 100,000 attendees in 2005 to 200,000 attendees in 2007.

All of the initial research and drawing for this book was completed by students at the University of Cincinnati using resources available through the U.S. Department of Energy Solar Decathlon website and the websites of the 20 teams. The content was studied and interpreted by myself with the participation of students, graduate assistants, consultants, colleagues and Solar Decathlon team members. The results are intended to offer insight and inspiration to students, educators, builders, researchers, owners or anyone who has an interest in the competition, an interest in zero-energy or "sustainable" design, or an interest in the integration of architecture and passive design.

The Contests

A building is a result of thousands of decisions based on ever-changing criteria. In a typical design solution, these would, at a minimum, include: location and climate, building program, budget, building-code and zoning requirements, available materials, resources and labor skills, spatial intentions and aesthetic or contextual expressions – as well as innumerable other factors. What is unique about the Solar Decathlon competition is that these 20 houses each had to respond to the same set of criteria and present their solutions collectively. All 20 teams arrived at profoundly different decisions, but what they had in common was a goal of being competitive in the ten contests.

The ten contests for 2007, with available points in parentheses:

1. Architecture (200)
2. Engineering (150)
3. Market Viability (150)
4. Communications (100)
5. Comfort Zone (100)
6. Appliances (100)
7. Hot Water (100)
8. Lighting (100)
9. Energy Balance (100)
10. Getting Around (100)

The Solar Decathlon competition integrates the architecture and engineering disciplines in these ten contests, which range from the purely quantitative to the purely qualitative. Engineering is an inherently quantitative discipline, while architecture commonly resists purely quantitative assessment. In 2007, there were a total of 1,200 points available, of which 600 were intended to be objective and 600 to be subjective. According to the *U.S. Department of Energy* (DOE), contests 1–4 were intended to be subjective while contests 5–10 were considered objective.[4]

Other authors writing on the Decathlon have discussed the disparity between the objective and subjective assessment of the ten contests.[5] The "*subjective*" competitions included the *Architecture Contest*. The sections of the Architecture Contest were tallied quantitatively with 50 points each for "*firmness, commodity*" and "*delight*" as well as for the construction documentation. It is unusual to quantify the subjective characteristics of a field such as architecture, yet even those contests that were seemingly objective, such as the Engineering Contest, showed inherent disparities between how data was measured, read and interpreted from house to house.[6] As a result, this book resists the tendency to offer steadfast conclusions and instead provides diagrams, comparisons and analyses that begin to illuminate relationships and trends.

This research focuses exclusively on the Architecture and *Comfort Zone Contest* rankings as well as the overall competition rankings. The diagrams, comparisons and analyses address the subjective subtleties of the architecture and comfort zone decisions that were made by each team. The overlap of decisions that spanned architecture and comfort zone conditions was a springboard towards the investigation into the role of passive design and its impact on architecture.

Architecture Contest – Solar Decathlon 2007 (200 points)

The Architecture Contest was judged by a team of "*esteemed Architects,*" who were asked to assess the designs from the perspective of the Vitruvian triad of "*firmness, commodity and delight"* – interpreted in a new way for this competition. Firmness was defined as "*the house's strength, suitability, and appropriateness of materials for the building"*. Commodity was defined as *"ease of entry into the house and circulation among the public and private zones; architectural strategy used to accommodate the technologies required to run the house; and generosity and sufficiency of space in the house"*. Delight became *"surprises, unusual use of ordinary materials, or use of extraordinary materials in the house"*.[7]

The Architecture Contest is further described in the 2007 Rules and Regulations, thus:

> *To be architecturally sound, a home's design must not only satisfy human comfort needs, it must also be well organized and visually pleasing both inside and out. The Architecture contest is intended to demonstrate that solar-powered, energy-efficient homes can be designed to meet enduring architectural standards.*[8]

The architectural diagrams, comparisons and analyses within are intended to delve deeply into the architectural decisions that the teams made and provide a basis for comparison from the perspective of a design (as opposed to an engineering) analysis.

The description of the Architecture Contest states that each team must design "*high performance houses that integrate solar and energy efficiency technologies seamlessly into home design*".[9] There were some teams who addressed this concept of integration vigorously – such as Madrid, which included a thickened roof that was covered in *photovoltaic* (PV) panels and south-facing sliding glass doors that had *Building Integrated Photovoltaics* (BIPV). The Cornell team's architectural concept involved two separate entities: a box for the indoor living space and a framework for the ever-changing external technologies. Other teams, such as Santa Clara University, added photovoltaics that effectively collected energy, but without an evident attempt to integrate them into the design.

There is a clear distinction between the definition of a "*good house"* from the perspectives of a designer, a builder or a homeowner. Conversations overheard in the Solar Village attested to the belief that attendees preferred the houses that looked more traditional as compared to those that were overtly expressive with technology (such as Cornell). This disparity is reflected in the Market Viability Contest (an evaluation of whether or not the house has "market appeal").

Part of the success of the Darmstadt design was that it appealed to the designer and homeowner alike; it was not overly expressive technologically and was predominantly clad in wood, creating a sense of warmth and welcome. However, the design was still extremely refined, minimal and elegant.

Comfort Zone Contest – Solar Decathlon 2007 (100 points)

The Comfort Zone Contest tests the capacity of a team to design a house that remains within an extremely limited temperature and humidity threshold throughout the competition, as measured at 15-minute intervals during specific times when the house is closed to the public. This is measured through a continuously monitored data logger that is placed in the house by the competition jury. The collected data is part of the complete scores, which are publicly available online through the DOE Solar Decathlon website.

Though this measurement doesn't distinguish between temperature and relative humidity levels that were achieved as a result of active mechanical systems and those reached under non-mechanized passive systems, the reality of the weather conditions in 2007 created a scenario in which teams that utilized integrated passive strategies were able to keep their homes within the comfort zone much more effectively than those that didn't.

Typically, designers in the United States design for a comfort zone that varies from approximately 68 to 78° Fahrenheit and 20 to 80% relative humidity. Within this temperature and humidity range, most individuals (while at rest) tend to feel thermally comfortable. In the Comfort Zone Contest, "[*F*]*ull points are awarded for maintaining narrow temperature (72° F/22.2° C to 76° F/24.4° C) and relative humidity (40% to 55%) ranges inside the houses.*"[10] Clearly, the range being measured in this contest is significantly narrower than what is typical in most US buildings. This challenge is exacerbated by the energy requirements of the Decathlon: each team must produce their own power exclusively through solar energy.

One relevant comparison is between Santa Clara University and the University of Cincinnati – the only two teams that utilized absorption chilling for their mechanical cooling. Both teams independently designed this innovative approach to cooling, but they barely used their absorption chiller during the competition. The Cincinnati team had technical problems with their air handler which caused them to shut down the system, while Santa Clara simply didn't need their system because of their capacity to remain within the comfort zone as a result of the thermal storage integrated into their exterior walls. This thermal storage strategy proved very effective for the Santa Clara team, who ranked fifth in the Comfort Zone Contest. The Cincinnati house had initially been designed to allow for the free flow of natural ventilation through its single, open space but the final construction included fixed windows on the north, east and west. As a result, on days when the weather made natural ventilation an ideal strategy this house was sweltering. The Cincinnati house placed twentieth in the Comfort Zone Contest.

In most examples of contemporary construction, the designers can plug into a local utility grid and rely on non-renewable energy sources to provide power and water to their project; the energy performance of the building is rarely assessed or considered by the designers. The combination of performance and design, however, is the hallmark of "green" design, and this competition offers much to learn for those practicing within this industry.

Why Isolate the Architecture and Comfort Zone Contests?

> *As we refine the technical performance of passive design prototypes, it is important that the designers begin to explore their latent formal content.*[11]

Though many of the contests reflect decisions that overlap and impact other contests, this book addresses the decisions made by teams that impacted only two contests – Architecture and Comfort Zone. This focus developed from an interest in understanding what factors impacted the specifically architectural decisions that each team made. In a typical design project, the array of decisions made evolves as a response to the site, the architectural program, the client needs, the budget or the designer's aesthetic. *Thermal comfort* unfortunately plays a minimal role in the conceptual design phase of most architectural projects. In this competition, every project must attempt to be a notable expression of *Architecture* without having the option of ignoring thermal comfort requirements. Furthermore, every architectural decision must not only address firmness, commodity and delight, but must respond to and augment the need to attain thermal comfort with a minimal use of energy.

The first step in reducing energy needs is to incorporate passive design strategies for reducing heating and cooling loads. However, the Solar Decathlon contests highlight and reward solar technology much more than passive design strategies. The relationship between the overall competition and the Architecture and Comfort Zone Contests results provides input as to how well each house performed from the perspective of architecture and passive design.

This book investigates the 20 houses of the 2007 Solar Decathlon competition with the goal of uncovering how these designers balanced the architectural and thermal comfort requirements with the technical agenda of the competition guidelines.

This investigation is predominantly a comparative and subjective analysis based on diagrams of architecture and passive design strategies, combined with quantitative data about each design. While there is a tremendous amount of technical data available, the focus here is not on engineering. The current analyses address the multiplicity of design decisions that anyone interested in zero-energy design faces in considering the relationship between architecture, passive design and zero-energy design. For anyone wishing to gain more insight into the technical performance of any house, there is data available through the U.S. Department of Energy as well as on each team's website.

Placing the Solar Decathlon competition in History

Passive solar design has been practiced for centuries by cultures seeking to utilize the *insolation*, or natural energy available from the sun, in order to keep their spaces comfortable. Principles such as shading, ventilation and daylighting were inherent aspects of design until the advent of the Industrial Revolution. The invention of electric lighting and air conditioning has allowed humans to inhabit interior space without any connection to the exterior. There have been a multitude of environmental movements that have addressed this over the last two centuries, but the latest has some unique characteristics.

The late twentieth century brought issues of the environment and climate change to the forefront of the world's attention. The last significant environmental movement at this scale was the *solar architecture* movement of the 1960s and 70s. These two movements developed from different sources; whereas rising energy costs led to the solar architecture movement, global climate change and its impacts have led to the current green building movement. This recent *green design* momentum is differentiated from that of the 1970s in many ways, but chiefly by its widespread public acceptance. The creation and success of the Solar Decathlon competition on the National Mall represents a powerful public expression of the contemporary green movement.

From the perspective of the public, the solar architecture movement of the 1970s defined energy conservation as a form of suffering – denying one's self what one really wanted. This movement was short-lived for several reasons (including decreasing energy costs), however one clear reason for its shortened life span was the distinction made between energy performance and architecture. The architectural media ignored solar architecture, and the public response to this movement was predominantly negative. Solar architecture was quickly replaced in the media by other postmodern styles of the 1980s and '90s.

It wasn't until the mid-1990s that environmentally responsive design began to receive positive public and media attention. This was the result of many cultural and environmental factors. Groups such as the US Green Building Council formed in the early 1990s, with the explicit goal of bringing about a *market transformation* to reflect a *green economy*. Architects such as William McDonough graced the pages of *Time* magazine, and business luminaries such as Paul Hawken and Ray Anderson drew thousands to their public presentations on the new economy of "Natural Capitalism". It was this trajectory that led to the creation of a public competition to showcase zero-energy design, held at one of the most significant civic spaces in the United States – the National Mall at the U.S. Capitol.

Richard King of the Department of Energy is the brainchild behind the Solar Decathlon. In conjunction with the *National Renewable Energy Lab* (NREL), he has crafted the competition requirements, put together the juries and organized everything for these events since their inception.

The first Solar Decathlon competition was held in 2002, and planning began early in 2000. The 2002 Decathlon had 14 competitors, the 2005 competition had 18 and the 2007 event reached 20 competitors. The upcoming 2009 competition received more than 20 submittals, and 20 teams were chosen to compete in Washington, DC on October 8 - 21, 2009. There are Solar Decathlon competitions scheduled for 2011, 2013 and 2015.

Sustainable Design: Active Solar Technology or Passive Solar Design?

It is a common U.S. response to attempt to solve problems with technology. It is undeniable that invaluable innovation has occurred as a result of this inclination, but we must also recognize that the future of sustainable design will not lie with technology alone. There are simple decisions that can be made in the design process, which can significantly decrease the energy use and overall impact of any building. Passive-energy design decisions concerning site orientation, shading, daylighting, ventilation and insulation can decrease the energy impacts of a given building by 50% without adding any additional cost to a project.

The utilization of active renewable-energy technologies to offset energy usage is only an option for projects with a significant budget. The teams that competed in the Solar Decathlon had no limit on their budget and were required to incorporate photovoltaic technology for electrical energy. Each team was required to design a house that could produce all of its own electricity, and some chose to incorporate principles of passive design. The investigation of the successes and failures of that incorporation will hopefully entice others to utilize these strategies in the future.

Passive Design and the Solar Decathlon

> *Rather than relying on HVAC equipment and the consumption of hidden energy to provide comfort, passive design seeks to use the form and envelope of a building to act as mediator between climate and people, providing comfort by natural energy flows.*[12]

The Solar Decathlon competition has been criticized for not incorporating passive-design assessment in its competition evaluations.[13] Passive design strategies are low-cost or free, and represent the most easily reproducible options for homeowners hoping to decrease their own energy costs.

Passive design is impacted by weather, and the weather in Washington, DC for the 2007 Decathlon was ideal for passive cooling. The days were warm and humid, but the weather cooled down in the evenings. Those houses that incorporated strategies such as daylighting, shading, natural ventilation or high levels of insulation had decidedly better results in the Comfort Zone Contest. The house that won the overall competition – from Darmstadt, Germany – did so by combining shading and natural ventilation with an extremely well insulated shell and the largest number of photovoltaic panels of any house in the competition.

While there are explicit rules against certain forms of alternative energy (as will be described later), there is nothing in the competition brief that limits the use of passive design opportunities to reduce energy needs and increase thermal comfort. Yet only a minority of the teams effectively incorporated these principles, while others seemed to ignore the potential benefits of passive design completely.

Seeking the source for this disparity, one quickly learns that the Decathlon is clearly not designed to focus upon passive solar design. As mentioned earlier, the competition guiding principles provided by the sponsor, the U.S. Department of Energy, state that energy must be supplied by global solar radiation, that houses must exemplify *"good design principles"* and that the competition should stimulate renewable-energy research and development. But the rules and regulations barely mention *daylight*, shading, *thermal mass* or non-mechanical ventilation – and there is no mention of passive design. What is explicit is the intention of increasing exposure to solar photovoltaic (PV) technology and, according to the analysis in Part 3, there is a direct correlation between *overall rankings* and increased size of PV array.

In Part 3 of this book there is a series of analyses that address the passive strategies of daylighting, shading, natural ventilation, thermal storage and *thermal conductance (or transfer)* through the *building envelope* as well as addressing the relationship between the layout of the mechanical system and the layout of the building. These analyses provide opportunities to see relationships and trends that emerge between architectural strategies and passive design strategies, and to assess how these strategies impact rankings overall and in the Architecture and Comfort Zone Contests specifically. These strategic relationships are relevant to all designers who are incorporating passive design into their projects.

Orientation and the Solar Angle

> *Can a solar building ignore the differentiation between north and south?*[14]

For the majority of contemporary houses built in the United States, the geometry of the sun's path in the sky doesn't have a major impact on the form of the building. To maximize efficiency, any project that is utilizing solar photovoltaic technology should have a shape that is designed to maximize its exposure to solar energy when it is available. As a result, all of the houses in the 2007 competition had a form that is roughly an east - west orientation, thereby maximizing their southern (equator-facing) exposure.

In passive solar design, a thin house exposed towards the equator (south in the US) is common, in order to maximize passive heat gain through glazing as well as to limit its exposures on the east and west. Northern exposures in moderate climates tend to have reduced glazing to minimize heat loss. However, in this competition the amounts of glazing on all four sides of some houses suggested that the principles of passive solar design were not always the main forces behind the orientation decisions.

Had it not been for the reliance on solar energy, it is unlikely that 20 teams would have designed houses with such similar form and orientation. Every team faced critical design decisions that had to address contradictory architectural intentions with the requirements of solar orientation and form. This was the basis of the analyses that compared the rankings of the houses on the north side of Main Street with those on the south.

All teams were required to utilize photovoltaics (PV) and a PV array is most efficient when it faces towards the equator. To maximize the available energy from the sun, the collector should be perpendicular to the angle of the sun when the most solar energy is available. In Washington, DC in October, the ideal *solar angle* is approximately 42° off the horizontal, facing south. Each degree off of this angle decreases the effectiveness of the solar array.

The angle of the PV arrays on the houses varied from flat to 62° off the horizontal, with a few having variable angles and one (Georgia Tech) fixed at 42°. Darmstadt challenged the assumptions about orientation by creating a square house plan that had no angle to its roof. Instead, the team utilized a massive amount of photovoltaics to cover the roof, south, east and west facades with solar collectors. Though the angle was far from ideal, the sheer amount of photovoltaics offset the inefficiencies.

Active Solar Technology

When designing a building that must produce all of its own energy, there are many potential strategies for reducing energy needs – passive design as well as sources for the production of renewable energy. These renewable energy sources extend beyond solar photovoltaics to include *solar thermal*, wind-driven energy, geothermal (geo-exchange), biofuels and others. This competition is explicit about the role that technologies other than solar photovoltaic can play.

From the 2007 Solar Decathlon Rules and Regulations:

> ***12.6 Energy Collection***
>
> *During the Solar-Only Period, the only source of energy with which houses shall operate and tasks shall be performed is global solar radiation received by the house without artificial external augmentation.*
>
> *The competition is clearly titled 'The Solar Decathlon'. It is not the 'alternative' or 'renewable energy' decathlon. There are in the Rules and Regulations, some allowances for renewable energy alternatives - thermal mass and evaporation.*[15]

The Texas A&M house actually had wind turbines on its deck, but they had to remain locked throughout the competition. Most teams used solar thermal to some degree for hot water heating, and the Santa Clara and Cincinnati teams utilized this heated water for cooling through absorption chilling.

Stretching the Comfort Zone

> *Thermal Comfort - that condition of mind which expresses satisfaction with the thermal environment.*[16]

The thermal comfort zone is the range of temperature and humidity in which most people feel comfortable. This zone adjusts, based on activity level, location, cultural expectation and other factors. Generally, within the United States, we assume a typical thermal comfort zone of 68 to 78°F with a relative humidity level from 20 to 80%.

In contemporary sustainable design, there is an ongoing debate about whether the solutions to our energy dependence lie more in changes in technology or changes in behavior. The projects that seek zero-energy targets throughout the world often achieve this decreased energy usage by challenging the commonly accepted thermal comfort zone. In contemporary European architecture, it is common to find temperature swings from 55° to 85° F. Occupants are encouraged to dress appropriately. Studies find that when people have control over their thermal environment (such as having the option to open a window) they are much more willing to stretch their own thermal comfort zone.[17]

The approach for the thermal comfort zone for the Solar Decathlon competition comes from a completely different perspective. For the competition, the comfort zone is defined as follows:

> ***Indoor Temperature Control (50 pts):***
>
> *Teams earn the maximum number of points per scored 15-minute interval by keeping the time averaged interior dry-bulb temperature between 72.0°F (22.2°C) and 76.0°F (24.4°C).*
>
> ***Indoor Humidity Control (50 pts):***
>
> *Teams earn a maximum number of points per scored 15-minute interval by keeping the time-averaged interior relative humidity between 40.0% and 55.0%.*[18]

Keeping any house within a 5-degree temperature range and 15% relative humidity range is difficult under any conditions. Designing a house that will use minimal energy for the variability of weather possibilities in Washington, DC in October is even more challenging. The weather in the U.S. capital in autumn can vary from crisp, cool temperatures to muggy, tropical nights. The design teams must be prepared to adjust for weather conditions that have historically varied significantly. In 2005 the weather during the competition was consistently rainy and humid, while in 2007 there were warm days with cool nights.

The comfort zone data is collected by measuring temperature and humidity within the house with a single sensor placed in a location chosen by the judges. Some teams stated that their sensor location unfairly impacted their performance. In some cases, the sensor may have been receiving direct solar gain, thereby increasing its own temperature reading.

In addition, each person adds sensible and latent heat to a space, leading to an increase in temperature and humidity levels. Therefore, the more visitors coming through a house, the greater the potential heat gain. There is no way to assess how many visitors each house will have, but in 2007 there were hundreds to thousands of people moving through the homes each day. Given these conditions, this minimal range of temperature and humidity control represents a significant, and potentially unrealistic, challenge. The measurements were taken when the houses were empty, but those houses that could disperse this sensible and latent heat effectively through natural ventilation or effective use of thermal storage were clearly at an advantage.

The Challenge of Thermal Storage

Thermal mass storage is an effective way to mediate temperature swings of typical autumn weather in a climate such as that of Washington, DC. In 2007, only seven teams utilized thermal storage in any capacity – four teams internally and four externally (with Penn State being the only one to use internal and external thermal mass). There is an intrinsic contradiction between the desire to mitigate heat gain through thermal mass and the realities of designing a transportable house. The houses are supposed to inspire homeowners to use these strategies to reduce their energy usage, but this effective strategy was unquestionably underutilized in 2007.

Under certain temperature conditions, thermal storage can be an especially effective way to significantly reduce interior temperatures at night, thereby delaying the build-up of internal heat during the day. The thermal mass strategy of night ventilation would have been ideal for the climate in Washington, DC in October 2007 – warm days and cool nights. However, as Peña points out "[B]ecause temperatures and relative humidity are monitored in the 'Thermal Comfort' contest all night long, any use of nighttime cooling of thermal mass is discouraged."[19]

Weather Variability

The inaugural Solar Decathlon competition was held September 19 to October 6, 2002. The next competition took place September 29–October 19, 2005 and the 2007 Decathlon occurred October 3 to 22, 2007. In 2005, it rained nearly every day and the temperature ranged from 56° to 68° F with 68–93% relative humidity. In 2007, there was little rain, the temperature ranged from 52° to 82° F and relative humidity was 28–90% (decreasing significantly at night).[20]

The abundant moisture conditions of 2005 significantly impacted the teams' performances in several contests during the competition. In 2007, the decreased humidity levels provided an advantage to houses that had significant amounts of *operable glazing*, whether windows or doors (such as that of Maryland). These same houses may have suffered in more extreme weather conditions. If the 2007 houses had experienced the weather of 2005, they would have ranked differently in the Comfort Zone Contest (among others). This is one paradox of the Decathlon.

A deeper paradox is that each of these houses is being measured for their performance in Washington, DC for two weeks in October, while their lifespan could be spent in a different climate (in some cases a radically different one, such as Puerto Rico or Montréal). Should a house be able to perform well in any climate? Preferably, a house should perform well in the climate In which it will "reside," though this issue is not addressed in the competition.

Site Specificity and Transportation

An environmentally responsive design must respond to the specificities of climate, topography, geology and countless other characteristics of its location. The Decathlon challenges this by having the teams design and build the project at their home location with the understanding that, other than for the days of the competition, the house will likely remain at or near its original location. The competition is based on testing that occurs in Washington, DC, though no house remains there.

While passive design strategies tend to lead to increasingly site-specific design responses, the overall competition winner, the Darmstadt house, is a design that is less influenced by site specificity than many others. The house doesn't engage daylight extensively in its design, and in the overheated periods performs well by being sealed. Darmstadt approaches thermal comfort based upon the German principles of the "Passivhaus," which is a tight-building strategy of minimal apertures and very high insulation levels. This team enclosed their house in an outer shell that covered the roof and three walls with photovoltaic panels. The house had significantly more photovoltaic surface than any team previously in the competition and their energy production was significant, but the Darmstadt house placed tenth in the Comfort Zone Contest.

In contrast, the Maryland house (the team located closest to the site of the Decathlon) placed second in the overall competition and fourth in the Comfort Zone Contest. They made extensive use of passive strategies – incorporating a south-facing louvered wall that offered several different opportunities for natural ventilation. The Maryland team also took an innovative approach to shading and daylighting by angling these louvers to afford shade when needed while increasing the amount of daylight reaching the interior. The project additionally showcased a desiccant waterfall to decrease the humidity – a thermal challenge in this region. This was achieved by adding calcium chloride (a highly absorptive material) into the waterfall in order to capture moisture from the air.

Another challenge posed by running the event in a single location was that every team expended a significant amount of money, time and greenhouse gas emissions in the transportation of their house from its original location to the National Mall in Washington, DC. The Darmstadt team spent nearly $250,000 to ship their house overseas. Many critics have commented on this hypocrisy and called for future Decathlons to be held regionally.

Active Systems Design

Though an architect could design a comfortable house with no mechanical systems in Washington, DC for a typical October climate by stretching the thermal comfort zone and incorporating principles of passive solar design, every team in the competition had comprehensive, oversized mechanical systems to assure that they could compete in the Comfort Zone and the Engineering Contests (based largely upon oversized, and often redundant, mechanical systems). The energy for these systems came predominantly from photovoltaic panels, though in two houses (University of Cincinnati and Santa Clara University) solar thermal was the predominant source of heating and cooling energy.

In the active systems analyses in Part 3 of this book, the intent is to assess how the design decisions that addressed architectural principles were impacted by those decisions that were made for active mechanical systems. In these projects, the mechanical systems were extremely complex, oversized and, in several cases, never turned on during the competition. The infrastructure for these systems was extensive and its integration within the architectural design ideas varied immensely. These analyses are an attempt to shed light on the overlap of architecture and engineering decision-making. How can designers incorporate complex mechanical systems within a clear, elegant architectural design?

Budget

The overall project cost of the Solar Decathlon 2007 houses ranged from $270,000 to $1,378,000. For this competition, there was no limitation on budget. In the case of Darmstadt (the most expensive house), the money spent paid off well. In other cases, significant expenditures didn't lead to high rankings in the competition. Maryland placed second spending $448,470 while Lawrence Tech spent $672,000 and placed last in the overall competition.

For this competition, there was also no limitation on resources. The Darmstadt team had support from the German government as well as research and technological backup from an international leader in photovoltaic technology. This company invited members of the team to utilize their laboratories and research facilities for the development of their photovoltaic systems. The team made great use of this opportunity, and the result was a cutting-edge photovoltaic design for their louvered panels.

Other teams developed their designs without the benefit of million-dollar budgets or corporate research support. One must ask if this constitutes an even playing field for the competition. Critics have pushed to have a cap on spending for the houses, or some way of accounting that benefited those teams that produced their projects at a lower overall cost.

In the Market Viability Contest, each team states what the house would cost in regular production. From one perspective, these houses are part of the cost of regular research and development for contemporary residential solar photovoltaic design. The event introduces these concepts and technologies to thousands of visitors and inspires and educates students and faculty involved with the projects. These buildings bring media attention and, in some cases, prestige to the institutions that design and build them. From another perspective, if these designs are intended to impact the traditional homeowner, they should prioritize strategies and technologies that are more attainable and effective. At present, solar photovoltaic panels average approximately 15% efficiency. There are much more effective, lower-cost strategies.

The overall prices listed include all of the systems and materials that were purchased, outside labor and consultants, research and transportation as well as the cost for the competition entrants to stay in the US capital for the length of the competition. These prices are all well above standard construction on a square footage basis. Furthermore, what these numbers strikingly fail to include are the endless hours that students and faculty spent designing and building the houses.

Navigating the Book

This research is presented in three parts, which cover the architecture and passive design decisions of the 2007 Solar Decathlon competition:

Part 1: Drawings and Diagrams

Part 2: Comparisons

Part 3: Analysis

Part 1: Drawings and Diagrams

Part 1 describes the 20 houses of the 2007 Solar Decathlon competition in the order of their overall competition rankings. There are four pages for each house:

Page 1 includes an overall photograph and one or more detail photos, and a description of the concept taken from each team's own description of their design intentions as well as an interpretative conceptual diagram.

Page 2 includes the team's location, the orthographic drawings, competition rankings and total project cost.

Page 3 includes a site plan location that shows where the team was located on the National Mall, as well a series of architectural diagrams that explore abstract descriptions of specific architectural characteristics.

Page 4 includes diagrams of the passive and active system strategies for each house.

Part 2: Comparisons

Part 2 compares the diagrams from Part 1 in order of the ranking in the overall competition, the Architecture Contest and the Comfort Zone Contest.

Part 3: Analysis

Part 3 provides analysis of aspects of the competition that address the overlap of architecture and thermal comfort.

Terms that are listed in the Glossary are italicized in the body of the text.

Precedents in Zero-Energy Design On-line

There is a sample case study as well as "*The Matrix: An Interactive Database of the 2007 Solar Decathlon Diagrams*" available at the Routledge website:

http://www.routledge.com/9780415778756

All of the diagrams are organized in an interactive graphic database that allows the user to assess results for any contest or diagram in an ascending order based on the specific variable.

Precedents in Zero-Energy Design:

Part 1: Drawings and Diagrams

Introduction to Drawings and Diagrams

Drawings

The first step towards developing an understanding of a building is to draw it. The students enrolled in the winter 2008 seminar class, in which this original analysis began, were asked to study the construction drawings, photos and images of their two adopted houses and then produce a series of hand-drawn orthographic drawings. This included site plan and floor plans, transverse and longitudinal sections and a typical wall section.

This project would have been more efficient and consistent if all drawings had been done on a computer. However, the goals of this project were pedagogical. I feel strongly that students gain a greater understanding of a project by hand drawing.

There are discrepancies in the drawings as a result of having ten students' hands at work. As a result, the drawings reflect a variety of line weight, material hatches and other drawing conventions.

Following the completion of the orthographic drawings, students began an analysis of various aspects of the architectural characteristics of the houses. These were pinned up and discussed within the class weekly, and the results were a series of schematic architectural diagrams describing each house.

The analysis then shifted to one of passive and active systems design. The diagams went through multiple iterations and have been redrawn countless times.

Diagrams

According to *Merriam-Webster*, a diagram is "...a graphic design that explains rather than represents; *especially*: a drawing that shows arrangement and relations (as of parts)."

The diagram has been used throughout the history of architecture as a graphic means of communication. Master masons used diagrams to explain the design of Gothic churches well before we had the verbal vocabulary to describe complex three-dimensional relationships between disparate components.

In contemporary architecture and architectural education, the diagram has a pivotal role in communication between designer and client, designer and builder, designer and teacher or student. However, the diagram has an equally critical role as a tool for analysis of a given project, place or idea.

For this project, diagrams are used as a tool for differentiating and comparing specific aspects or components of these houses. These aspects were individually and collectively analyzed. Part 1 portrays the individual diagrams of each project, Part 2 compares them through the overall, Architecture and Comfort Zone rankings and Part 3 presents an analysis of the houses.

The intention was to develop diagrams that would facilitate an understanding of design decisions for three different aspects of the buildings: architecture, passive design and active mechanical system layout as it informed design decisions.

Diagrams – Architecture

These diagrams are analyses of specific architectural and organizational characteristics in each project. These are subjective interpretations that come from an in-depth study of the recurring similarities and differences between these characteristics.

site circulation – This site plan diagram shows the intended path of a visitor from the main street of the National Mall into and through the house.

outdoor spaces – This site plan diagram shows the outdoor spaces that are created for people to use around the exterior of the house.

construction module – This diagram shows the module that was used in transporting the house to the National Mall in Washington, DC.

structure – This diagram shows the load-bearing elements in solid lines and the spanning elements in dashed lines.

site plan location – This site plan identifies where the specific house was located in the Solar Village on the National Mall.

additive/subtractive – Beginning with the basic forms, this diagram demonstrates formal additions or subtractions.

public/private spaces – This diagram differentiates the public and private spaces. The "public" spaces (shaded) include the areas where an interaction might occur with a guest – living/dining space, study and kitchen. The "private" space includes the bathroom and bedroom.

parti – A *parti* is the basic scheme of an architectural design, which could be sketched in a few lines. This axonometric diagram represents the most distilled expression of the building form in three dimensions. This drawing is an overlay of the exterior form parti and the interior spatial parti.

interior spatial parti – This axonometric diagram represents the basic shape of the interior usable space in three dimensions.

interior spatial form – This diagram represents the basic shape of the usable interior space in two dimensions.

served/service – Louis Kahn introduced the concept of "served/service." "Service" space is that which is dedicated to supporting the "served" spaces. Served space (shaded) includes the living and bedroom spaces.

interior zones – This diagram differentiates the basic zones of the house by shades of grey:

- living space
- bedroom
- kitchen
- bathroom

geometry – This diagram shows basic geometric shapes evident in the plan. These include the square, "√2 rectangle" (1:1.414) and "golden rectangle" (1:1.618).

Diagrams – Passive Design

Although passive design wasn't the focus of the Solar Decathlon, every team considered these strategies to a greater or lesser degree. These diagrams distill what we perceive as the passive design intentions of each project.

daylighting – This diagram shows where direct and ambient daylight are entering the interior of the houses. Direct light is shown with a clearly bound white area while ambient daylight is shown with a light, gradient hatch that decreases as distance from the daylight source increases.

shading – The shading diagram shows where direct light can enter a given space and where direct light is kept out as a result of shading. It also shows the general shape of any physical shade element (overhang, sunshade, etc.).

natural ventilation – plan – This diagram shows how ventilation moves through a space in plan (cross ventilation).

natural ventilation – section – This diagram shows how ventilation moves through a space in section (stack ventilation).

solar angle – This diagram shows the predominant angle of solar panels on the given house. In some cases there are multiple angles. We are showing the angle that receives the majority of insolation.

thermal storage – Some teams utilized thermal mass as a passive strategy in the project. In these cases, we have noted where the thermal mass is located.

Diagrams – Active Mechanical Systems

The active mechanical system diagrams are not meant to describe how the systems work. What is being assessed is the relationship between the overall systems design and the architectural design.

core location – This diagram shows the location of the mechanical systems core within the building plan. In some cases, the distribution and services are also shown.

active heating – This diagram shows where the active heating system is deployed in the house.

active cooling – This diagram shows where the active cooling system is deployed in the house.

mechanical ventilation – This diagram shows where the ventilation system is deployed in the house. For forced-air systems, this is often the same as the heating and cooling diagrams.

hot water system – This diagram shows where the hot water is created, stored and distributed throughout the house.

mechanical parti overlay – this diagram combines the major components of the active mechanical system and the architectural parti in one drawing. This is done to assess the relationship between the Heating, Ventilation and Air Conditioning (HVAC) parti and the architectural parti.

01

The Team Darmstadt house is a series of overlapping, situationally-responsive layers. The outer layer contains all of the solar power equipment as well as acting as a breathable skin, providing security and helping to control privacy and transparency. The second layer is the thermal envelope, which acts as a weather barrier and the main container for the conditioned living spaces. The third layer is the mechanical core of the house, which contains the kitchen, bathroom and the building's mechanical systems.

Passivhaus

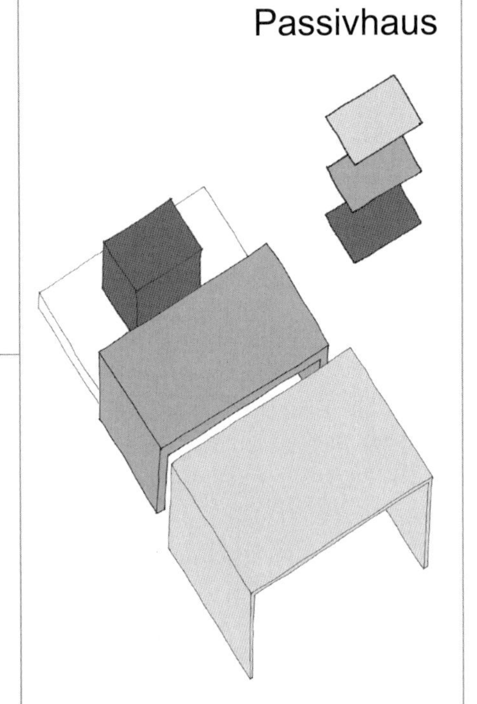

img #01

img #02

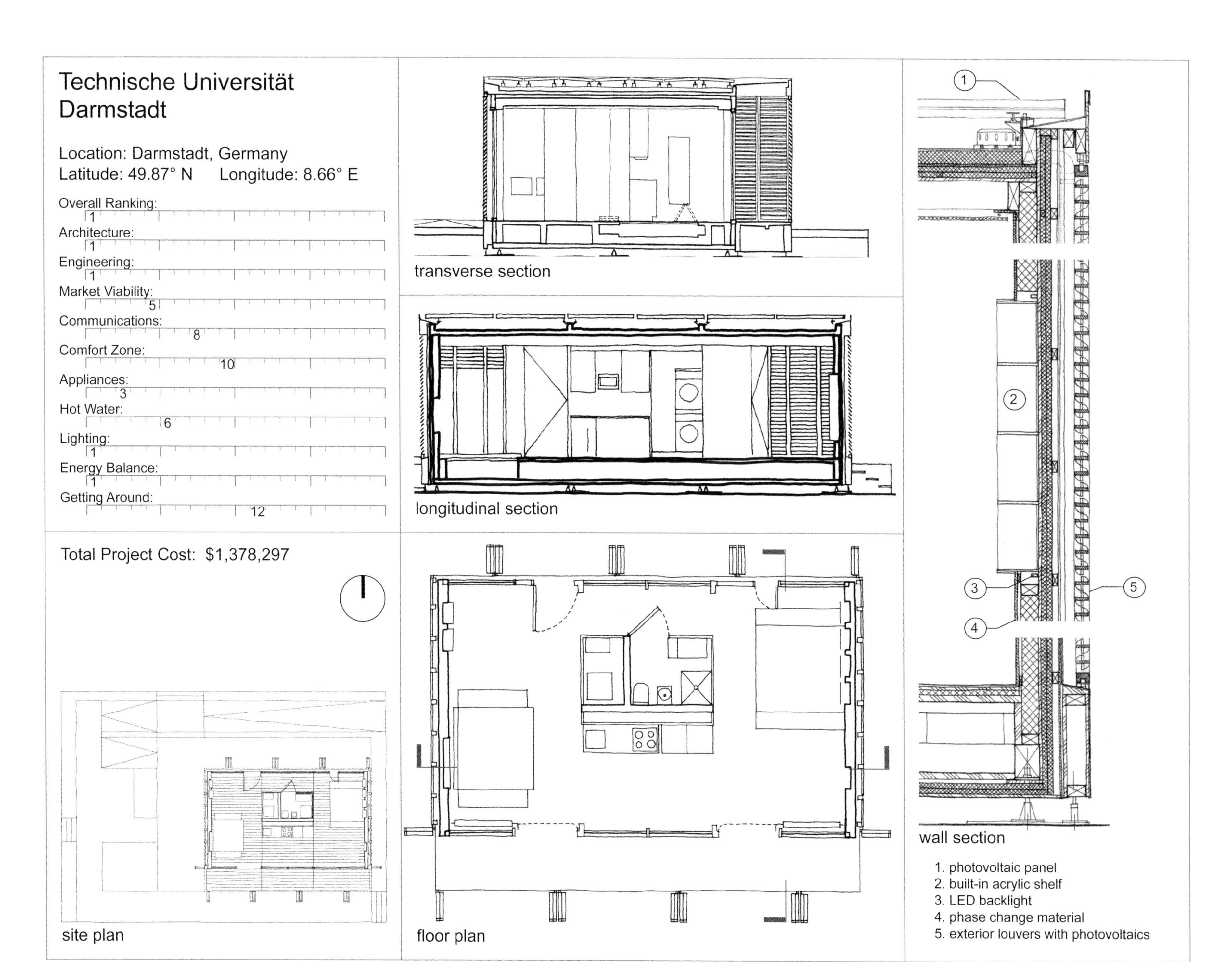

Technische Universität Darmstadt
Location: Darmstadt, Germany
Latitude: 49.87° N Longitude: 8.66° E
Overall Ranking:
1
Architecture:
1
Engineering:
1
Market Viability:
5
Communications:
8
Comfort Zone:
10
Appliances:
3
Hot Water:
6
Lighting:
1
Energy Balance:
1
Getting Around:
12
Total Project Cost: $1,378,297
site plan
transverse section
longitudinal section
floor plan
1
2
3
4
5
wall section
1. photovoltaic panel
2. built-in acrylic shelf
3. LED backlight
4. phase change material
5. exterior louvers with photovoltaics

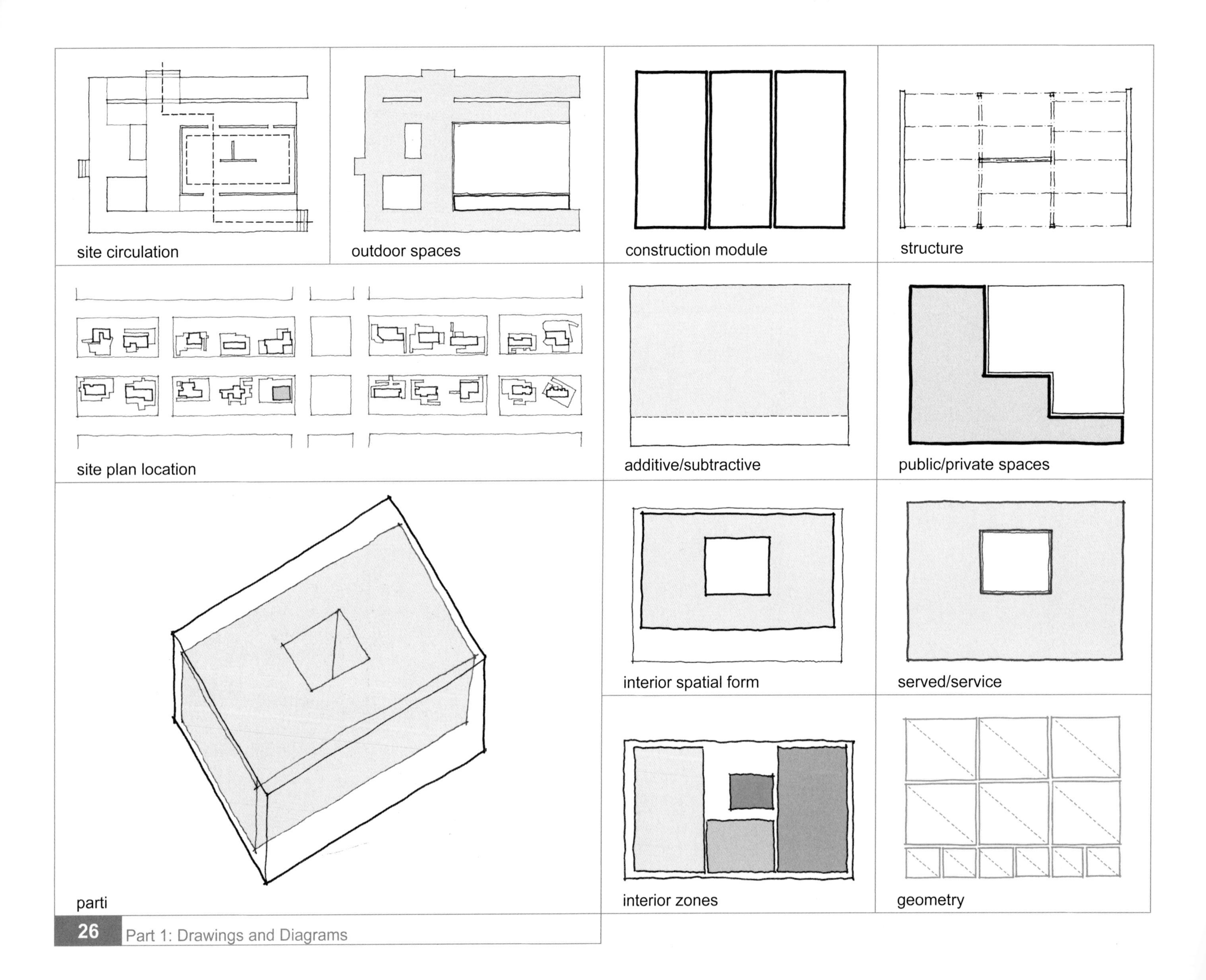
site circulation
outdoor spaces
construction module
structure
site plan location
additive/subtractive
public/private spaces
parti
interior spatial form
served/service
interior zones
geometry

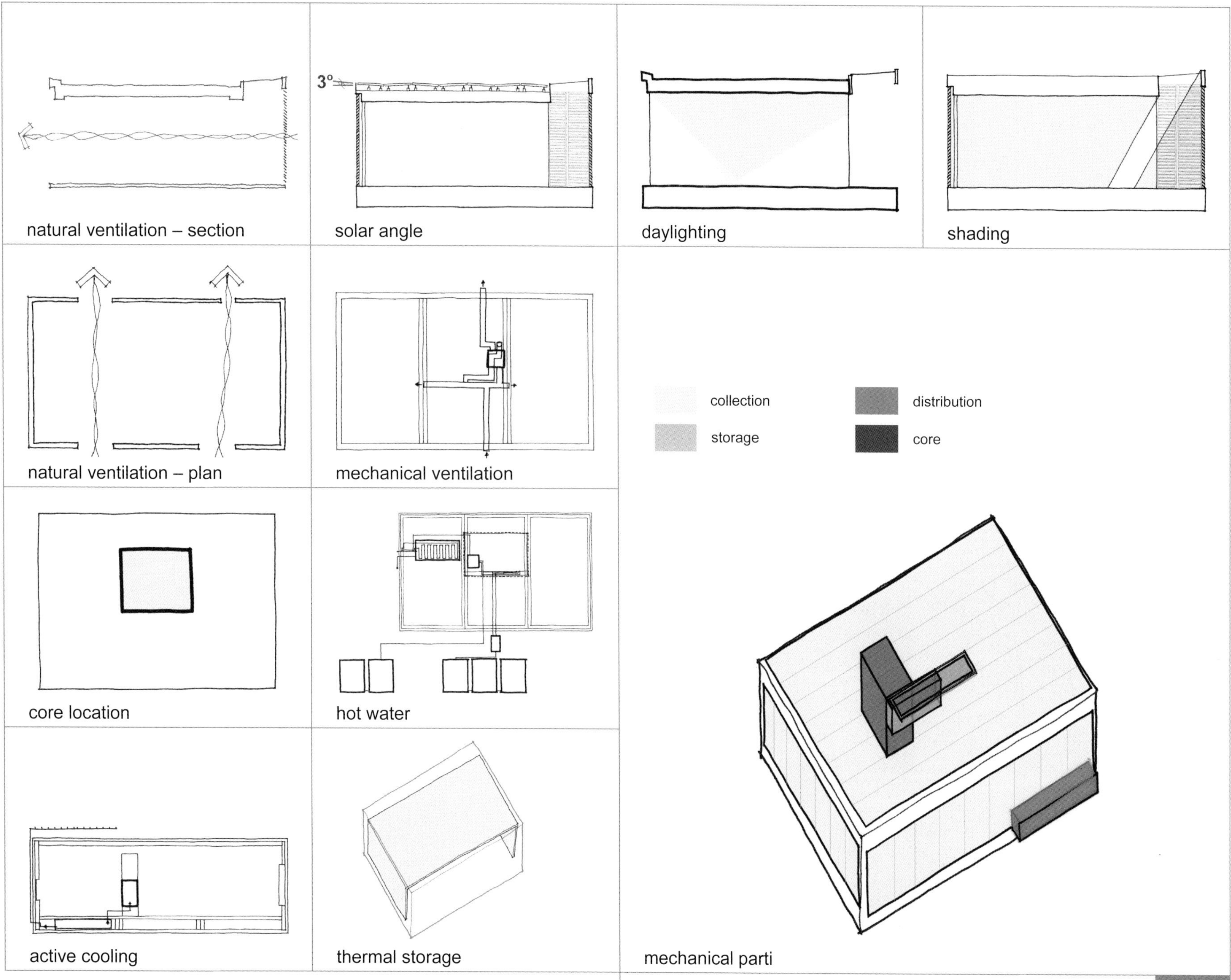
natural ventilation – section
3°
solar angle
daylighting
shading
natural ventilation – plan
mechanical ventilation
collection
distribution
storage
core
core location
hot water
active cooling
thermal storage
mechanical parti

02

LEAFhouse

The LEAFhouse (Leading Everyone to an Abundant Future) is an acronym that embodies the team's three goals: advance sustainable design and construction, use nature as an inspiration and mentor and demonstrate the potential of solar technology. The LEAF is flexible, provides comfortable space below it, allows views through and around it and produces energy for itself. Through the use of shading, solar panels and planting devices on the exterior of the house, the house produces energy, comfortable space and sources of food. The goals are to have a light touch on nature, engage local conditions and develop a contextually responsive design.

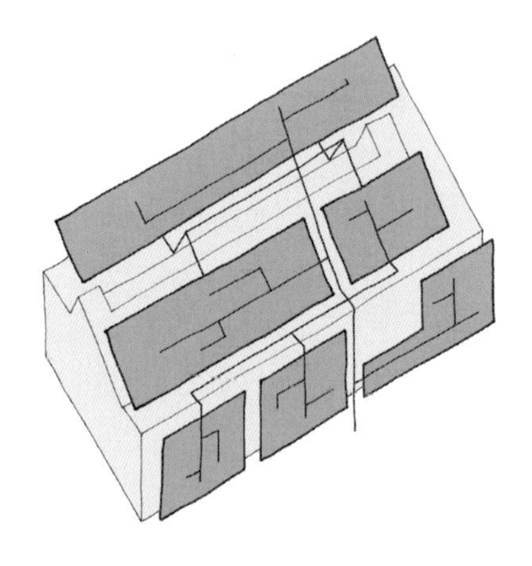

img #03

img #04

University of Maryland

Location: College Park, MD, USA
Latitude: 38.99° N   Longitude: 76.93° W

Overall Ranking: 2
Architecture: 2
Engineering: 6
Market Viability: 2
Communications: 1
Comfort Zone: 4
Appliances: 17
Hot Water: 8
Lighting: 2
Energy Balance: 1
Getting Around: 2

Total Project Cost: $448,470

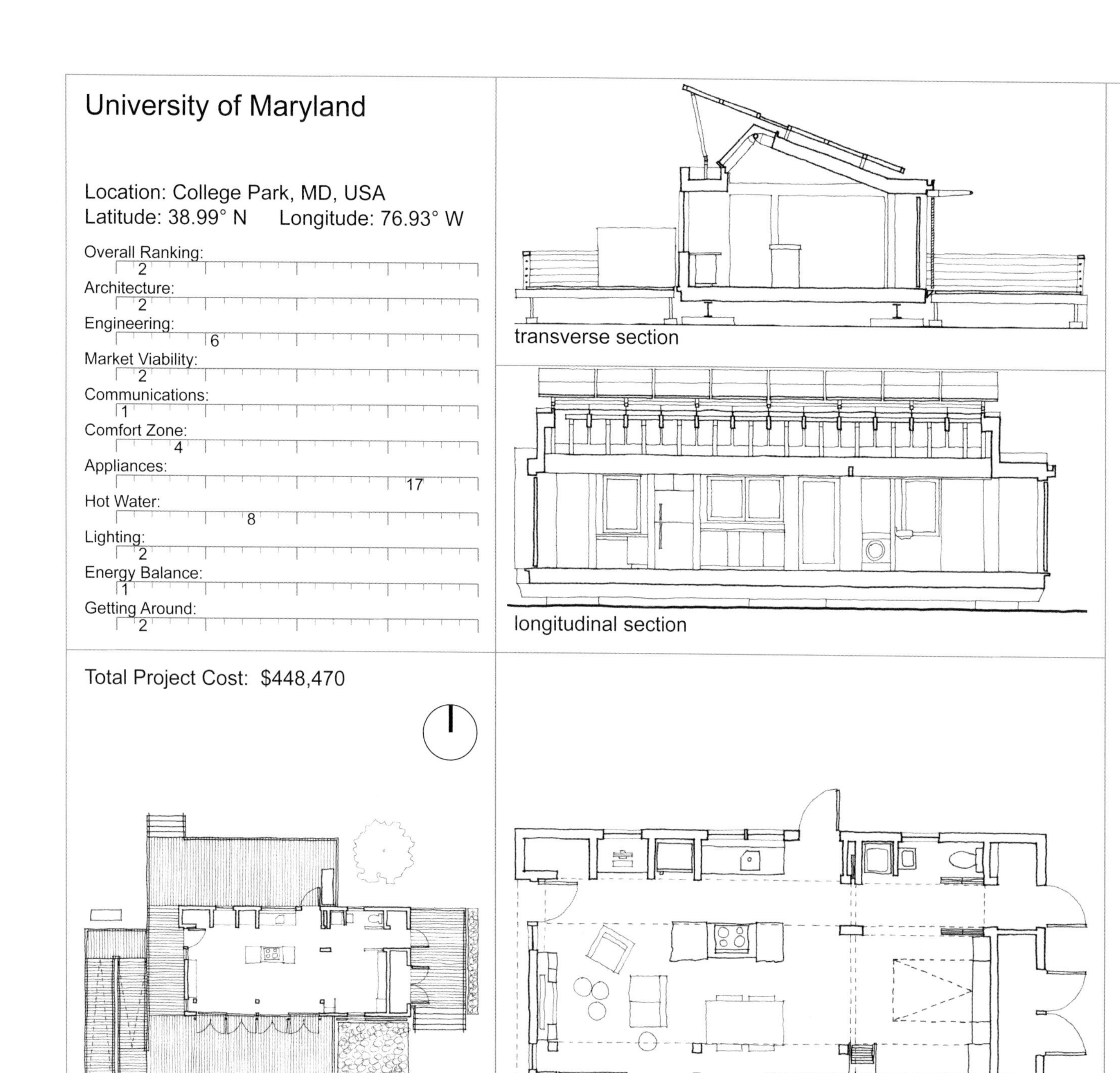

transverse section

longitudinal section

site plan

floor plan

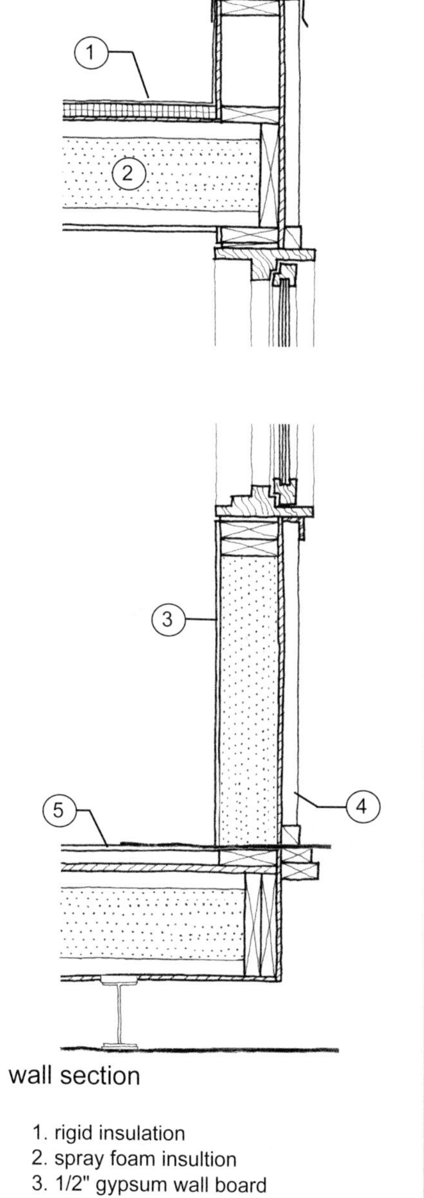

wall section

1. rigid insulation
2. spray foam insultion
3. 1/2" gypsum wall board
4. 1 1/2" corrugated metal cladding
5. 3/4" plywood subfloor

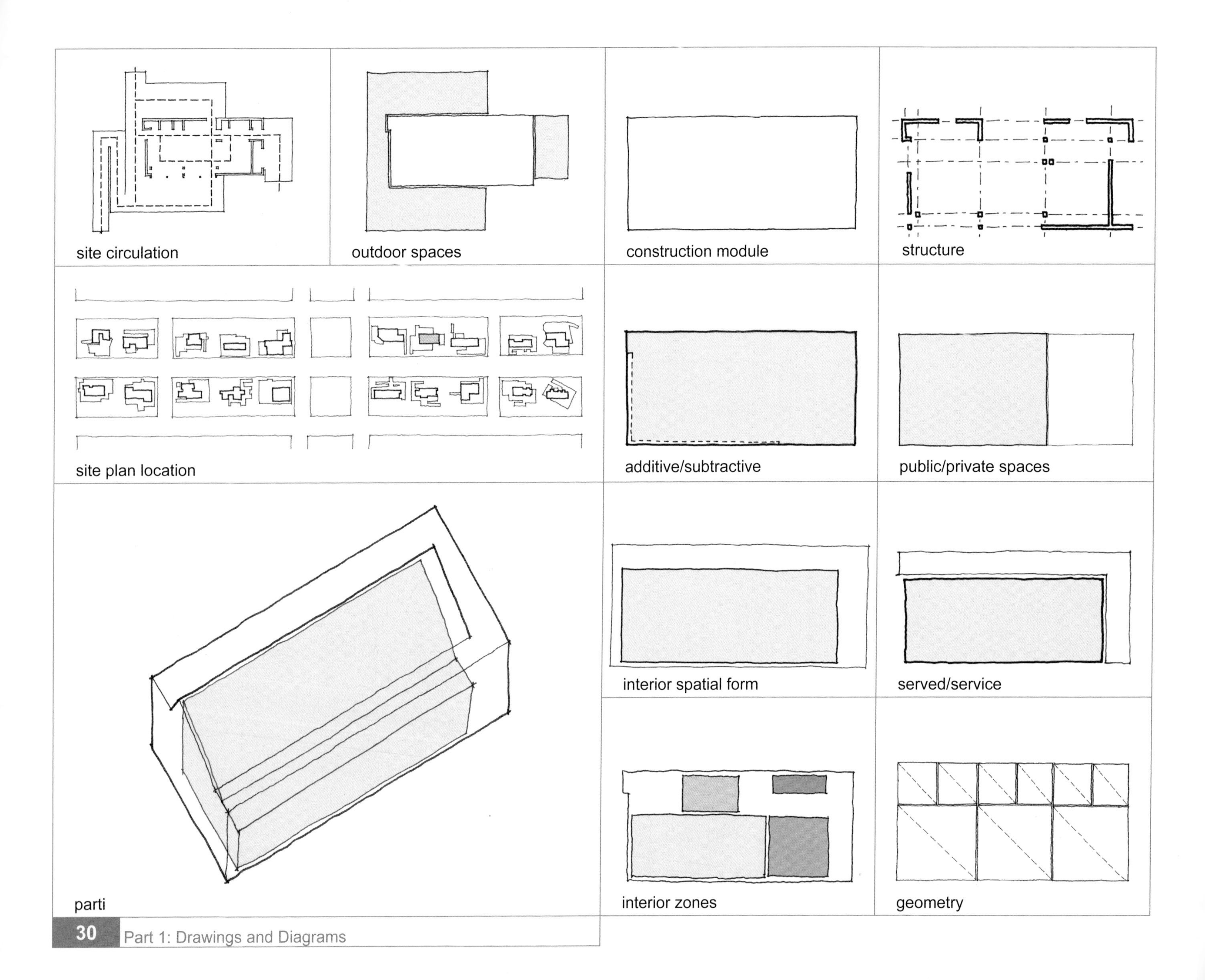
site circulation
outdoor spaces
construction module
structure
site plan location
additive/subtractive
public/private spaces
parti
interior spatial form
served/service
interior zones
geometry

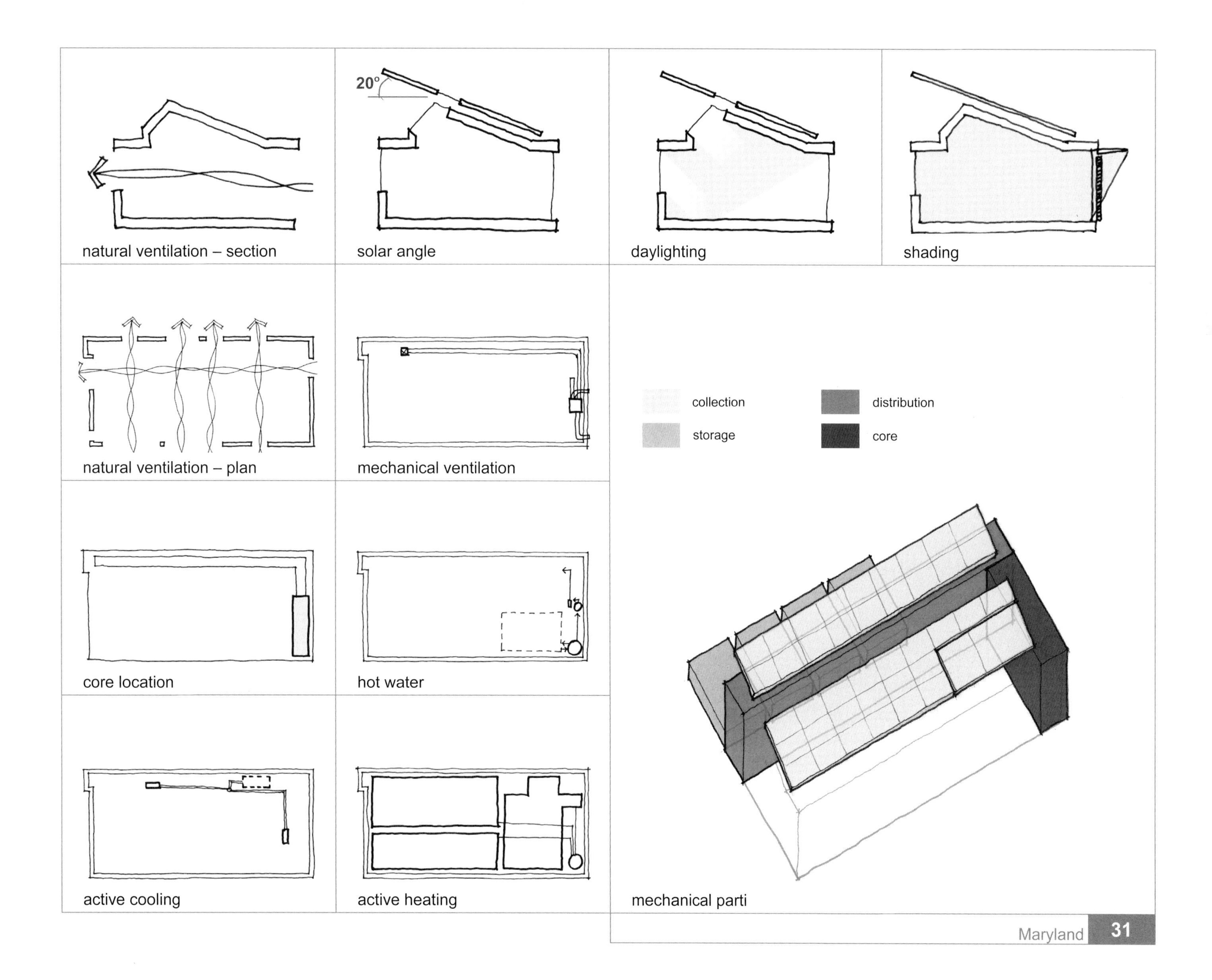
20°
natural ventilation – section
solar angle
daylighting
shading
natural ventilation – plan
mechanical ventilation
collection
distribution
storage
core
core location
hot water
active cooling
active heating
mechanical parti

03

Ripple Home

The Santa Clara house was designed to accomplish two things: embody the unique spirit of California and provide a sustainable and environmentally sound building while maintaining the look and feel of a traditional home. The design includes a large, insulated, translucent opening that extends the living space to the outdoors, as well as utilizing innovative bamboo ceiling joists and vernacular-influenced details. The design incorporated an architectural language on the interior and exterior that is similar to traditional California homes in an attempt to appeal to the largest homeowner market.

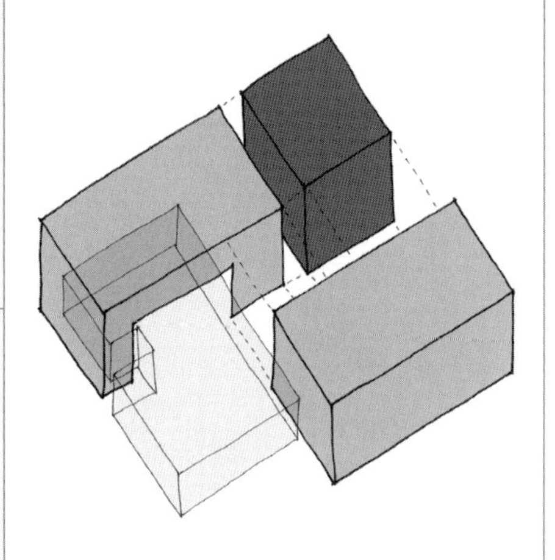

img #05

img #06

Santa Clara University

Location: Santa Clara, CA, USA
Latitude: 37.35° N Longitude: 121.95° W

- Overall Ranking: 3
- Architecture: 18
- Engineering: 10
- Market Viability: 6
- Communications: 2
- Comfort Zone: 5
- Appliances: 2
- Hot Water: 1
- Lighting: 8
- Energy Balance: 1
- Getting Around: 2

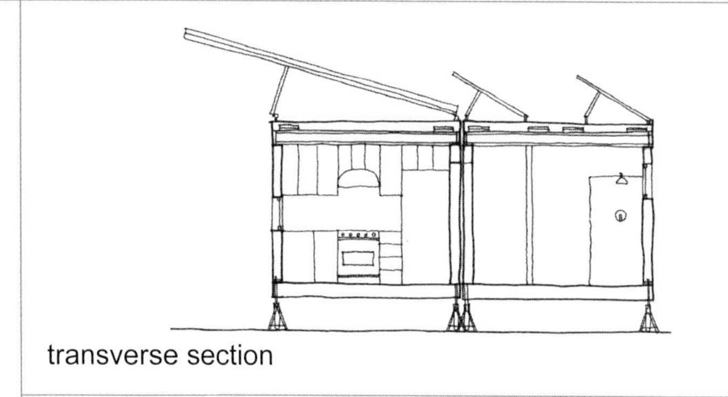

transverse section

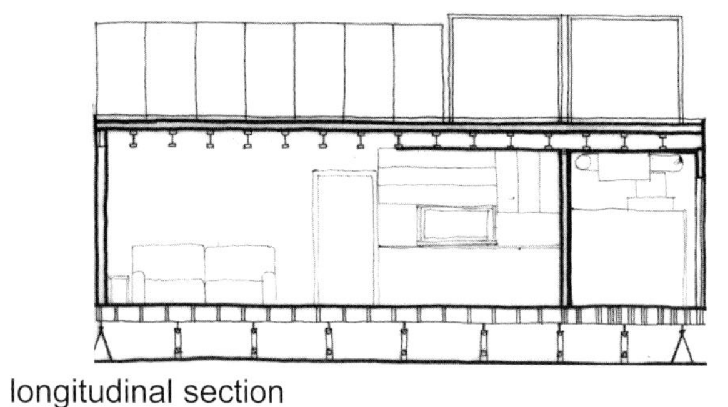

longitudinal section

Total Project Cost: $800,000

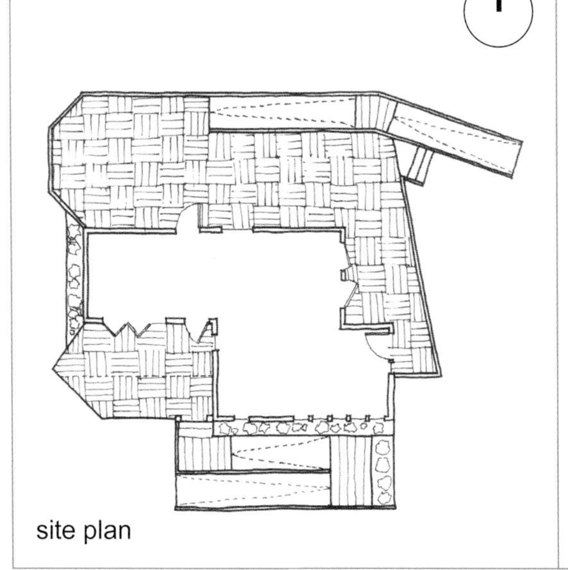

site plan

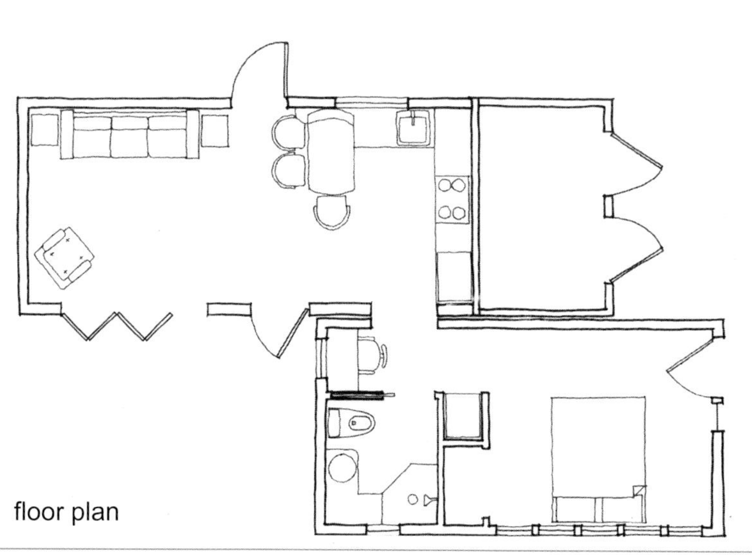

floor plan

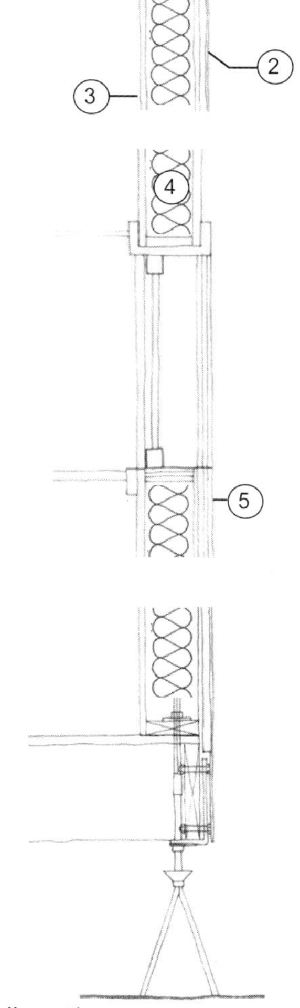

wall section

1. polyurethane spray-on insulation
2. 1/2" plywood sheathing
3. 5/8" paper-based sheet
4. cotton-batt insulation
5. fiber-cement board

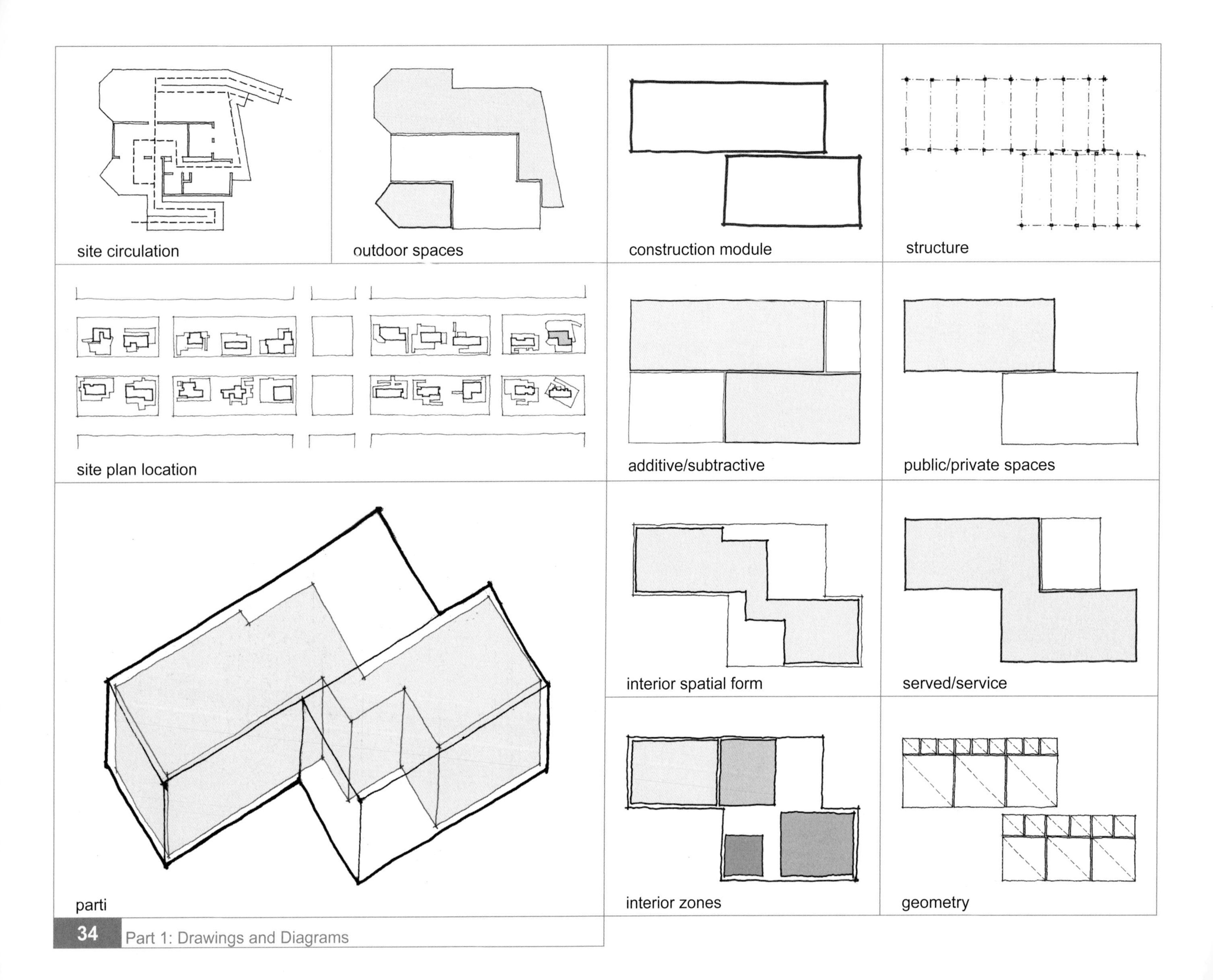
site circulation
outdoor spaces
construction module
structure
site plan location
additive/subtractive
public/private spaces
parti
interior spatial form
served/service
interior zones
geometry

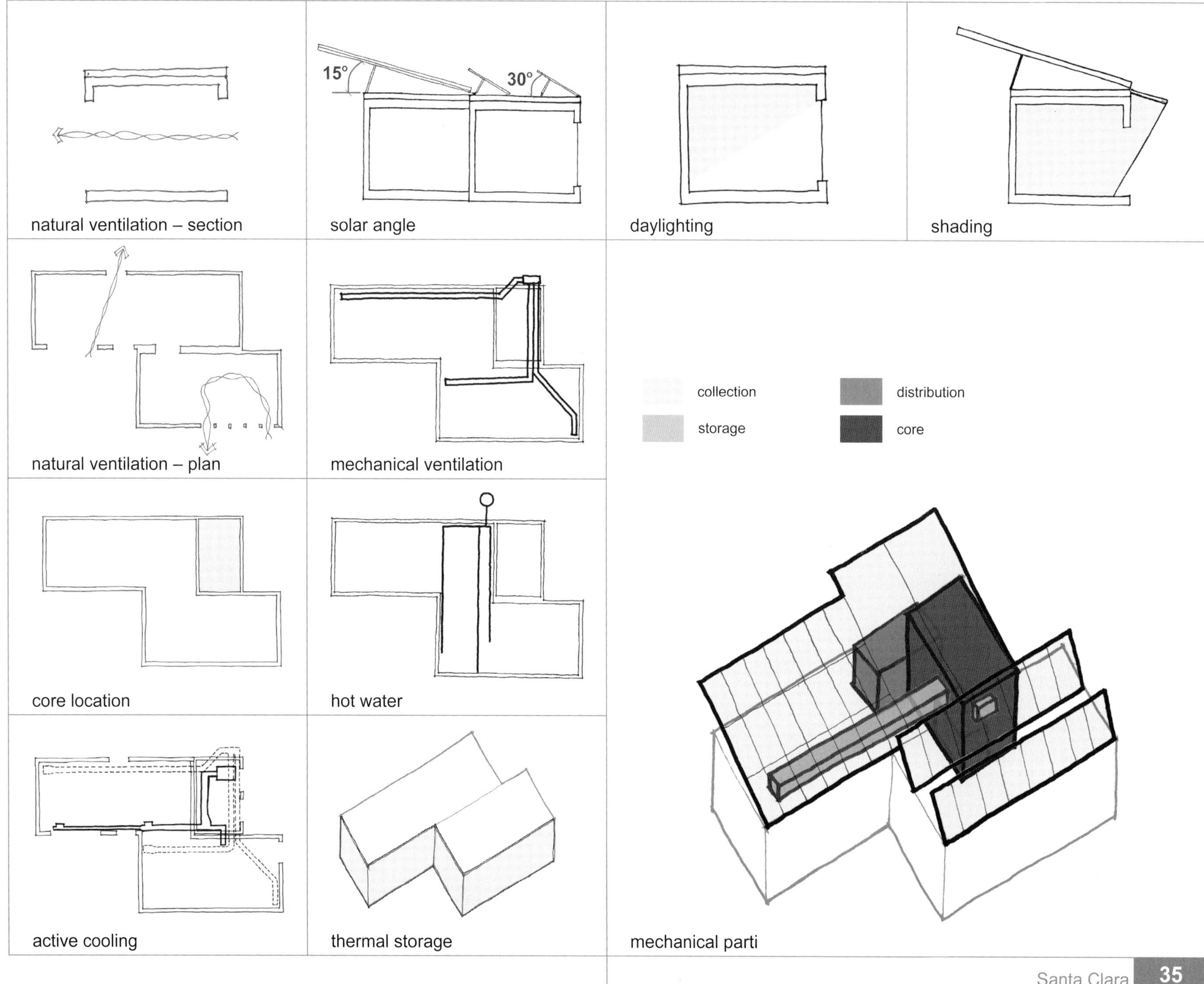

15°
30°
natural ventilation – section
solar angle
daylighting
shading
natural ventilation – plan
mechanical ventilation
collection
distribution
storage
core
core location
hot water
active cooling
thermal storage
mechanical parti

04

The Penn State house was designed to "sustain life" using three main components referencing the human body and its organs: the "brain", the "heart" and the "lungs." The "brain" of the house on the north contains all the services for the house, including the kitchen, bathroom and mechanical systems. The "heart" of the home contains the living space – an open-plan containing the living room and bedroom. The "lungs" of the house, which connect the "brain" and the "heart", act as both the entry and circulation area for the house. This space is bathed in light and is intended to ventilate the house.

MorningStar Home

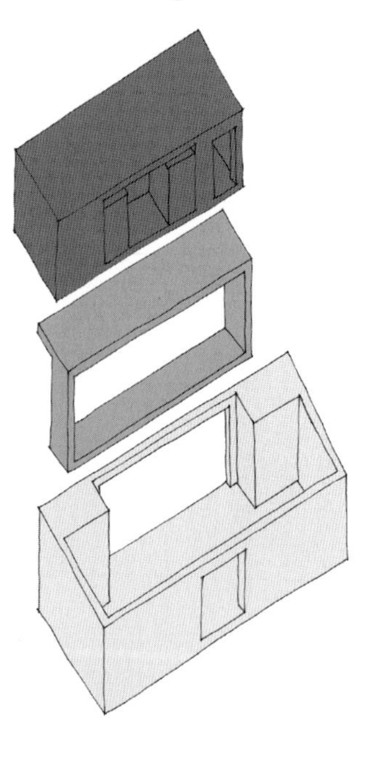

img #07

img #08

The Pennsylvania State University

Location: University Park, PA, USA
Latitude: 40.79° N Longitude: 77.86° W

Overall Ranking: 4
Architecture: 8
Engineering: 5
Market Viability: 3
Communications: 3
Comfort Zone: 13
Appliances: 7
Hot Water: 1
Lighting: 3
Energy Balance: 12
Getting Around: 14

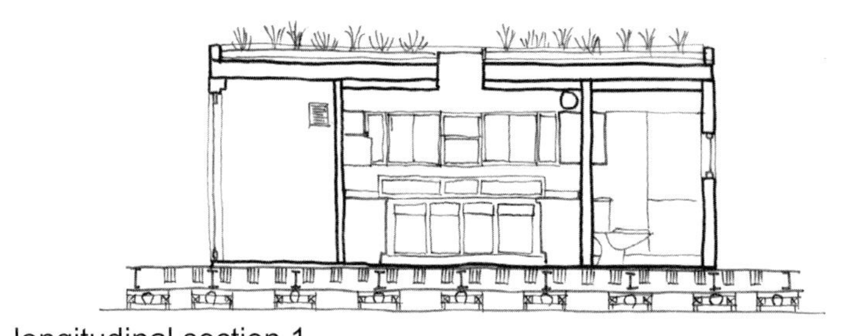
longitudinal section 1

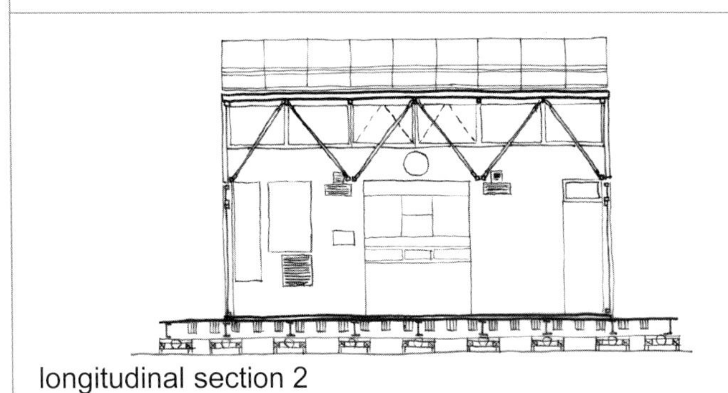
longitudinal section 2

Total Project Cost: $505,000

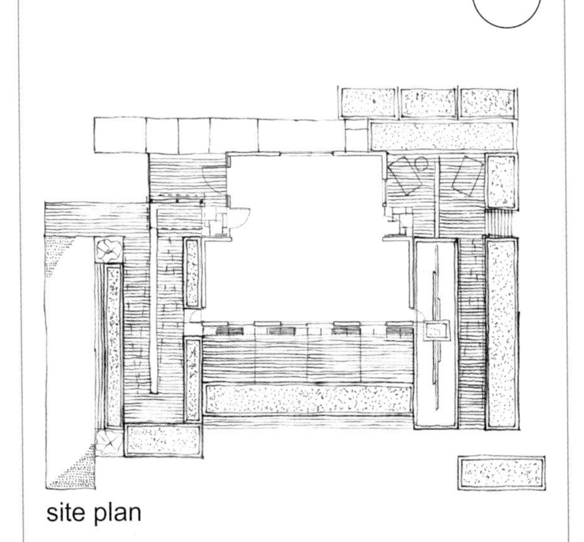
site plan

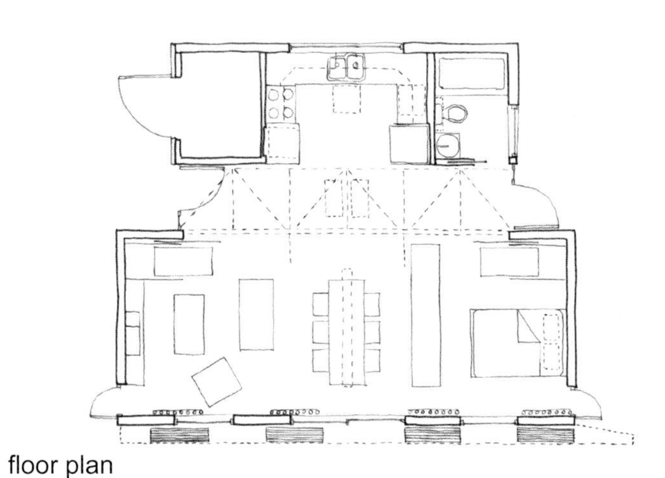
floor plan

wall section

1. spray-foam insulation
2. structural insulated panels
3. laminated veneer lumber
4. 1 1/2" x 1" wood slats
5. milk-bottle wall

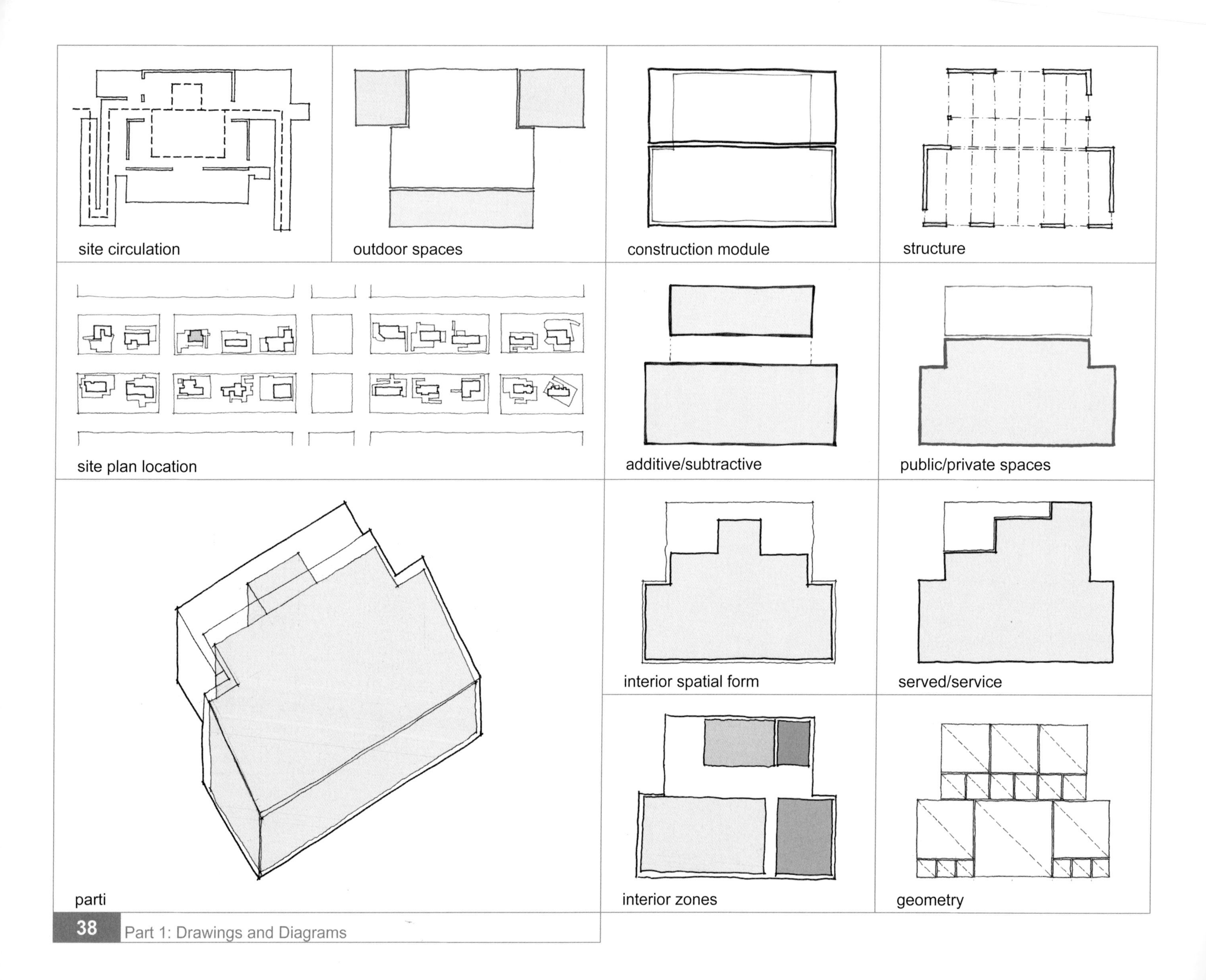
site circulation
outdoor spaces
construction module
structure
site plan location
additive/subtractive
public/private spaces
parti
interior spatial form
served/service
interior zones
geometry

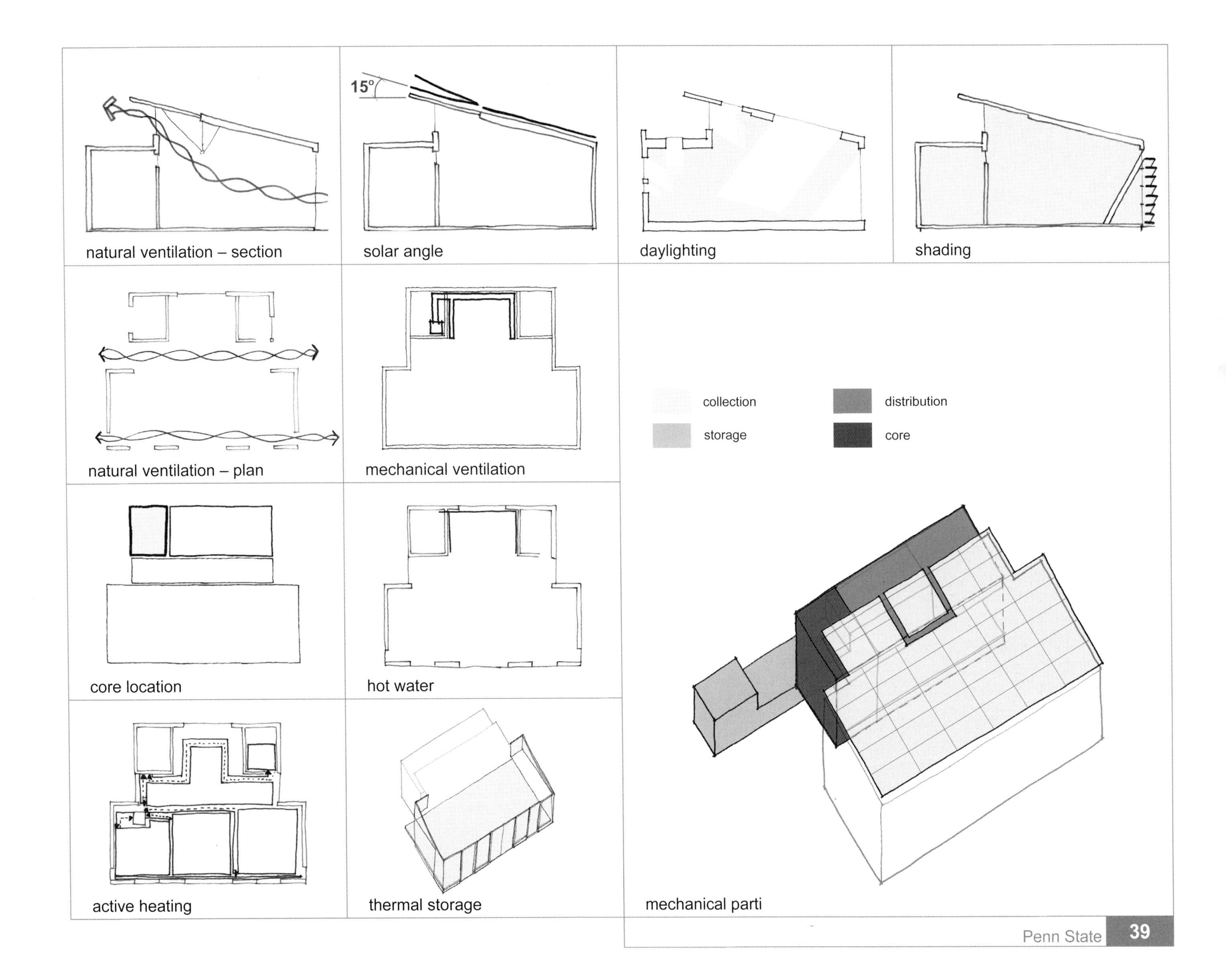
15°
natural ventilation – section
solar angle
daylighting
shading
natural ventilation – plan
mechanical ventilation
collection
distribution
storage
core
core location
hot water
active heating
thermal storage
mechanical parti

05

Casa Solar

The concept for the Madrid house combines three main components: roof, main space and tower. The roof is tilted to maximize solar energy and designed to appear floating just above the main space. The roof is optimized for the solar panels on its surface, with an overhang on all sides of the house providing shade. The main space is enclosed on three sides and opened on the south, admitting natural light and ventilation. Lastly, the mechanical tower plugs into and intersects both the main living space and the roof. The tower contains all the services for the building, structure to support the roof and a vegetated wall.

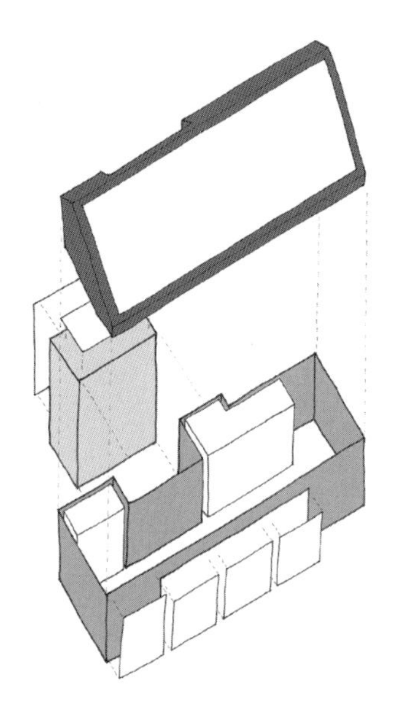

img #09

img #10

Universidad Politécnica de Madrid

Location: Madrid, Spain
Latitude: 40.48° N   Longitude: 3.87° W

Overall Ranking: 5
Architecture: 3
Engineering: 13
Market Viability: 17
Communications: 16
Comfort Zone: 7
Appliances: 4
Hot Water: 15
Lighting: 6
Energy Balance: 1
Getting Around: 5

Total Project Cost: $1,000,000

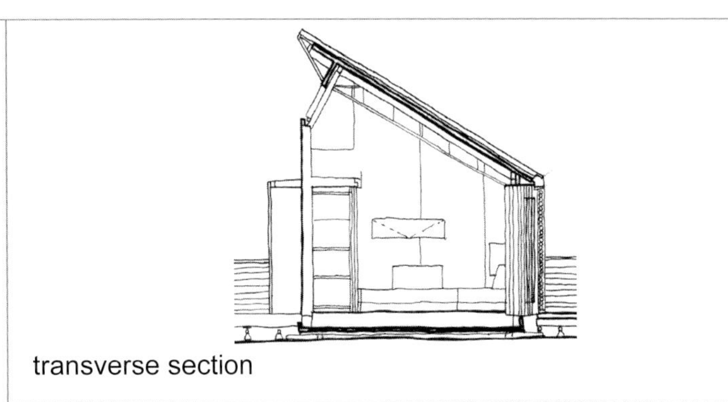

transverse section

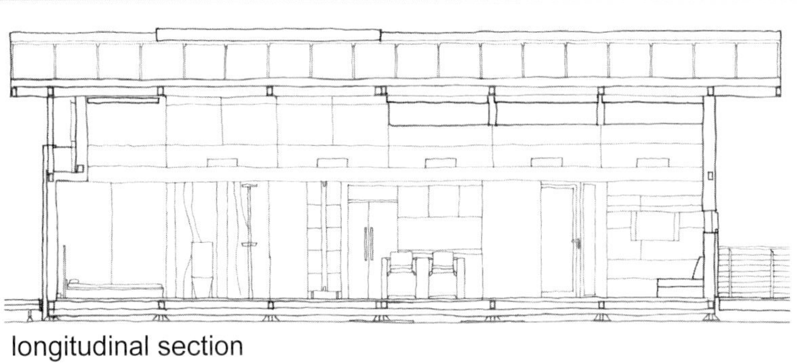

longitudinal section

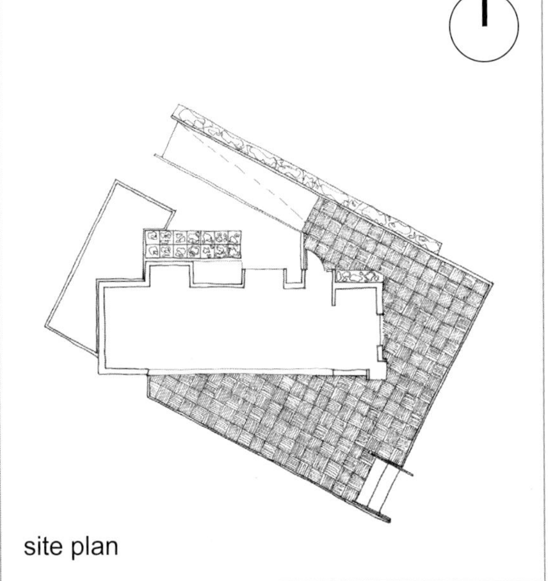

site plan

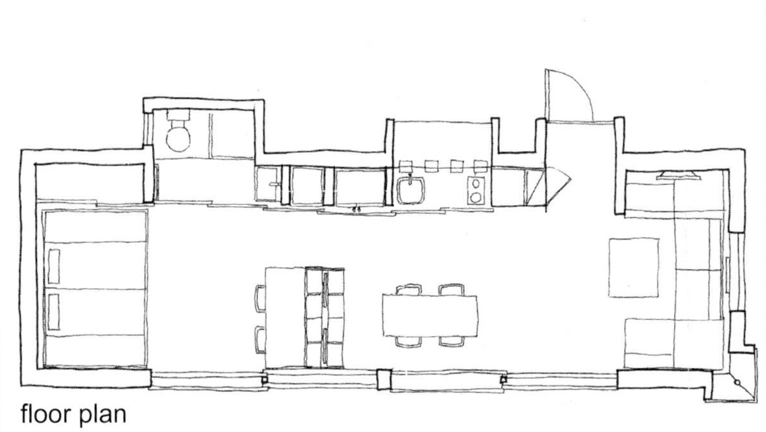

floor plan

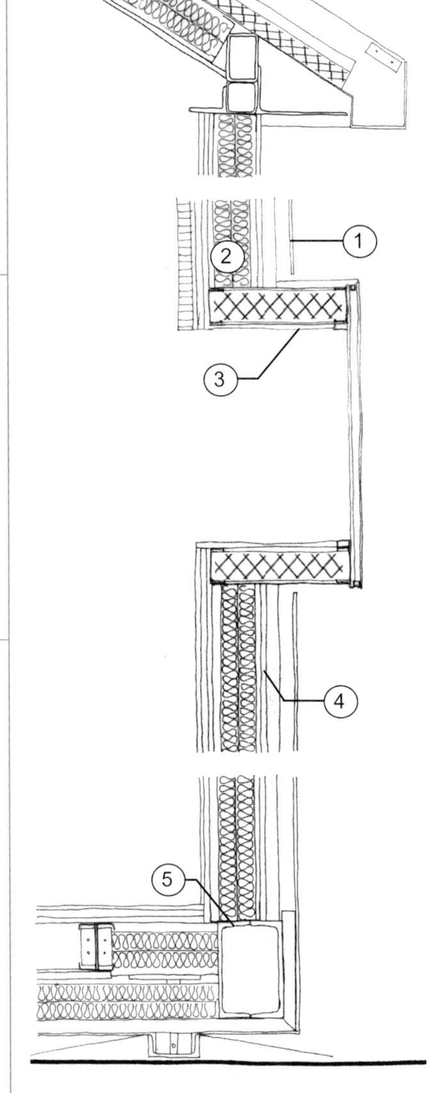

wall section

1. cellulose-fiber wood-faced panel
2. natural-fiber insulating batts
3. expanded-polyurethane rigid board
4. oriented strand board panel
5. steel frame C-profile

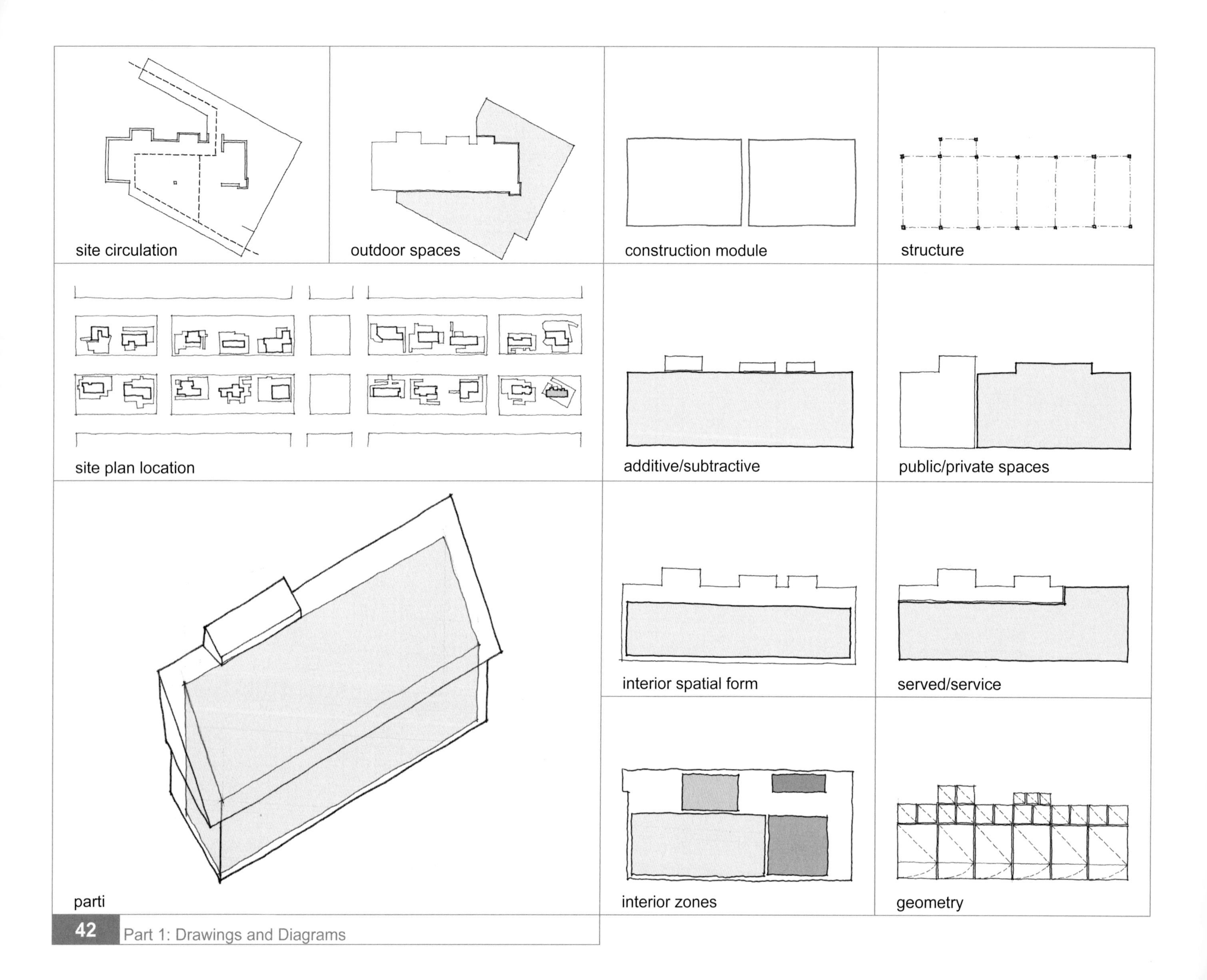
site circulation
outdoor spaces
construction module
structure
site plan location
additive/subtractive
public/private spaces
parti
interior spatial form
served/service
interior zones
geometry

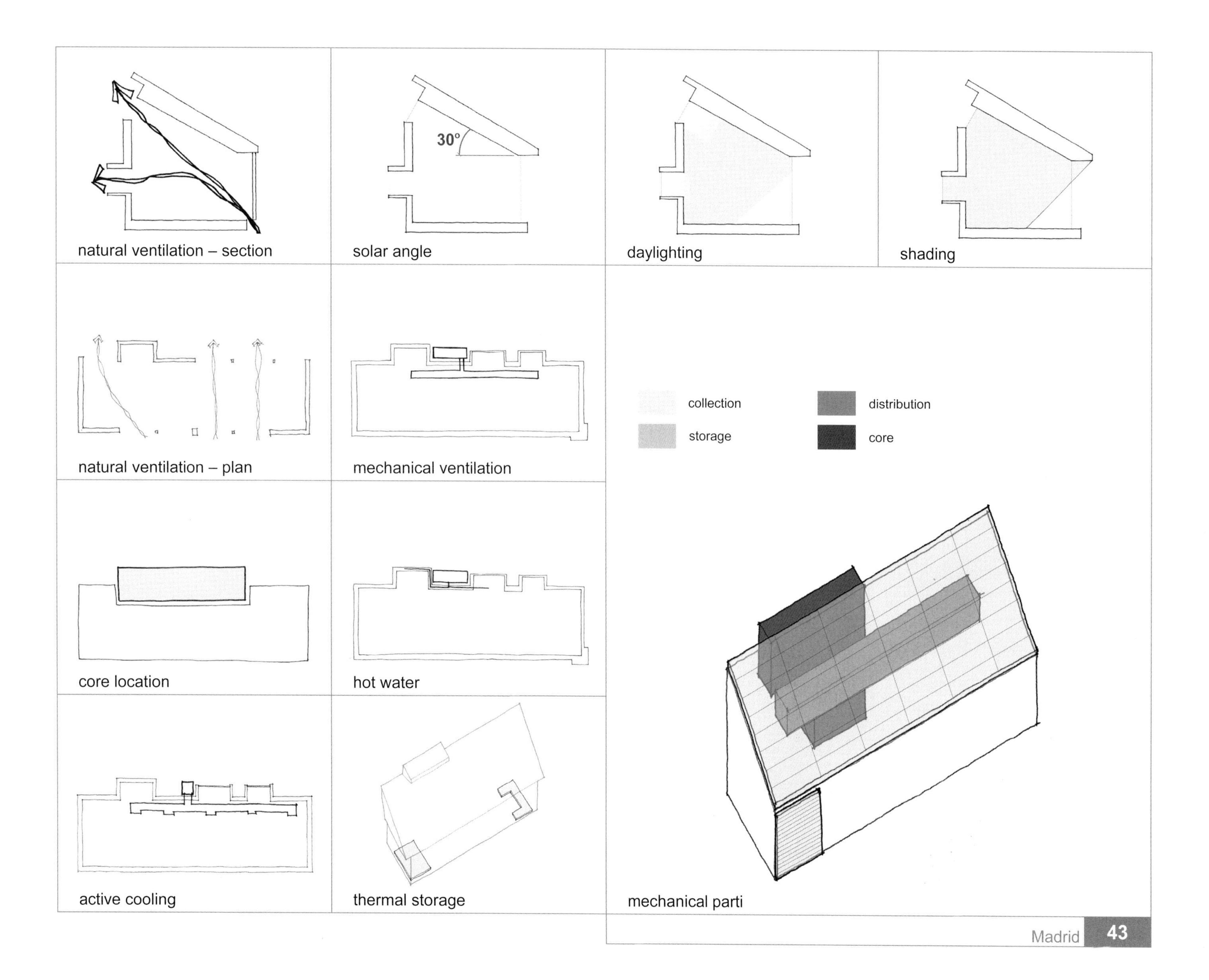
natural ventilation – section
solar angle
30°
daylighting
shading
natural ventilation – plan
mechanical ventilation
collection
storage
distribution
core
core location
hot water
active cooling
thermal storage
mechanical parti

06

The Icarus House by the Georgia Tech team was designed to "harness and celebrate the sun's power." The design explores the paradox between lightness and energy conservation. The house roof is a translucent assembly called a light pillow by its designers – a sandwich of plastic and insulation allowing the roof to be lightweight and to transmit light. Approximately one third of the south and east walls are translucent and made of a similar lightweight sandwich of materials, allowing additional light into the interior. The roof is shaded by "wing-like" solar panels with attached reflectors providing additional shade and light for power collection.

Icarus House

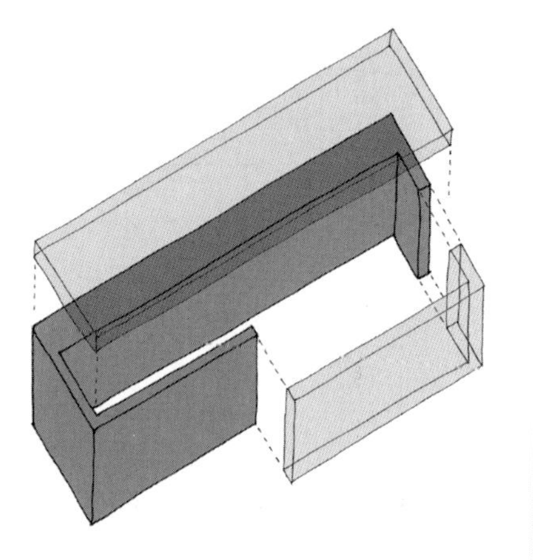

img #11

img #12

Georgia Institute of Technology

Location: Atlanta, GA, USA
Latitude: 33.76° N Longitude: 84.40° W

Overall Ranking: 6
Architecture: 4
Engineering: 12
Market Viability: 9
Communications: 5
Comfort Zone: 15
Appliances: 13
Hot Water: 12
Lighting: 20
Energy Balance: 11
Getting Around: 3

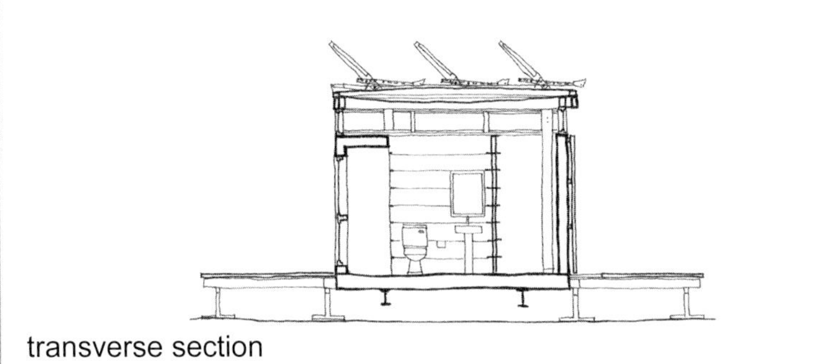

transverse section

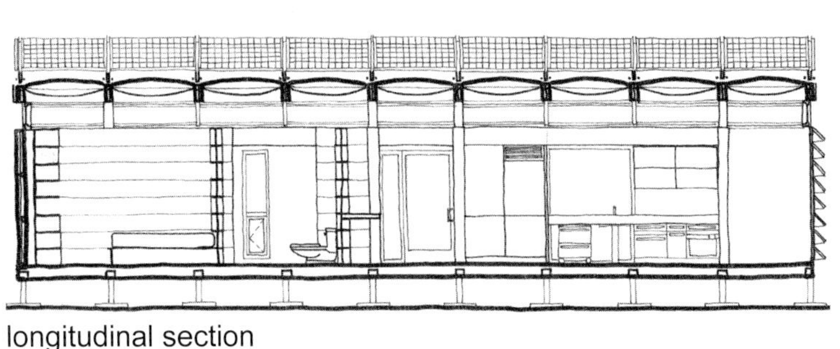

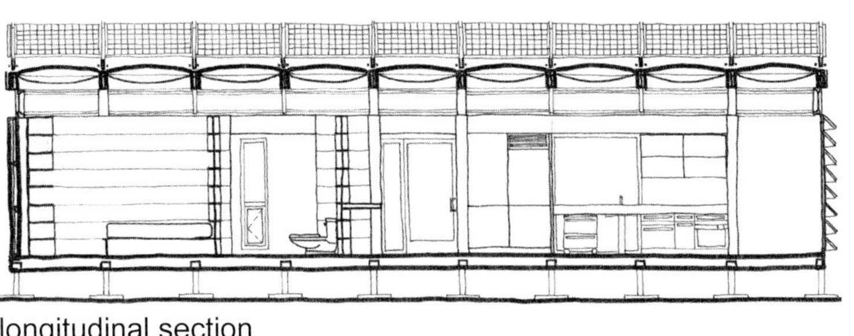

longitudinal section

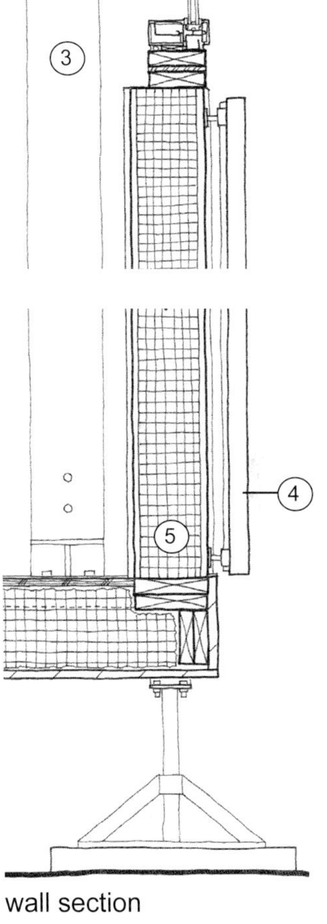

wall section

1. ethylene tetrafluoroethylene (ETFE) plastic roof membrane
2. aerogel layer
3. steel and wood-composite column
4. photovoltaic panel
5. structural insulated panel

Total Project Cost: $700,000

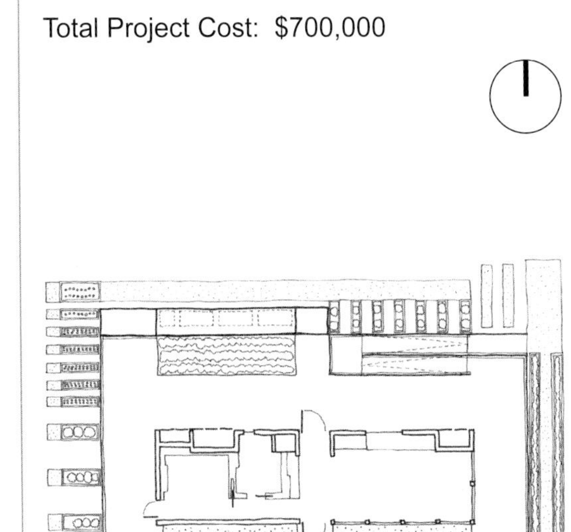

site plan

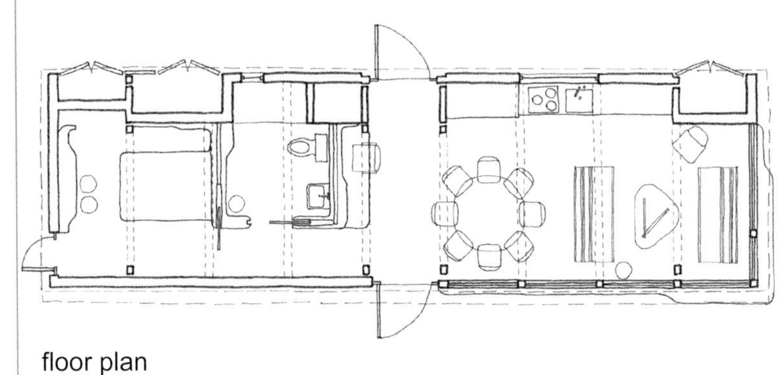

floor plan

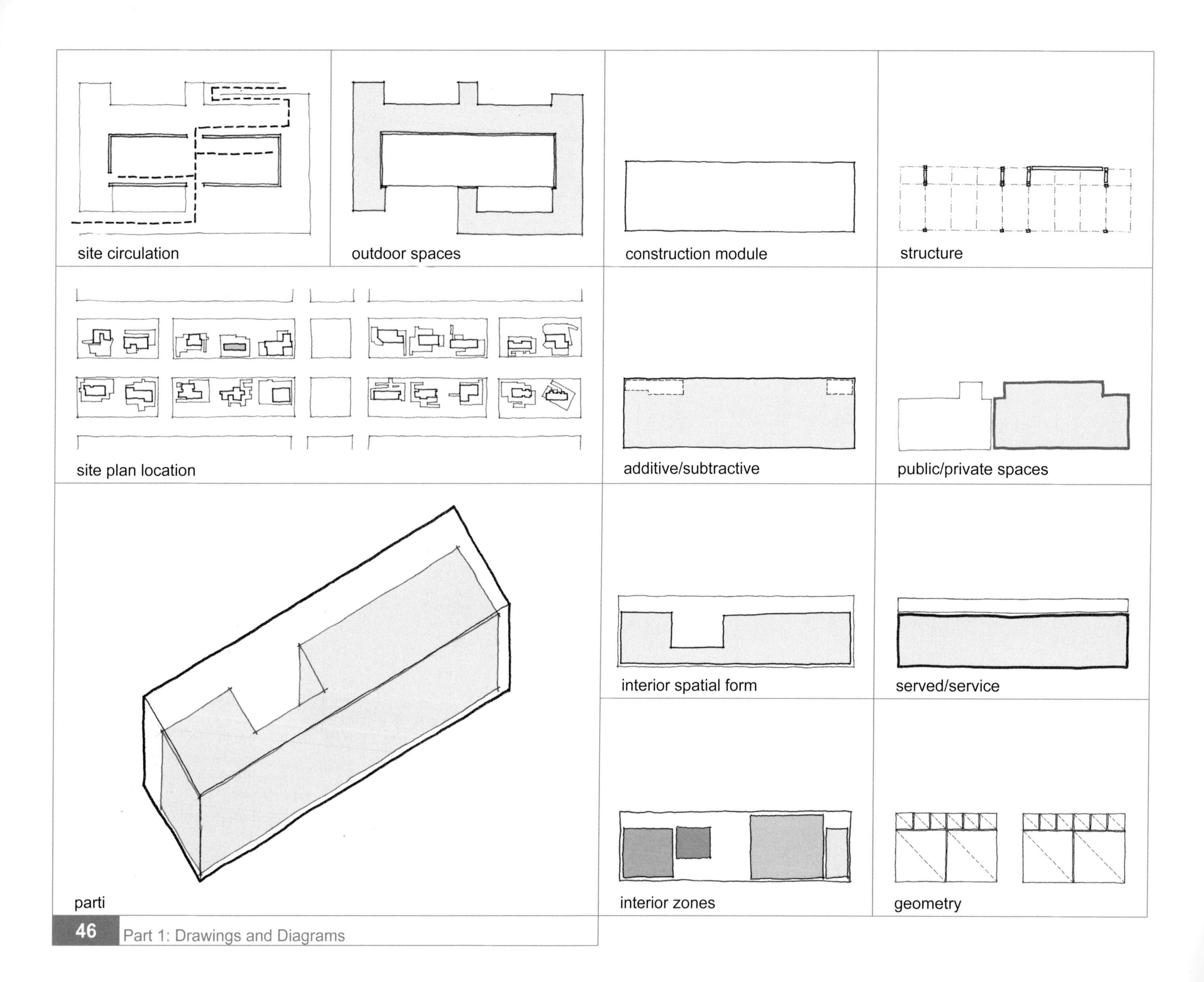
site circulation
outdoor spaces
construction module
structure
site plan location
additive/subtractive
public/private spaces
parti
interior spatial form
served/service
interior zones
geometry

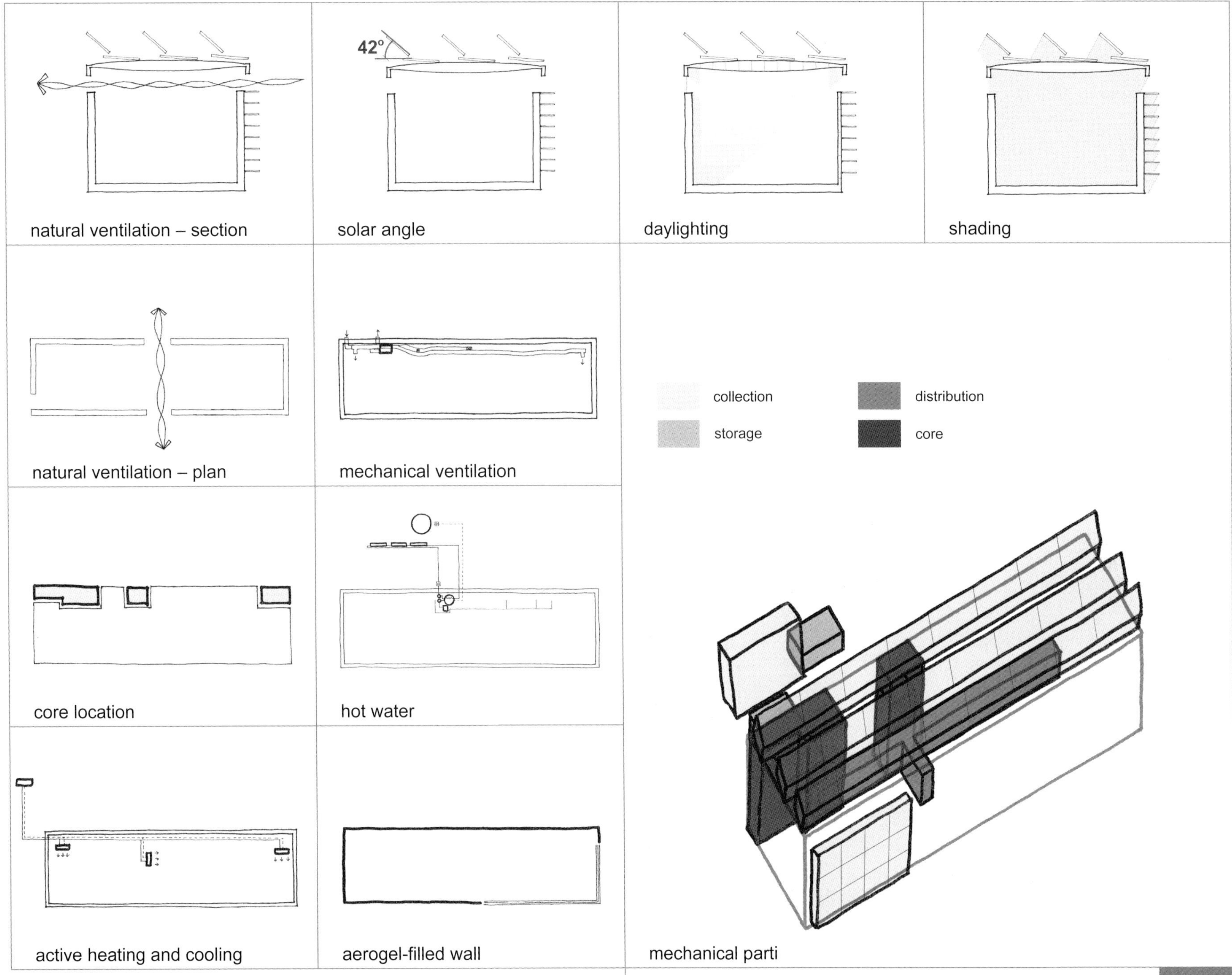
42°
natural ventilation – section
solar angle
daylighting
shading
natural ventilation – plan
mechanical ventilation
collection
distribution
storage
core
core location
hot water
active heating and cooling
aerogel-filled wall
mechanical parti

07

The concept for the Colorado house revolves around a central customizable "CORE", which contains all the mechanical and service functions of the home as well as the circulation space between the various parts of the home. Because these cores are customizable, factory-built and housed within a standard, industrial shipping container, they can be used in a variety of different applications. The intent is that the additional spaces built around the cores be influenced by local building materials, methods and aesthetics, furthering their breadth of design possibilities.

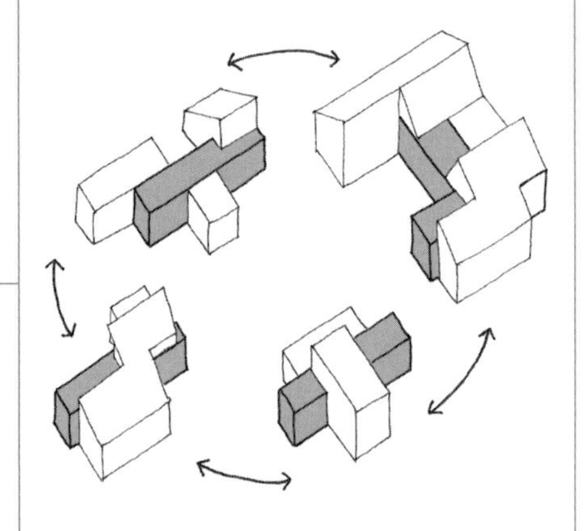

img #13

img #14

University of Colorado at Boulder

Location: Boulder, CO, USA
Latitude: 40.02° N Longitude: 105.25° W

Overall Ranking: 7

Architecture: 10

Engineering: 3

Market Viability: 7

Communications: 18

Comfort Zone: 19

Appliances: 15

Hot Water: 14

Lighting: 4

Energy Balance: 8

Getting Around: 1

Total Project Cost: $300,000

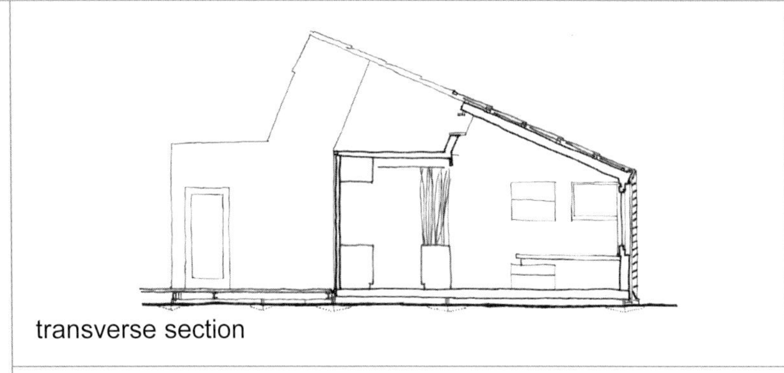

transverse section

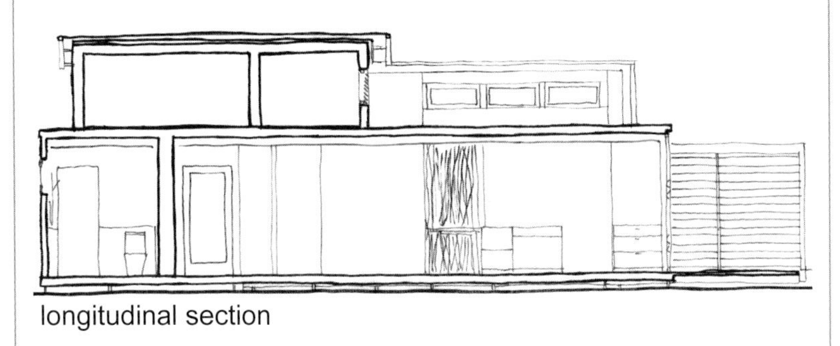

longitudinal section

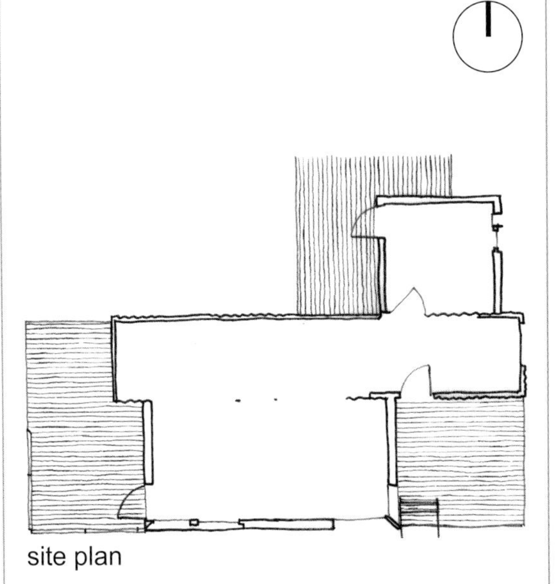

site plan

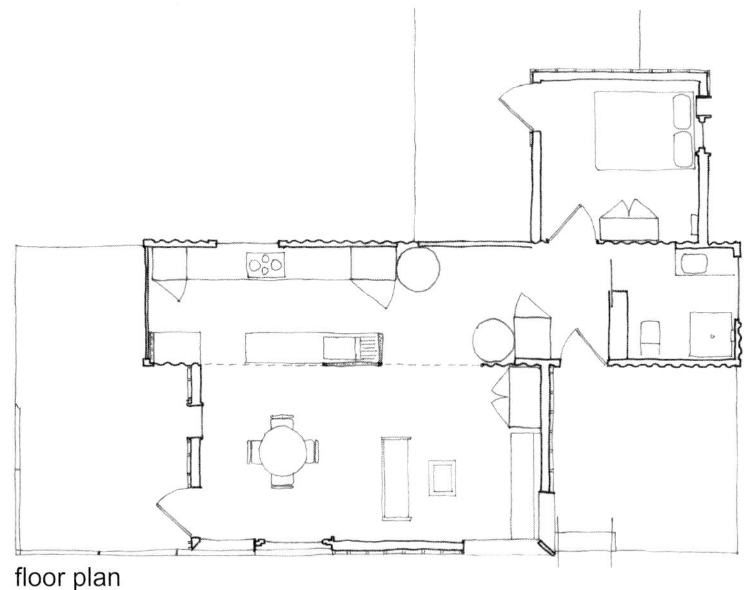

floor plan

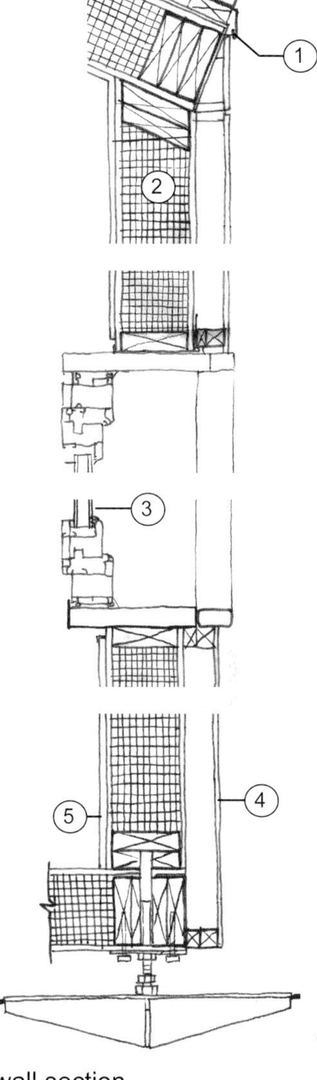

wall section

1. galvanized steel flashing
2. 6 5/8" structural insulated panel
3. double-glazed insulated unit
4. fiber-cement rainscreen cladding
5. 1/2" gypsum wall board

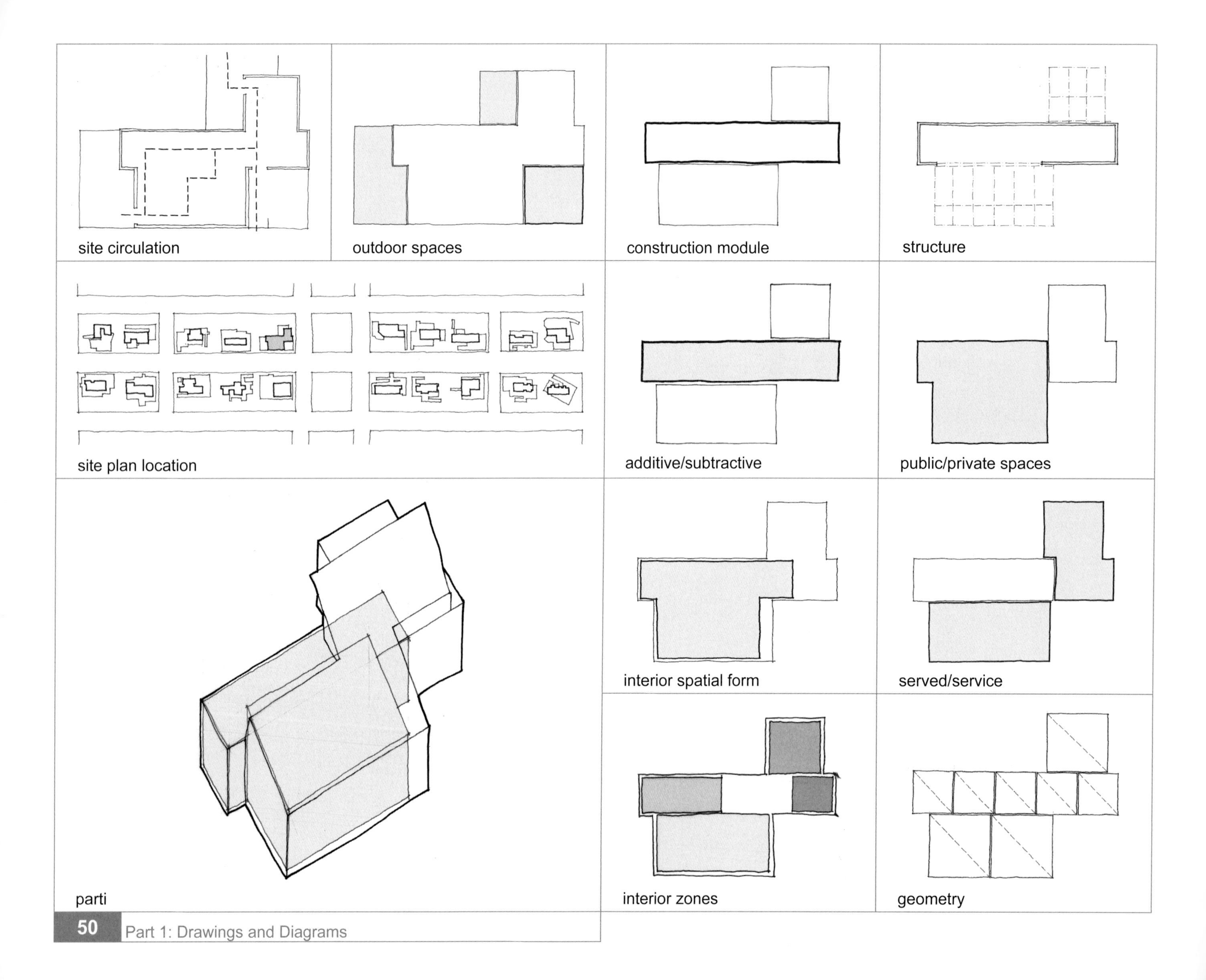
site circulation
outdoor spaces
construction module
structure
site plan location
additive/subtractive
public/private spaces
parti
interior spatial form
served/service
interior zones
geometry

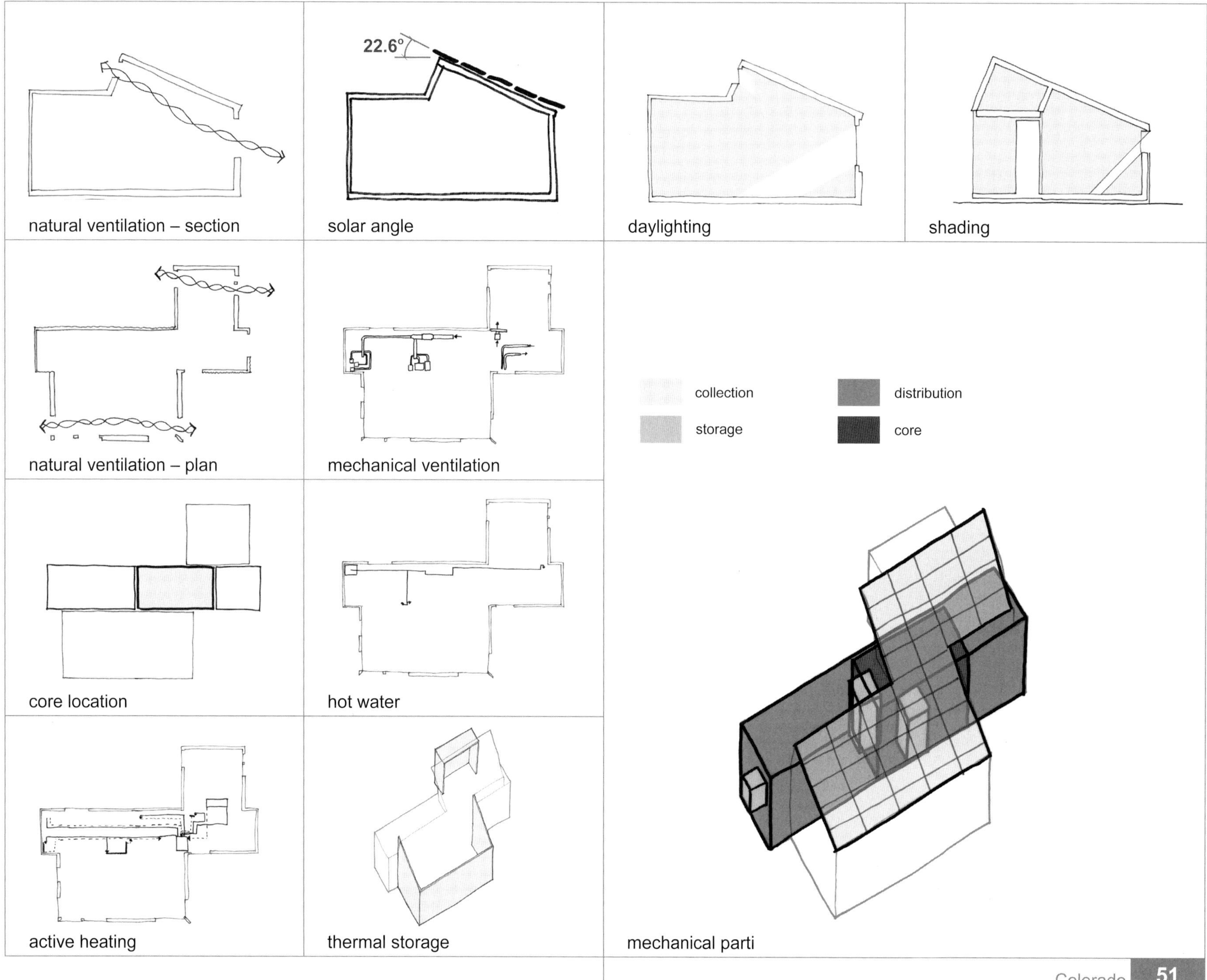
22.6°
natural ventilation – section
solar angle
daylighting
shading
natural ventilation – plan
mechanical ventilation
collection
distribution
storage
core
core location
hot water
active heating
thermal storage
mechanical parti

08

The intention of the Team Montréal house involves the incorporation of light as a constructive building element. The building envelope of this house wraps around the light element and service core, which plugs into and intersects these elements, distributing services to all areas of the house. The building envelope wraps the walls, floors and roof in its architectural language while the light element acts as the spatial volume, extending the design through the building envelope to the exterior. This house is predominantly daylit, and allows for natural ventilation and visual connectivity to the exterior.

Lumen Essence

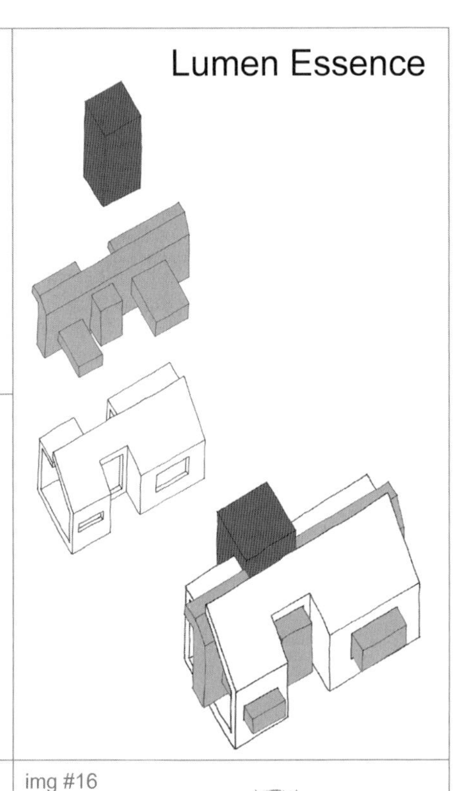

img #15

img #16

Team Montréal

École de Technologie Supérieure, McGill University and Université de Montréal

Location: Montréal, Québec, Canada
Latitude: 45.47° N Longitude: 73.75° W

Total Project Cost: $510,000

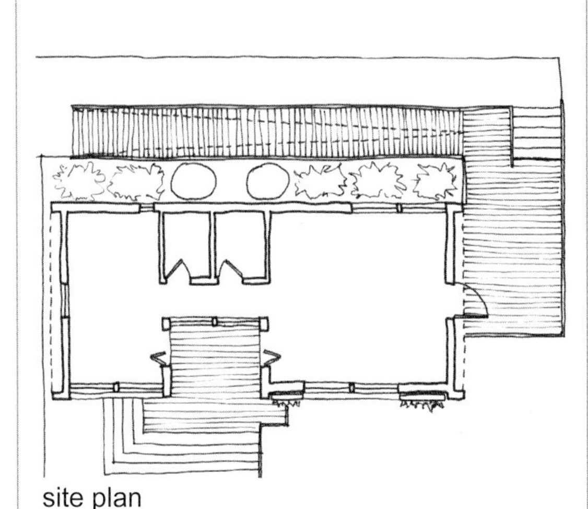

site plan

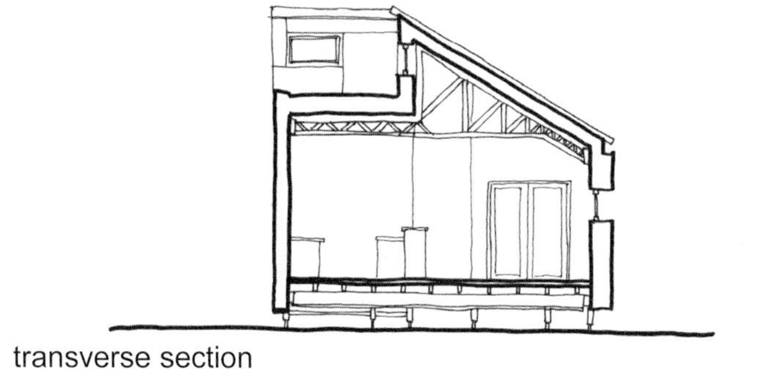

transverse section

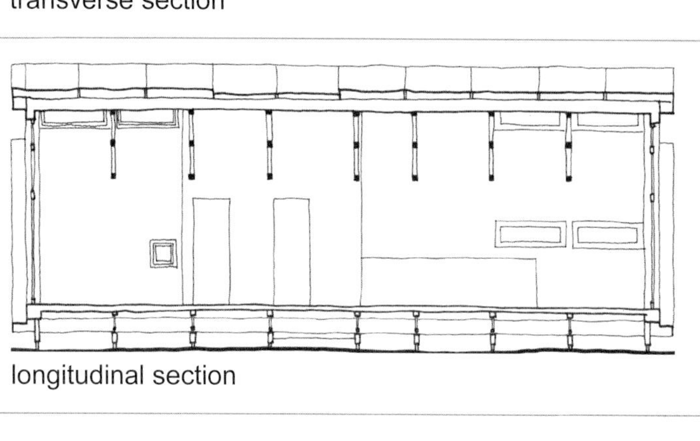

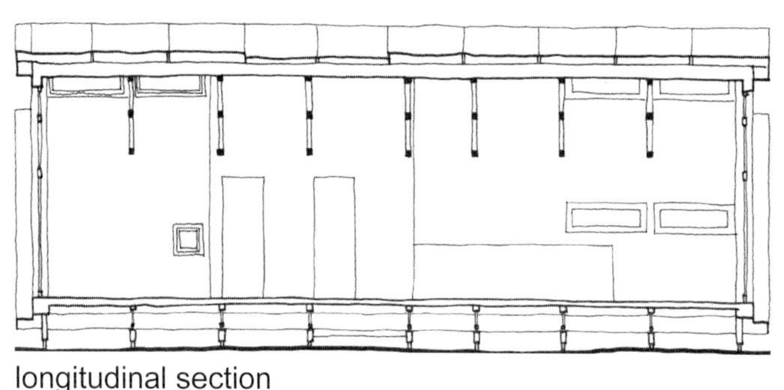

longitudinal section

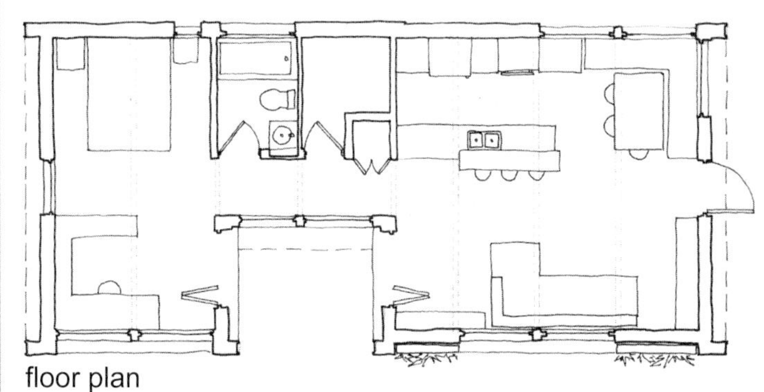

floor plan

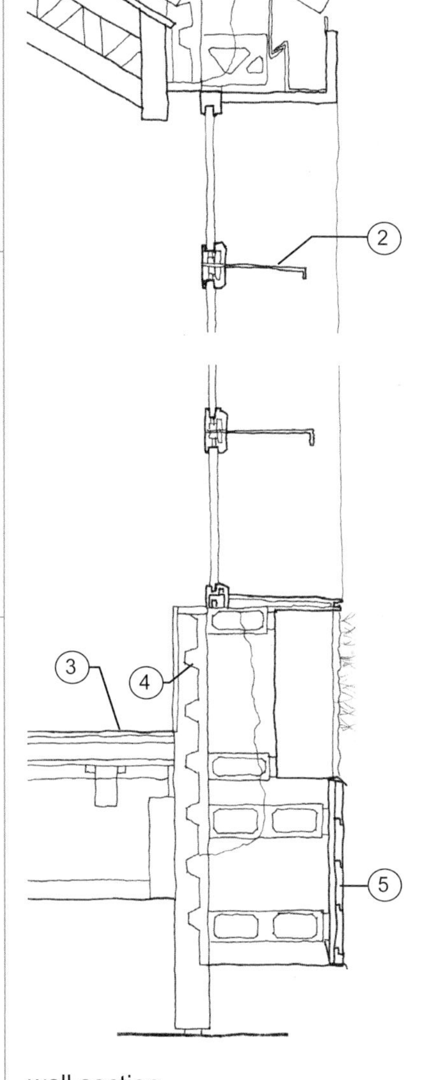

wall section

1. soy urethane insulation
2. shading devices: 1/8" steel channel
3. radiant heating system
4. steel deck
5. exterior wood cladding

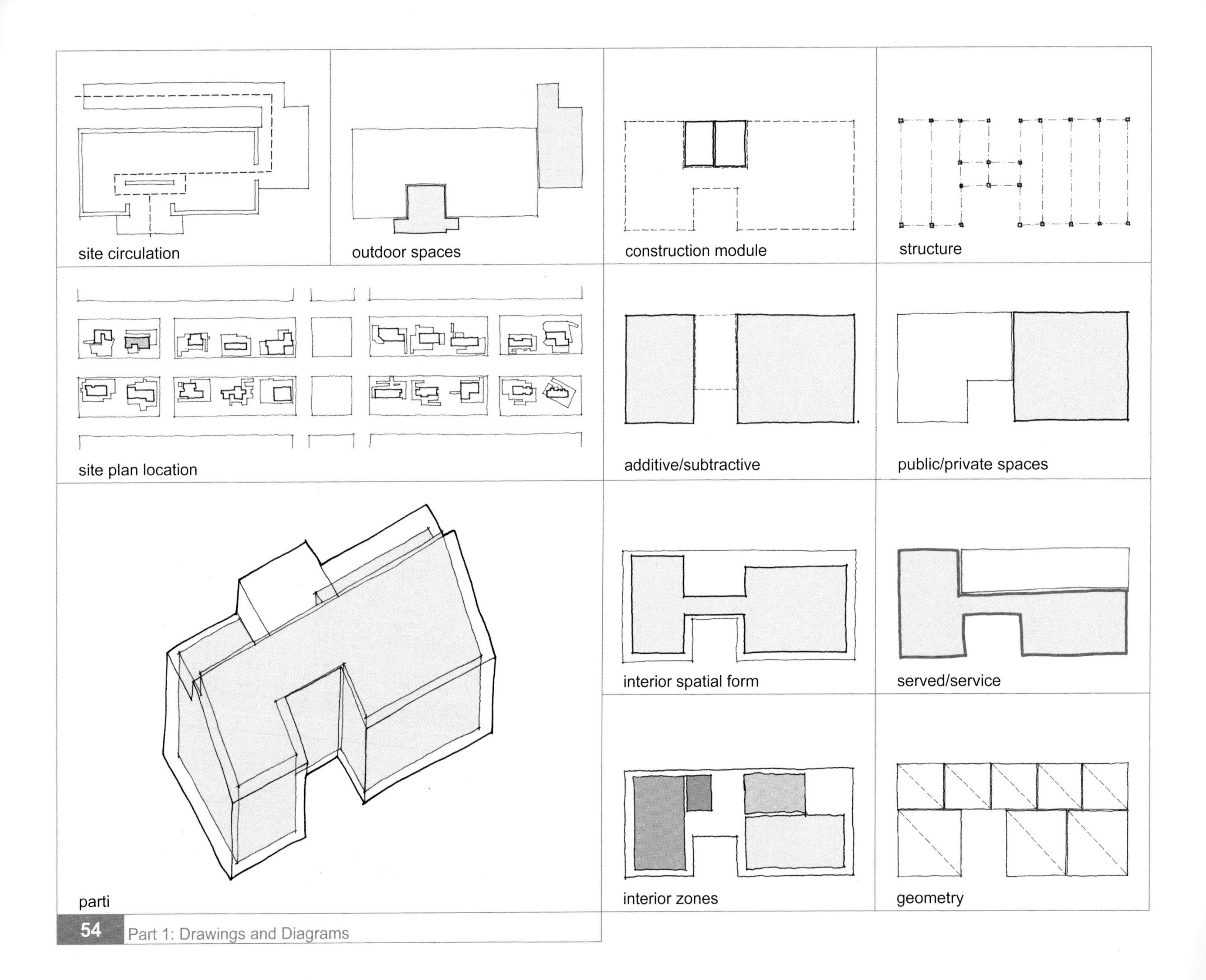
site circulation
outdoor spaces
construction module
structure
site plan location
additive/subtractive
public/private spaces
parti
interior spatial form
served/service
interior zones
geometry

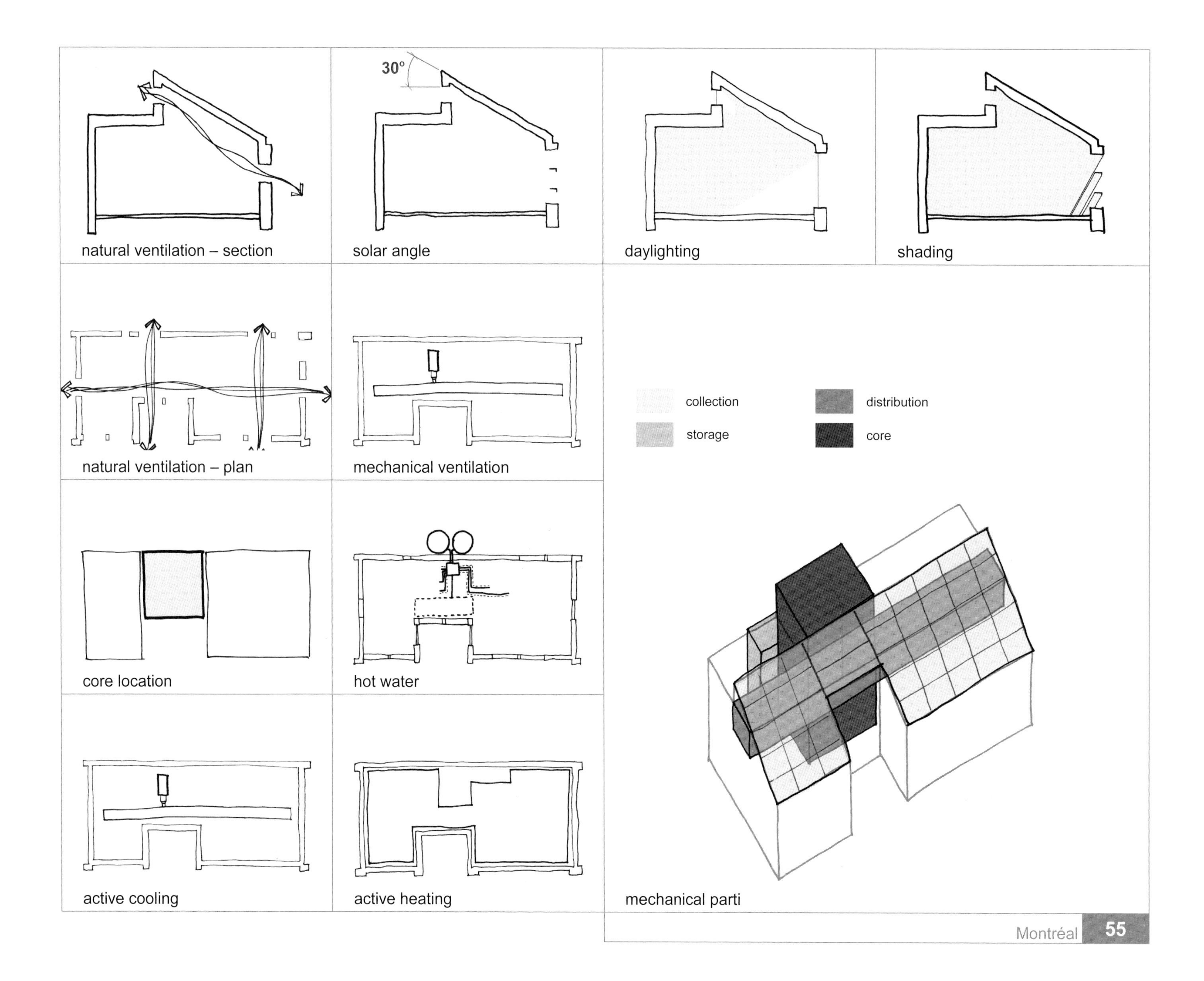

30°
natural ventilation – section
solar angle
daylighting
shading
natural ventilation – plan
mechanical ventilation
collection
storage
distribution
core
core location
hot water
active cooling
active heating
mechanical parti

09

Element House

The Illinois house design involves the use of a singular, repetitive, spatial "element" repeated in unlimited combinations to provide adaptable and flexible buildings. Each element contains its own living and service components, allowing it to be self-sufficient and adaptable on site to current and future configurations. The built prototype of this design includes three of these "elements" connected in an "S" configuration with one element containing the kitchen and entry, a second containing the bedroom and bathroom and a third acting as a living and office space. The intention is for these units to be mass produced and shipped to provide relief for sites in need of temporary housing, mobile offices and command centers.

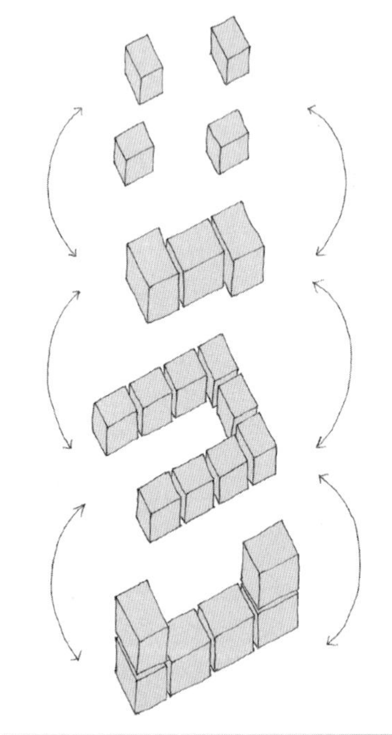

img #17

img #18

University of Illinois at Urbana-Champaign

Location: Urbana, IL, USA
Latitude: 40.11° N Longitude: 88.21° W

Overall Ranking: 9
Architecture: 14
Engineering: 11
Market Viability: 1
Communications: 10
Comfort Zone: 1
Appliances: 18
Hot Water: 19
Lighting: 7
Energy Balance: 10
Getting Around: 8

Total Project Cost: $622,000

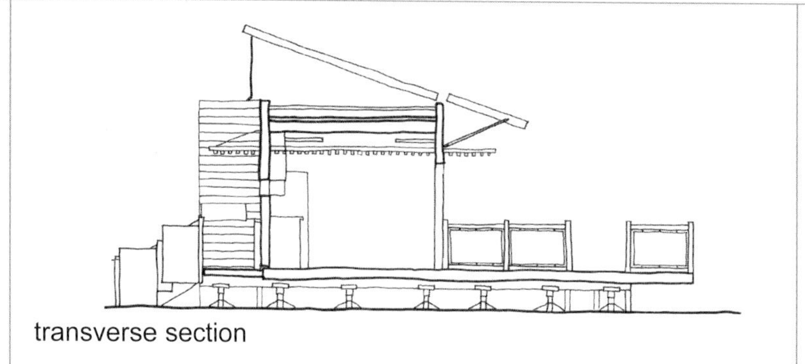
transverse section

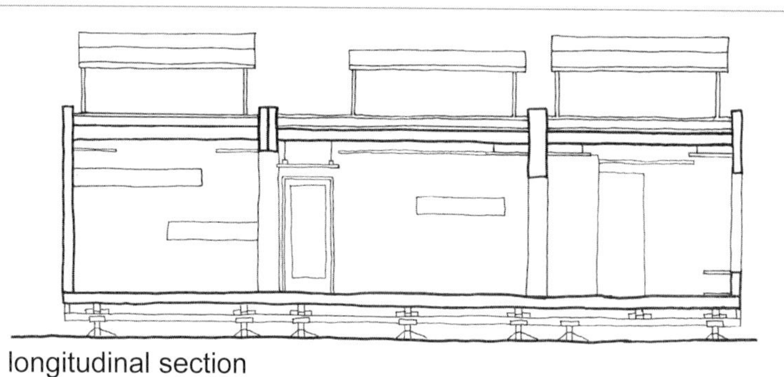
longitudinal section

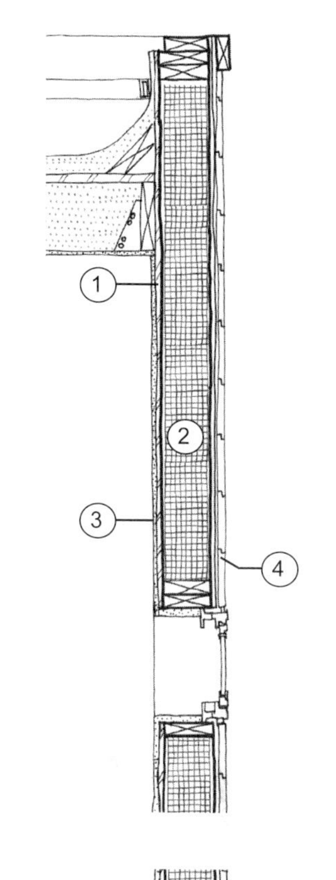

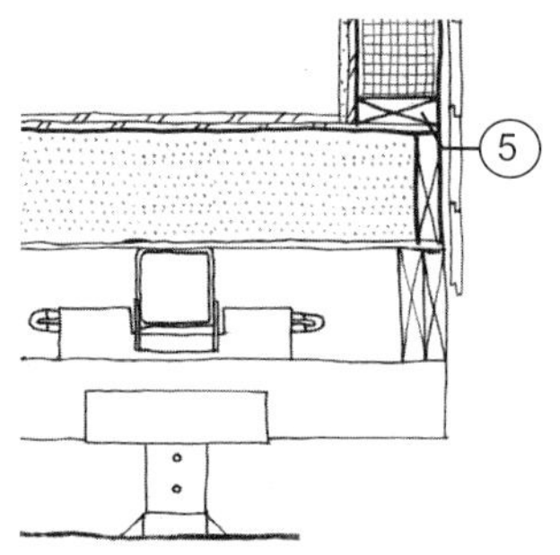

wall section

1. 1/2" oriented strand board
2. 4" foam insulation
3. 1/2" gypsum wall board
4. cedar lap siding
5. 2" x 6" studs

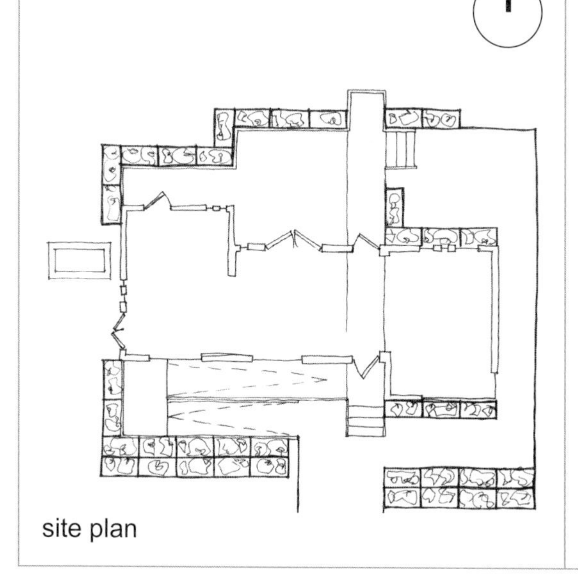
site plan

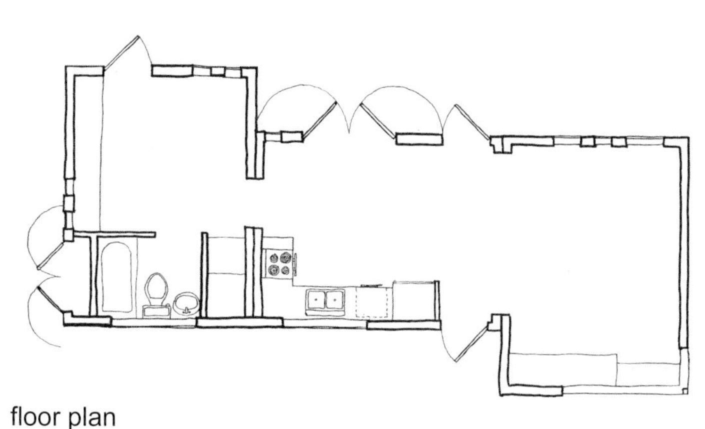
floor plan

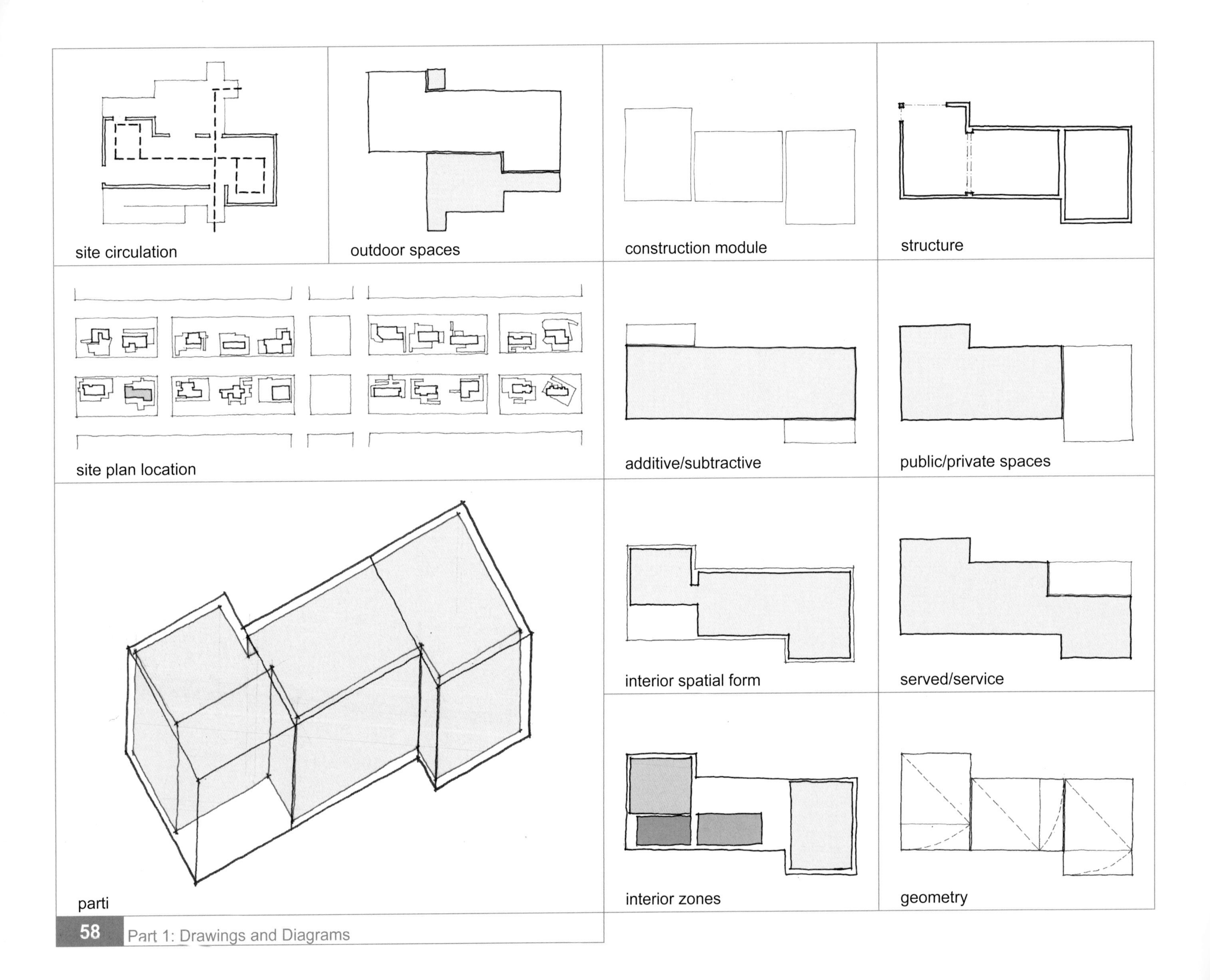
site circulation
outdoor spaces
construction module
structure
site plan location
additive/subtractive
public/private spaces
parti
interior spatial form
served/service
interior zones
geometry

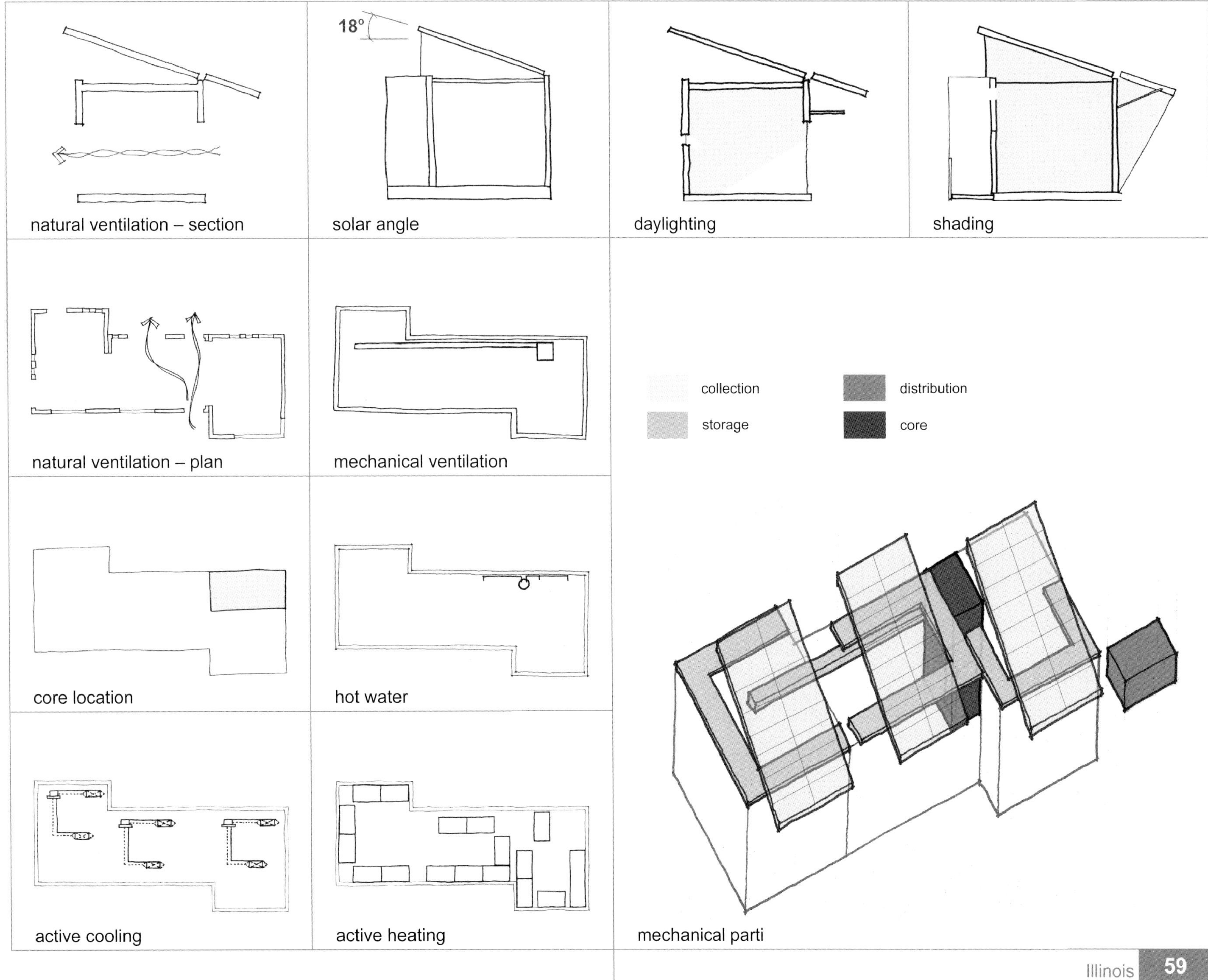
18°
natural ventilation – section
solar angle
daylighting
shading
natural ventilation – plan
mechanical ventilation
collection
distribution
storage
core
core location
hot water
active cooling
active heating
mechanical parti

10

BLOOMhouse

The nature of the UT Austin house is adaptability and dynamism. The house has an architectural "skin" that is flexible in terms of light, heat and ventilation. The outer walls of the north, south, and east facades are adjustable and operable. Windows are operable and shading is adjustable. The outer walls are constructed of adjustable, hinged flaps that can be raised when desired, allowing the house walls to "breathe". Alternatively, the east wall can slide open allowing more light to penetrate the interior. As the east wall opens, the house connects to an outdoor space with a solar-heated hot tub on the deck.

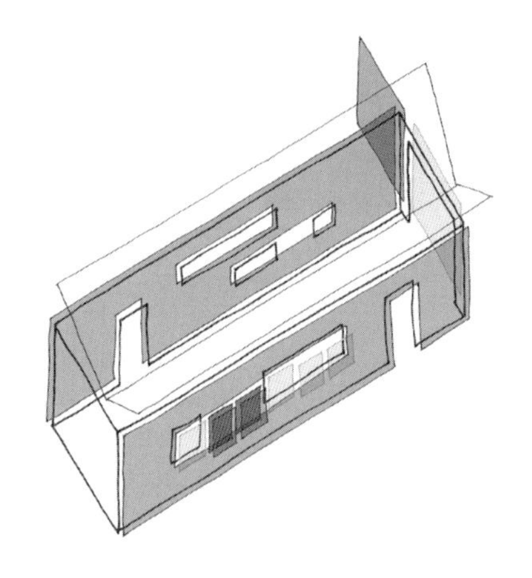

img #19

img #20

University of Texas at Austin

Location: Austin, TX, USA
Latitude: 30.32° N   Longitude: 97.77° W

Overall Ranking: 10
Architecture: 9
Engineering: 2
Market Viability: 10
Communications: 4
Comfort Zone: 3
Appliances: 12
Hot Water: 1
Lighting: 16
Energy Balance: 14
Getting Around: 18

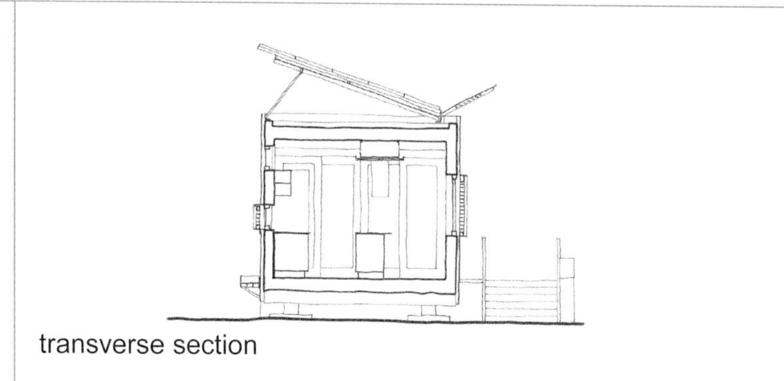

transverse section

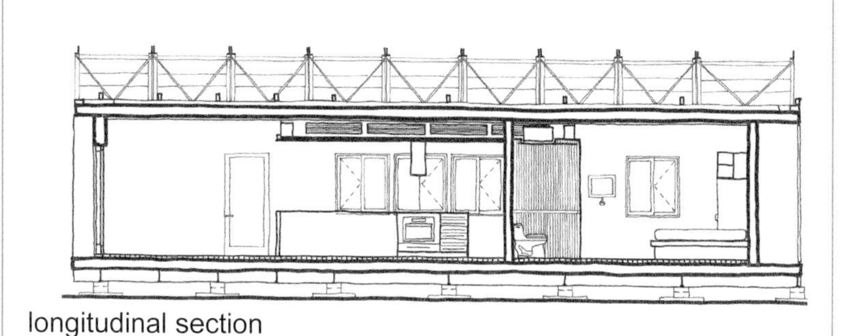

longitudinal section

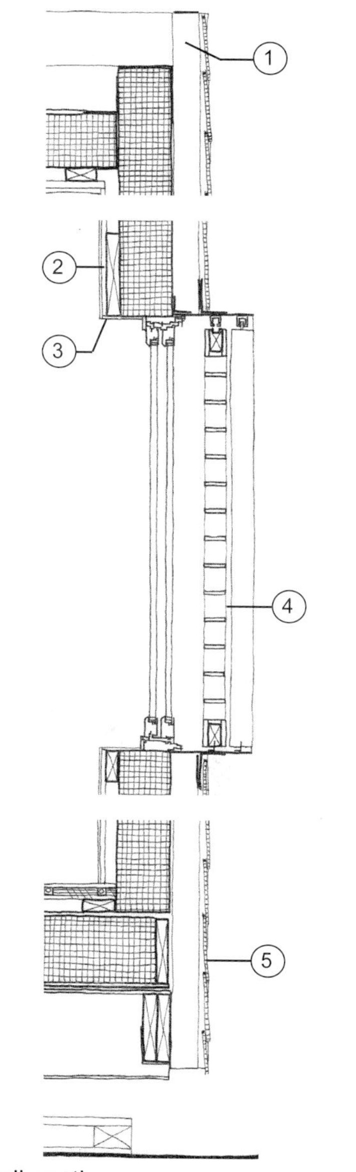

wall section

1. steel frame beyond
2. 1/2" gypsum wall board
3. 1/4" medium-density fiber board
4. sliding sunshade panel
5. polycarbonate panel

Total Project Cost: $450,000

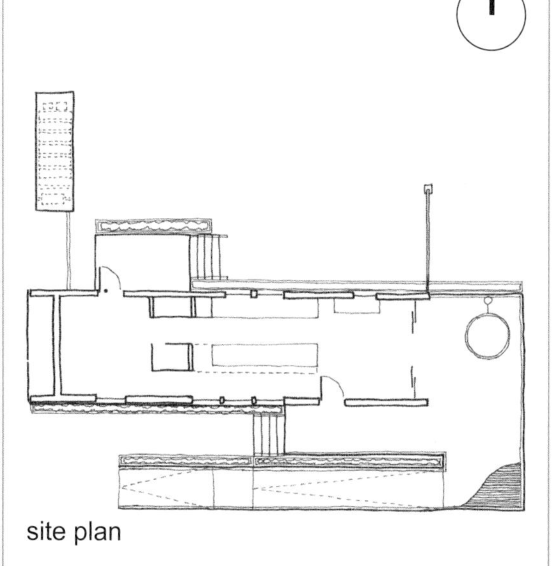

site plan

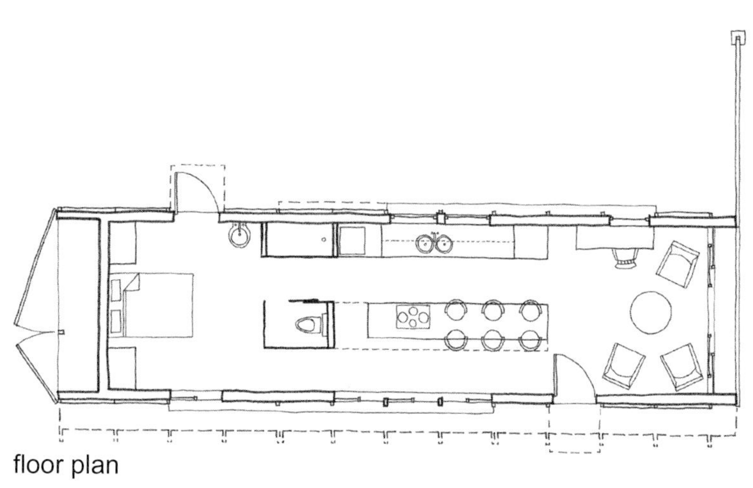

floor plan

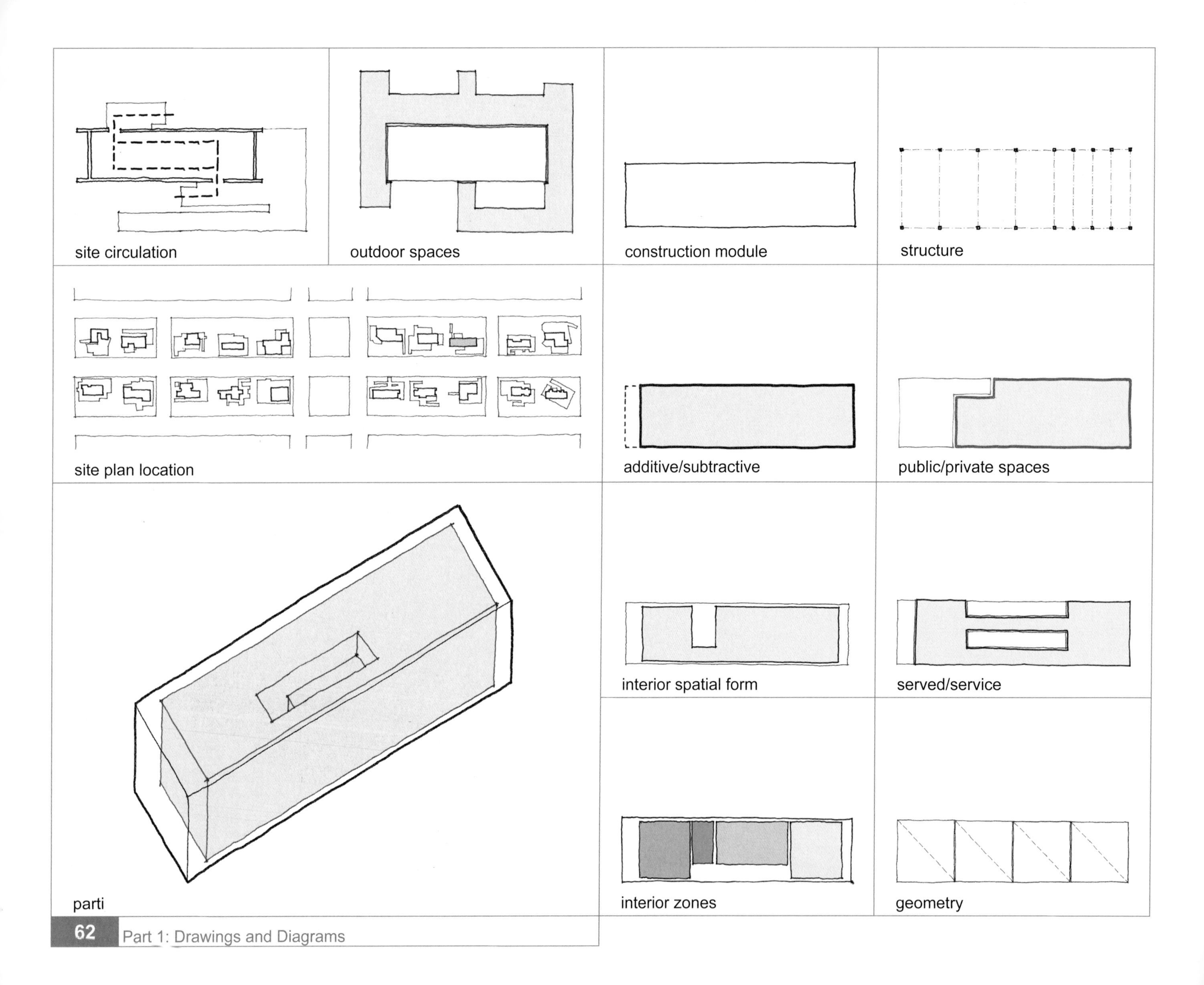
site circulation
outdoor spaces
construction module
structure
site plan location
additive/subtractive
public/private spaces
parti
interior spatial form
served/service
interior zones
geometry

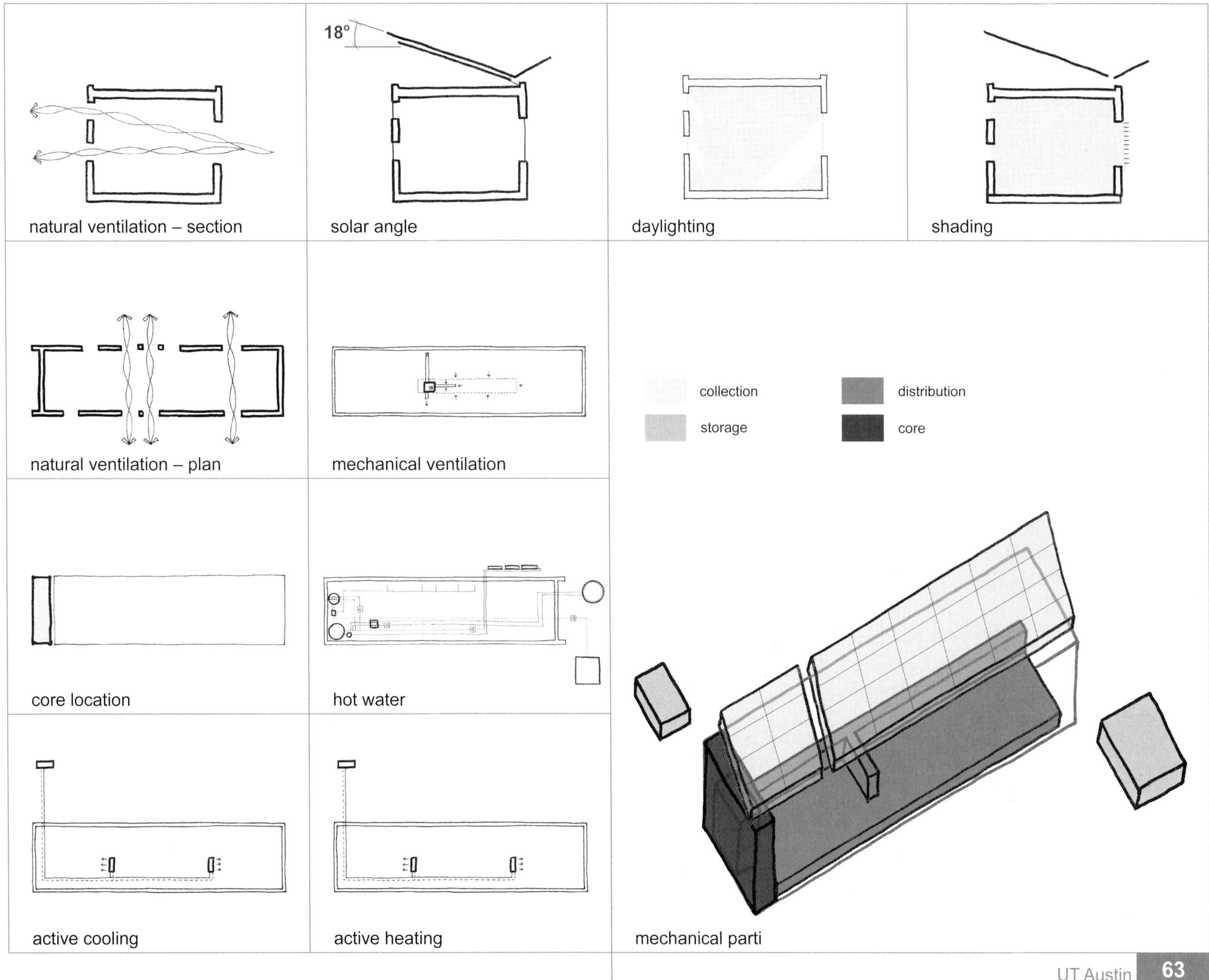
18°
natural ventilation – section
solar angle
daylighting
shading
natural ventilation – plan
mechanical ventilation
collection
distribution
storage
core
core location
hot water
active cooling
active heating
mechanical parti

11

Expanding Horizons

The design of the Missouri-Rolla house is intended to intersect indoor and outdoor spaces. The result of these overlaps includes the apparent intersection of the outdoors with the main living space and a private outdoor space connected directly to the bedroom. A service core divides the entire house into two halves and provides an efficient distribution of services to all areas of the home. For continuity, all of these intersected components of the house are wrapped on the exterior with an architectural "band" of cypress wood stained a crimson color – exemplifying the colors of the region.

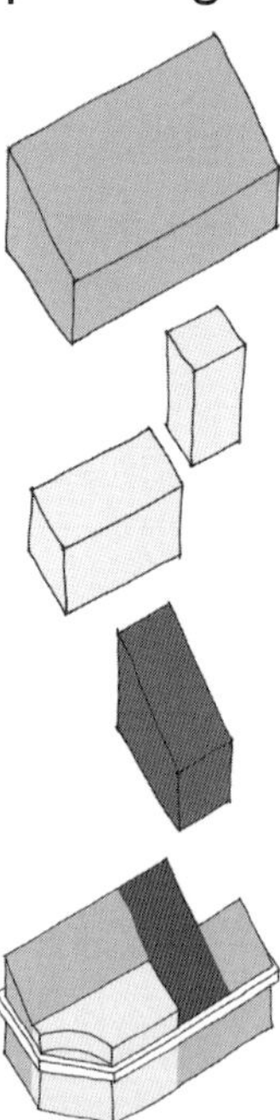

img #21

img #22

University of Missouri-Rolla

Location: Rolla, MO, USA
Latitude: 37.95° N   Longitude: 91.77° W

Overall Ranking: 11
Architecture: 20
Engineering: 14
Market Viability: 4
Communications: 13
Comfort Zone: 6
Appliances: 5
Hot Water: 7
Lighting: 9
Energy Balance: 9
Getting Around: 16

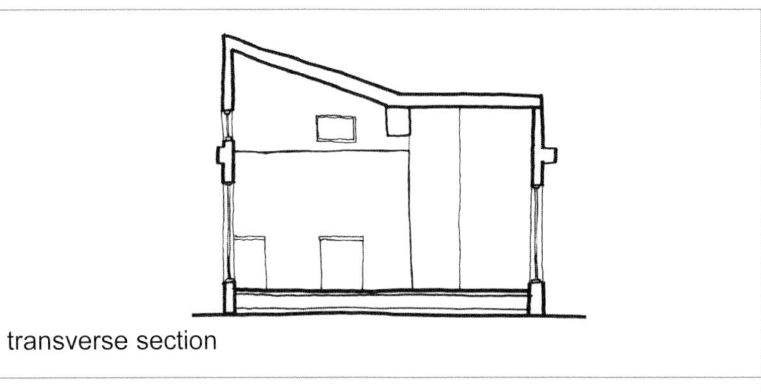

transverse section

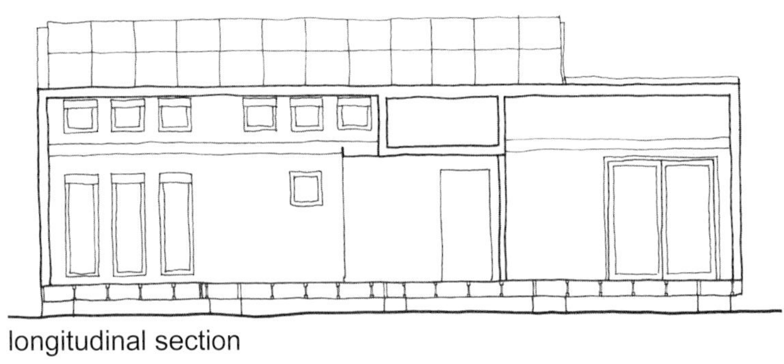

longitudinal section

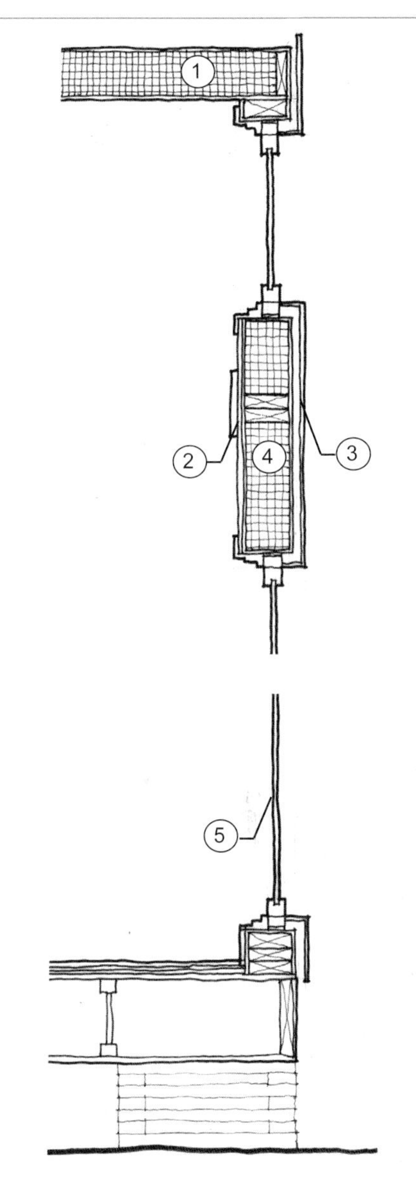

wall section

1. polyurethane foam
2. 1/2" gypsum wall board
3. paper-based exterior board
4. structural insulated panel
5. insulated double glazing

Total Project Cost: $425,000

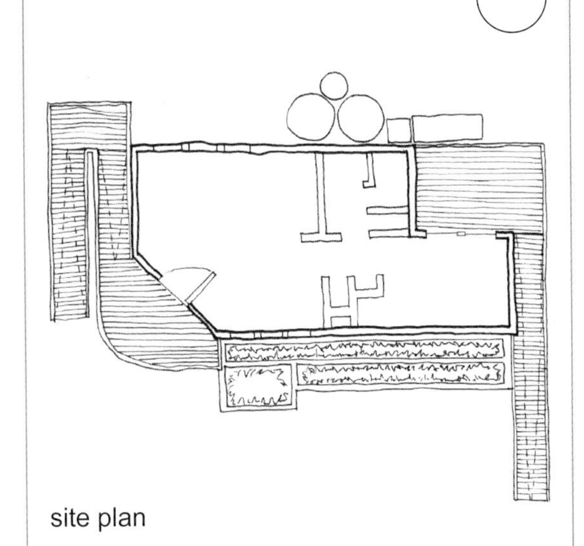

site plan

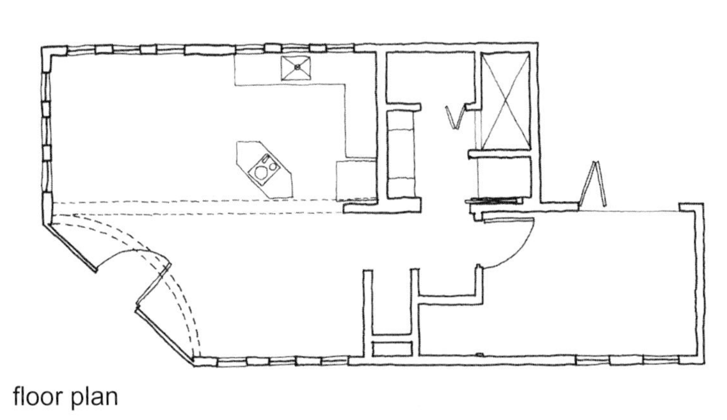

floor plan

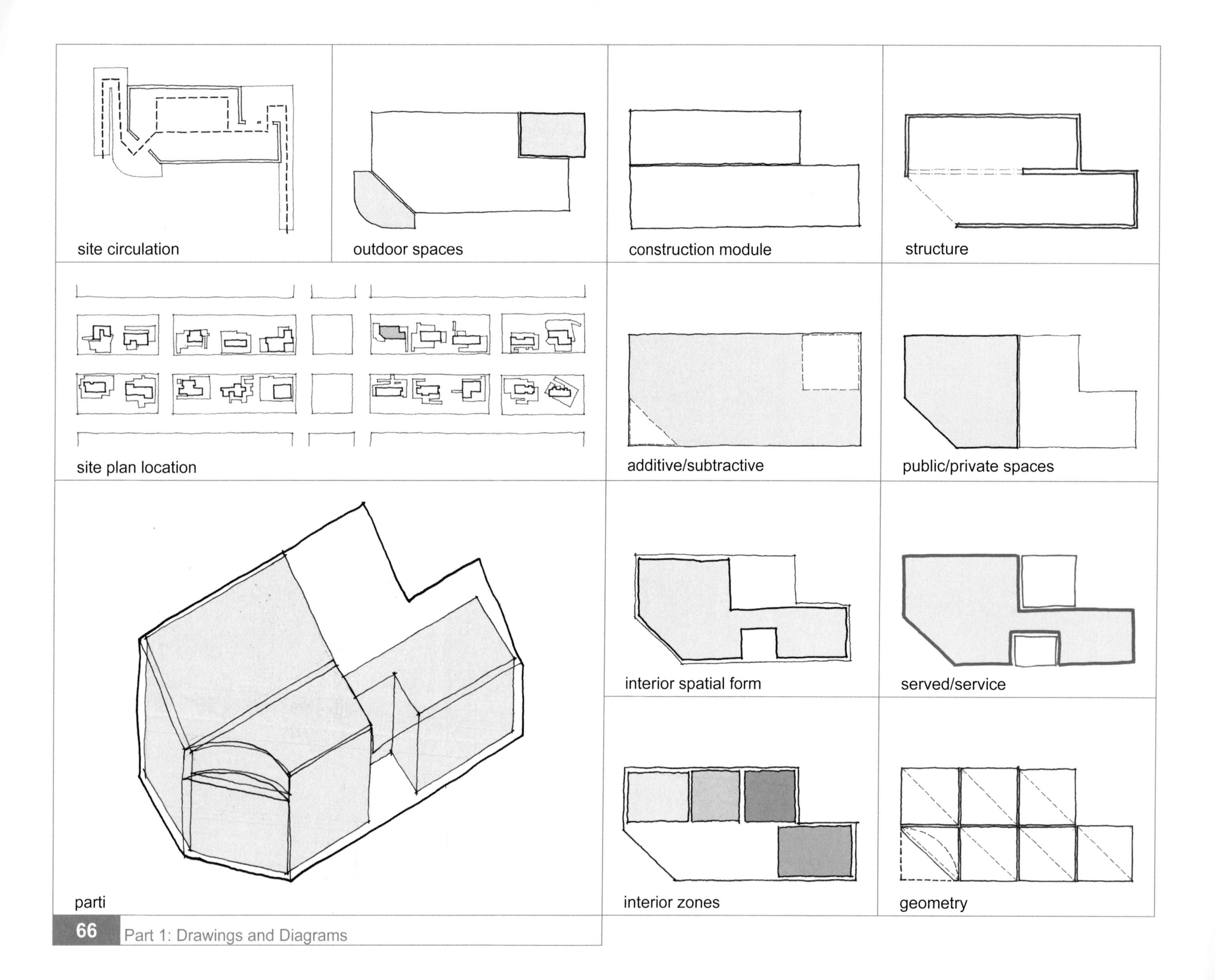
site circulation
outdoor spaces
construction module
structure
site plan location
additive/subtractive
public/private spaces
parti
interior spatial form
served/service
interior zones
geometry

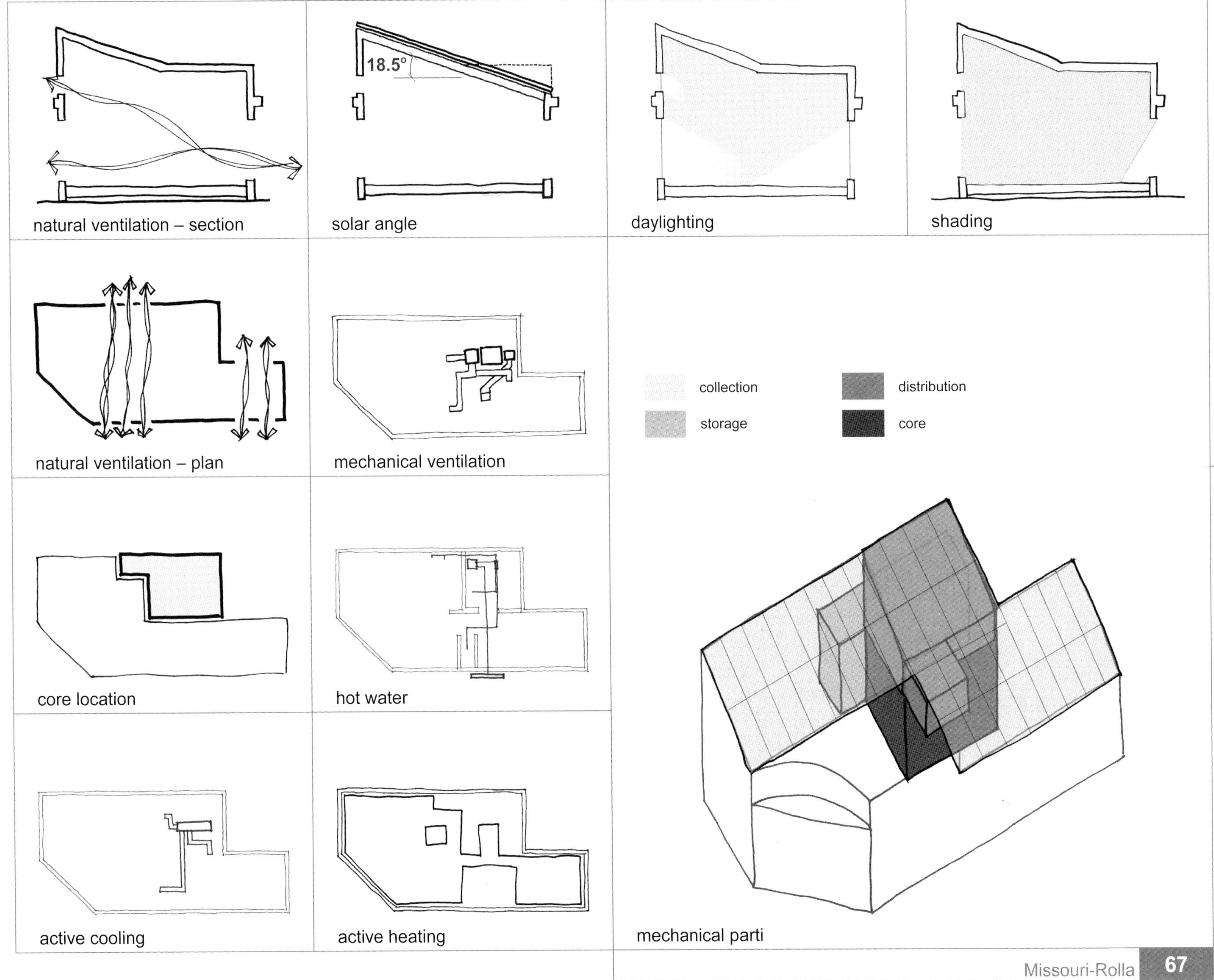
18.5°
natural ventilation – section
solar angle
daylighting
shading
natural ventilation – plan
mechanical ventilation
collection
distribution
storage
core
core location
hot water
active cooling
active heating
mechanical parti

12

OPEN House

The NYIT house combines two main ideas – the conceptual and the spatial. Conceptually, the house is considered open access or "open source" so that various parties can contribute to its construction and design, leading to an evolution through feedback. This occurs over time, as the house is intended to be an open and evolving experiment in sustainable design rather than a final solution. Second, the spatial organization includes two linear elements: one as its core and the other as a main living space that is a single, open area designed to encourage diverse and adaptable living situations.

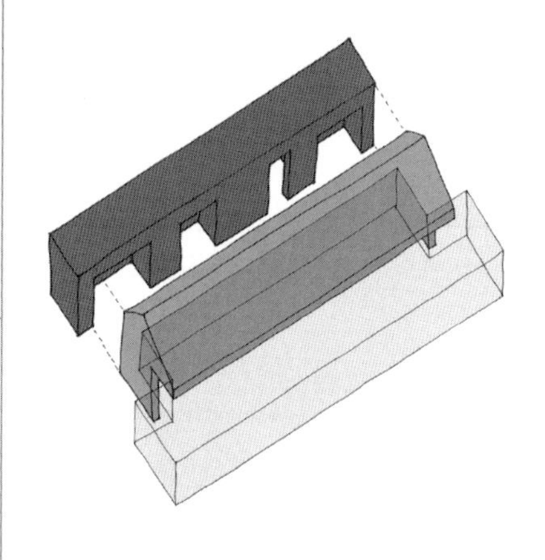

img #23

img #24

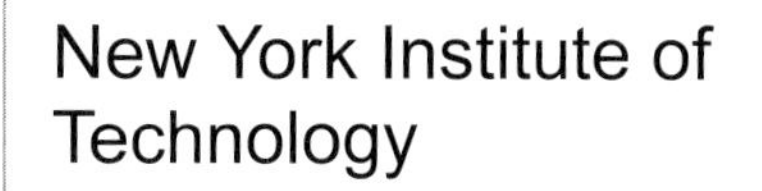

New York Institute of Technology

Location: New York, NY, USA
Latitude: 40.78° N Longitude: 73.97° W

Total Project Cost: $400,000

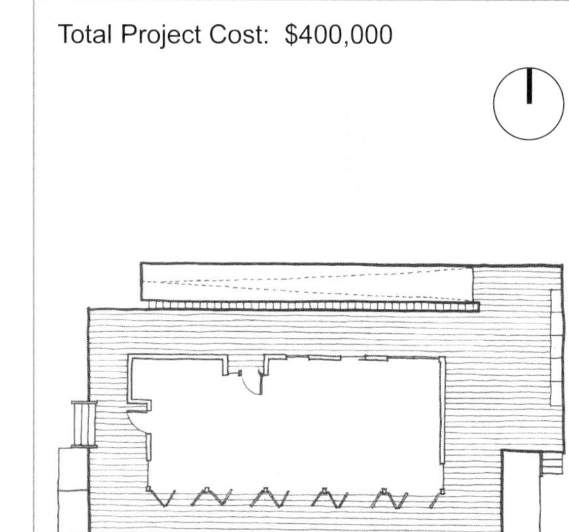

site plan

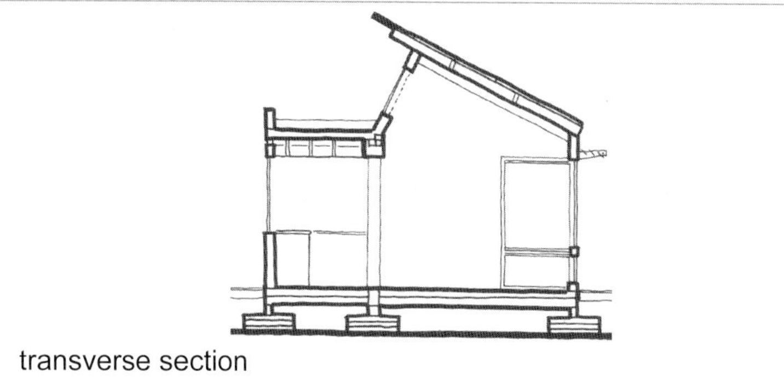

transverse section

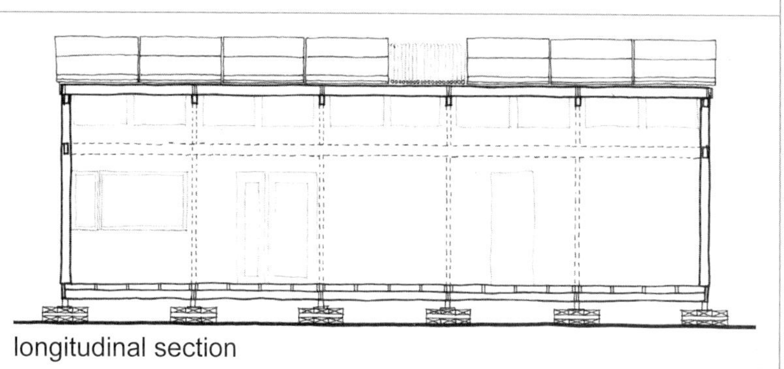

longitudinal section

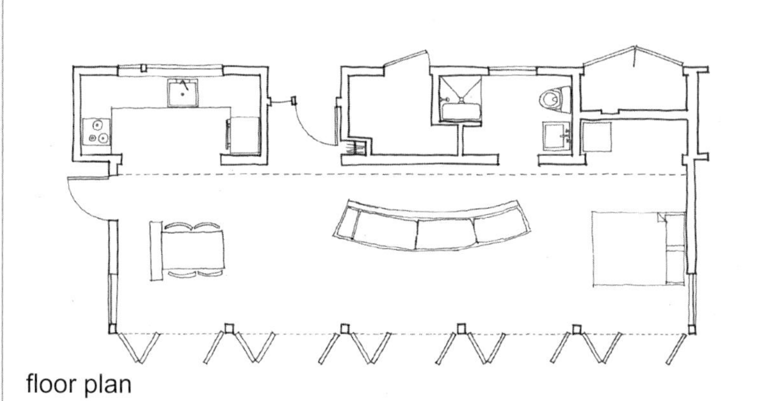

floor plan

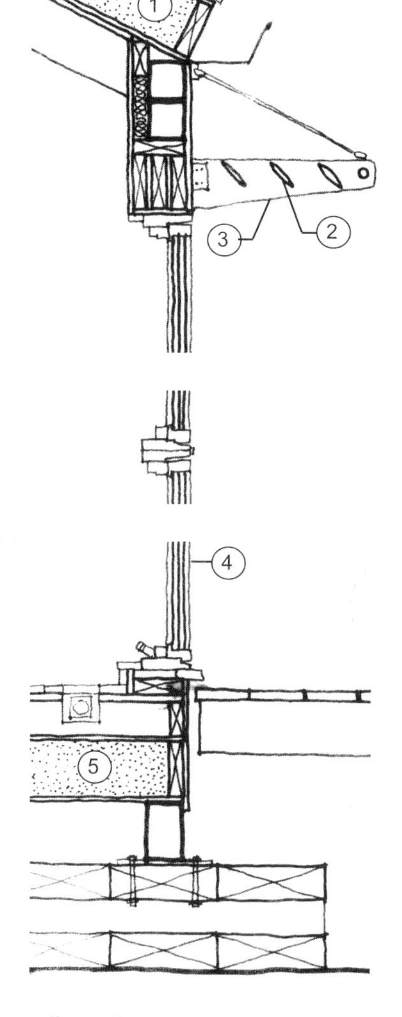

wall section

1. structural insulated panel
2. airfoil louver blade
3. aluminum-plate outrigger
4. operable glazed wall
5. structural insulated panel

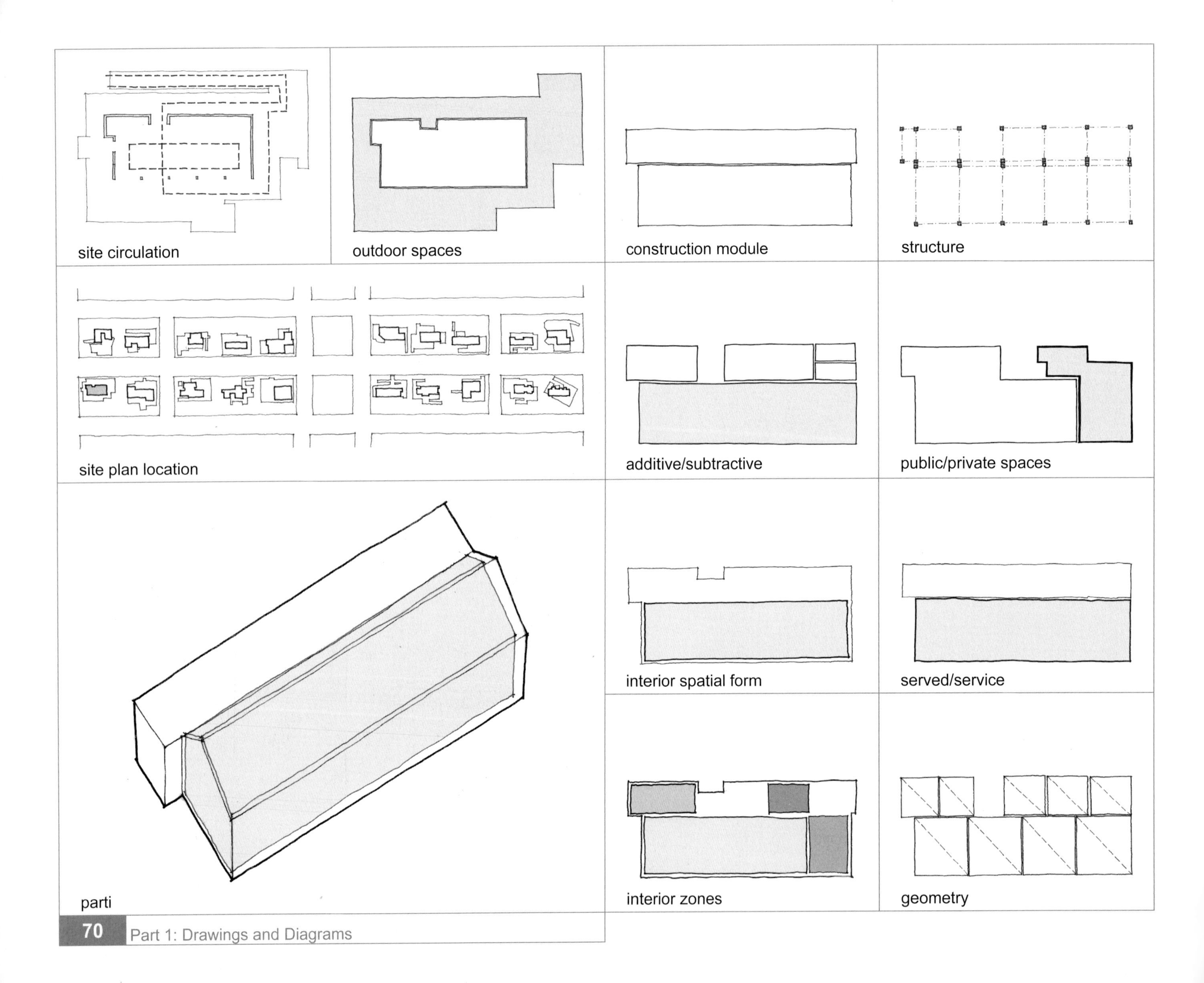
site circulation
outdoor spaces
construction module
structure
site plan location
additive/subtractive
public/private spaces
parti
interior spatial form
served/service
interior zones
geometry

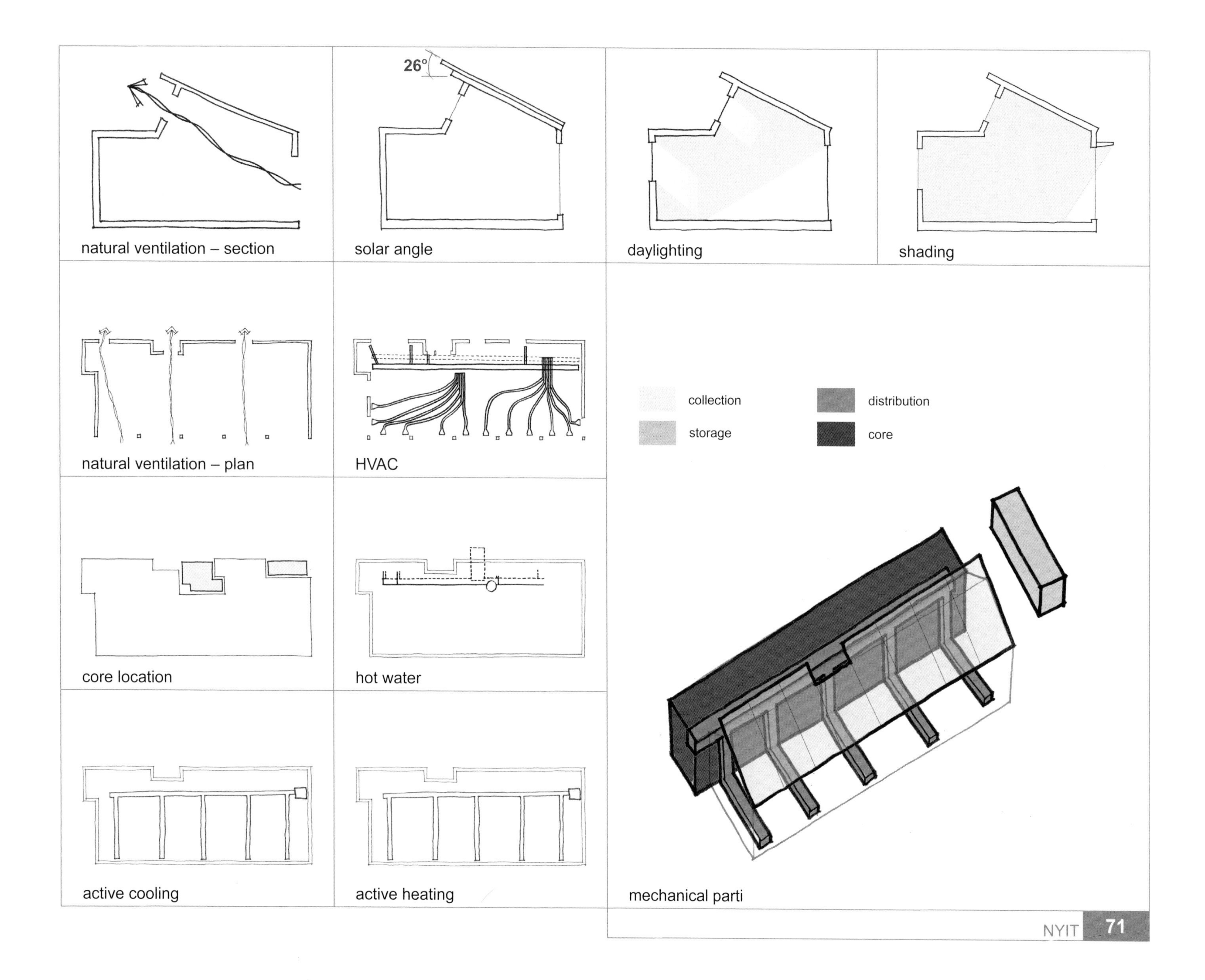
26°
natural ventilation – section
solar angle
daylighting
shading
natural ventilation – plan
HVAC
collection
distribution
storage
core
core location
hot water
active cooling
active heating
mechanical parti

13

The MIT SOLAR7 House is the seventh in a series of solar houses they have developed within the university. The team goals included seeking flexibility in the connection to the outdoors and the use of energy. There are multiple penetrations in the building envelope, allowing ventilation and natural light into the interior spaces. The design incorporates operable clerestory windows that aid in passively ventilating and lighting the interior. A single, large, operable wall in the living space invites the exterior into the interior.

SOLAR7

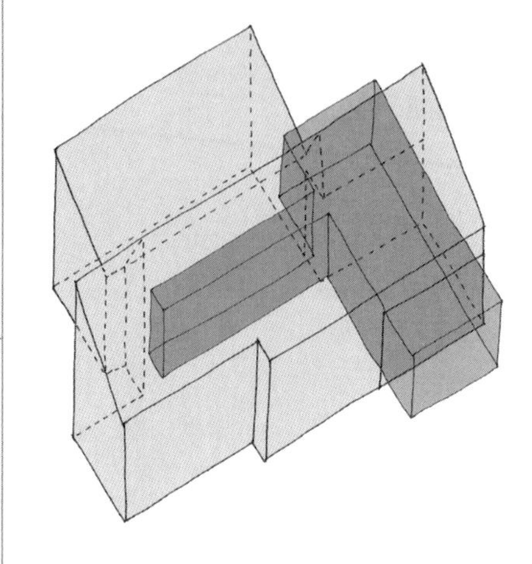

img #25

img #26

Massachusetts Institute of Technology

Location: Cambridge, MA, USA
Latitude: 42.38° N Longitude: 71.11° W

Overall Ranking:	13
Architecture:	19
Engineering:	8
Market Viability:	20
Communications:	19
Comfort Zone:	14
Appliances:	14
Hot Water:	13
Lighting:	18
Energy Balance:	1
Getting Around:	6

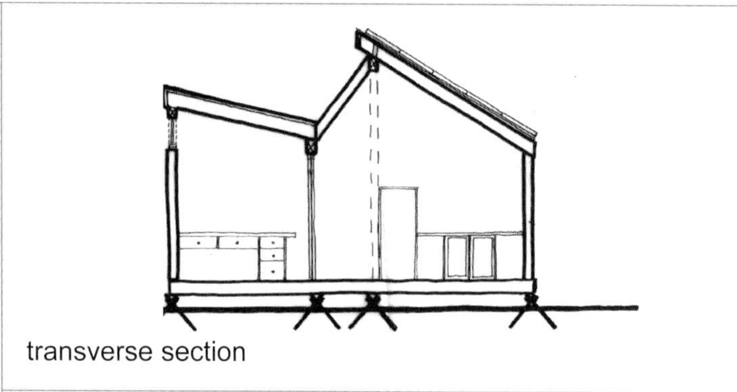

transverse section

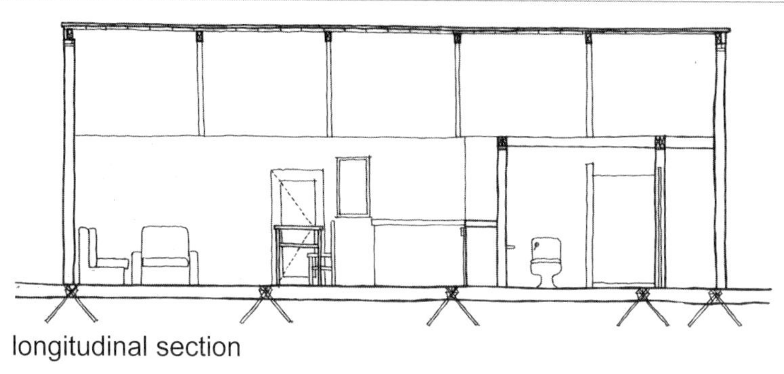

longitudinal section

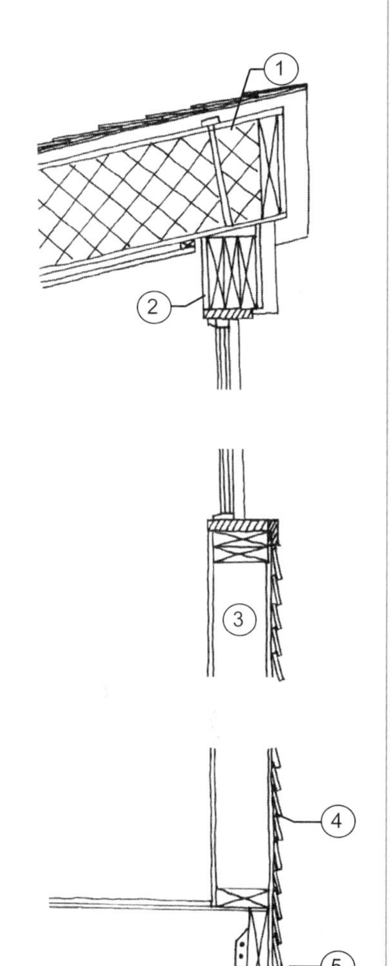

wall section

1. structural insulated roof panel
2. 1/2" gypsum wall board
3. structural insulated panel
4. exterior wood siding
5. joist hanger

Total Project Cost: $270,000

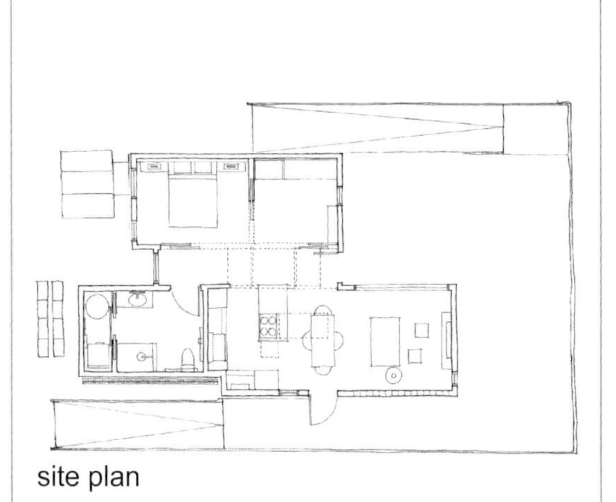

site plan

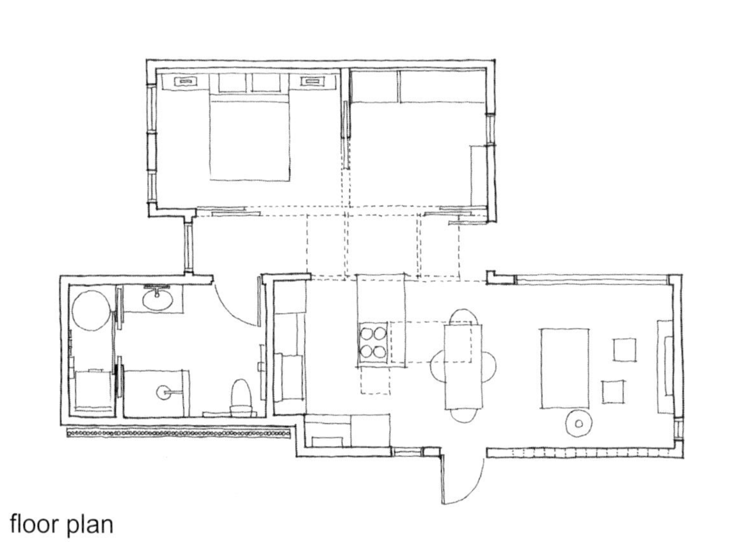

floor plan

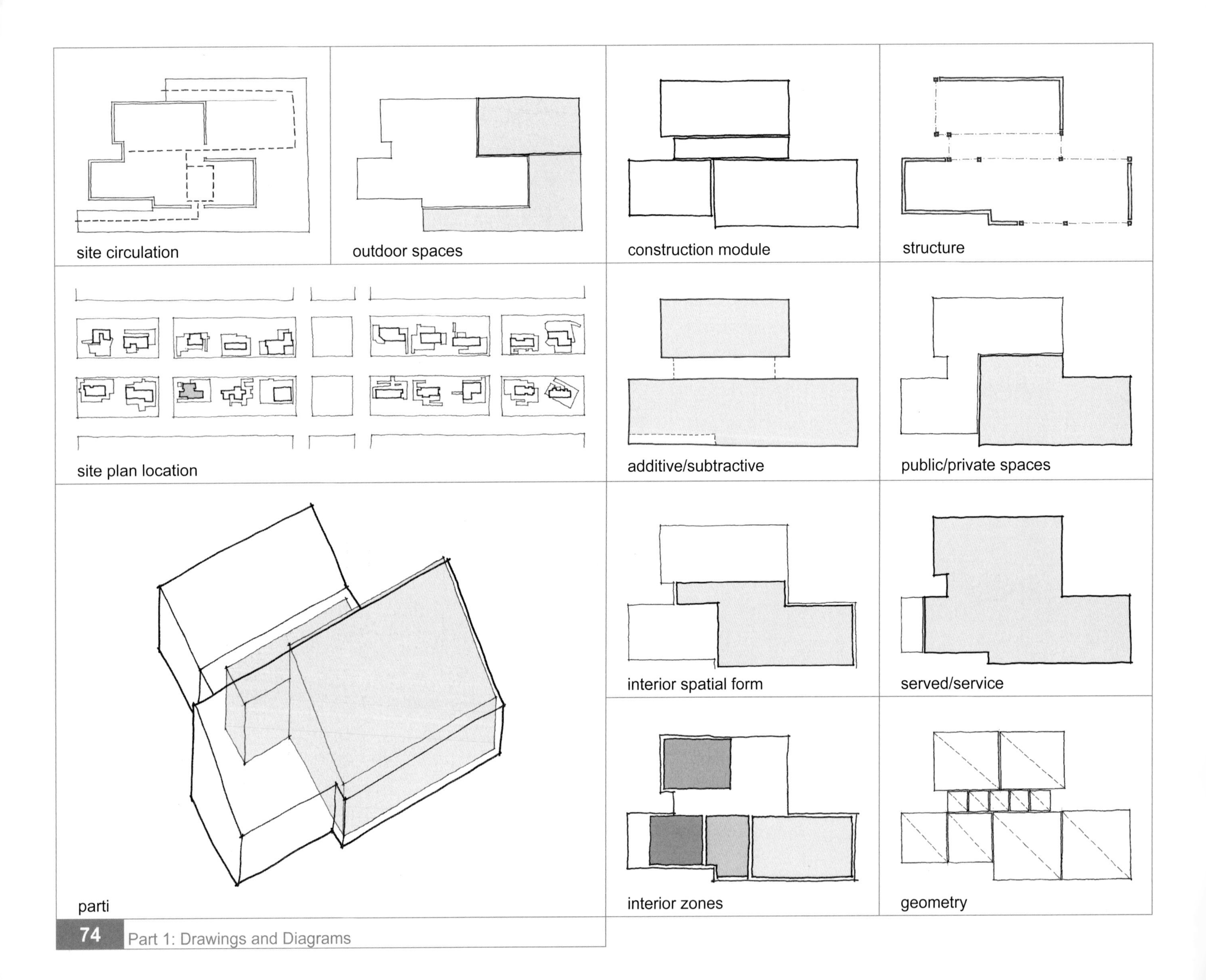
site circulation
outdoor spaces
construction module
structure
site plan location
additive/subtractive
public/private spaces
parti
interior spatial form
served/service
interior zones
geometry

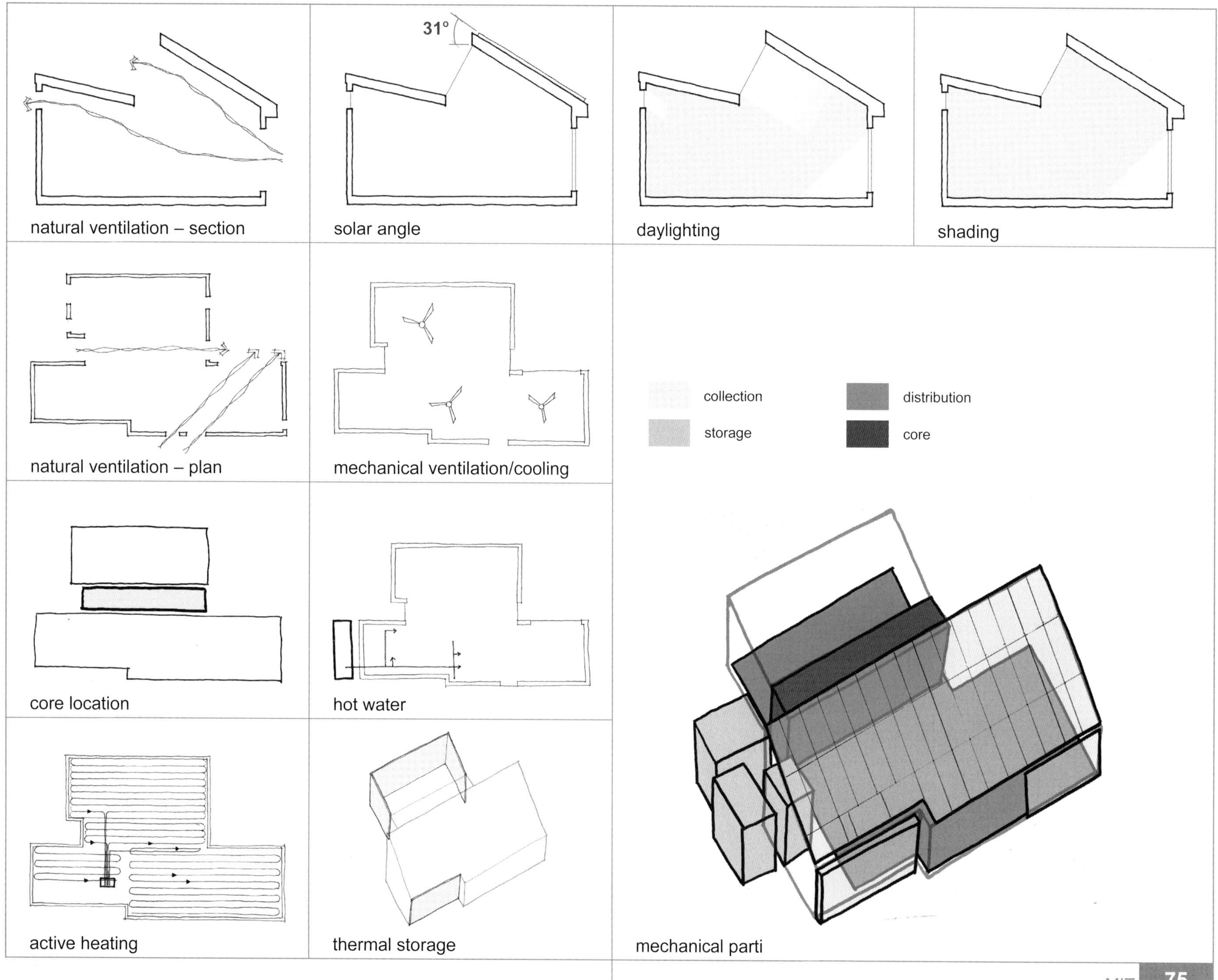
31°
natural ventilation – section
solar angle
daylighting
shading
natural ventilation – plan
mechanical ventilation/cooling
collection
distribution
storage
core
core location
hot water
active heating
thermal storage
mechanical parti

14

The Carnegie Mellon house is one iteration of a flexible modular system. The team adopted a "Plug 'n' Play" approach to design at various scales within all parts of the home. The intended flexibility is achieved with a central spine, which contains all core services for the home and allows different types of units to be "plugged" in to it in varying configurations. This core unit also houses alternative energy systems on its roof and allows for additional systems to be plugged in. Based on a single, consistent module, the house is able to adapt its spatial configuration, its fixtures and its materials and finishes to the users' preferences.

Tri-Pod

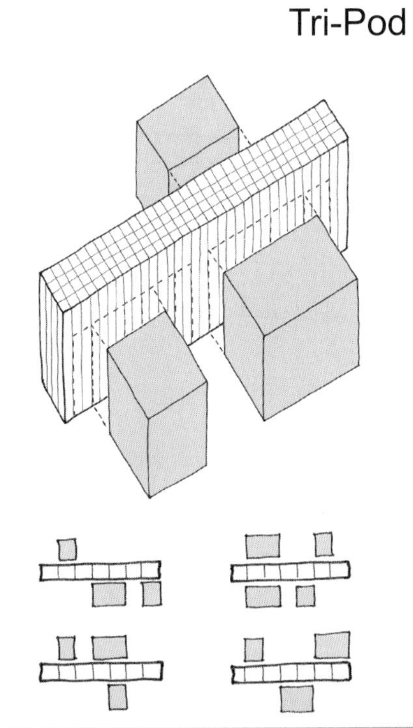

img #27

img #28

Carnegie Mellon University

Location: Pittsburgh, PA, USA
Latitude: 40.45° N Longitude: 79.95° W

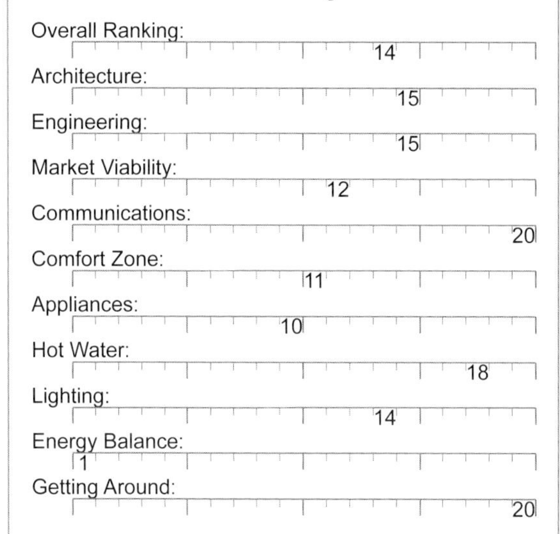

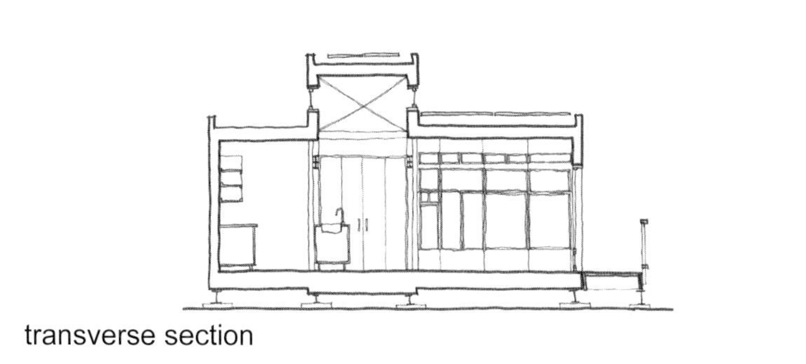

transverse section

longitudinal section

Total Project Cost: $410,000

site plan

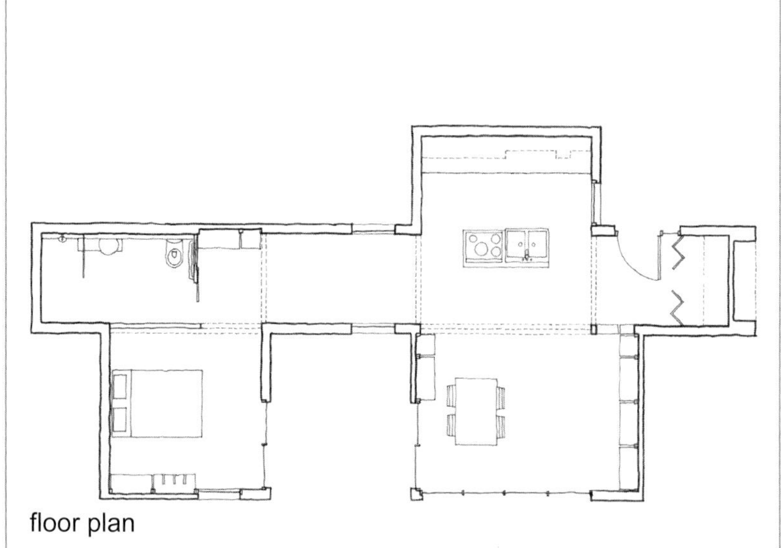

floor plan

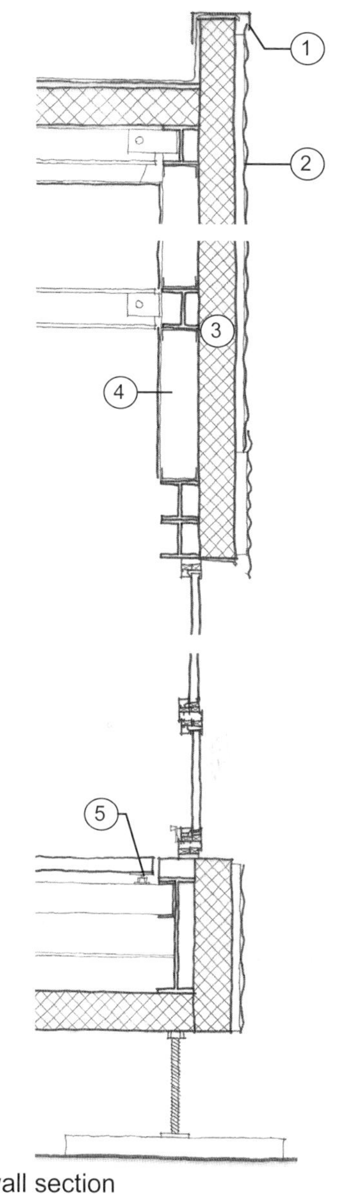

wall section

1. sheet-metal coping
2. rainscreen exterior
3. insulated metal panel
4. light-gauge metal frame
5. steel strut

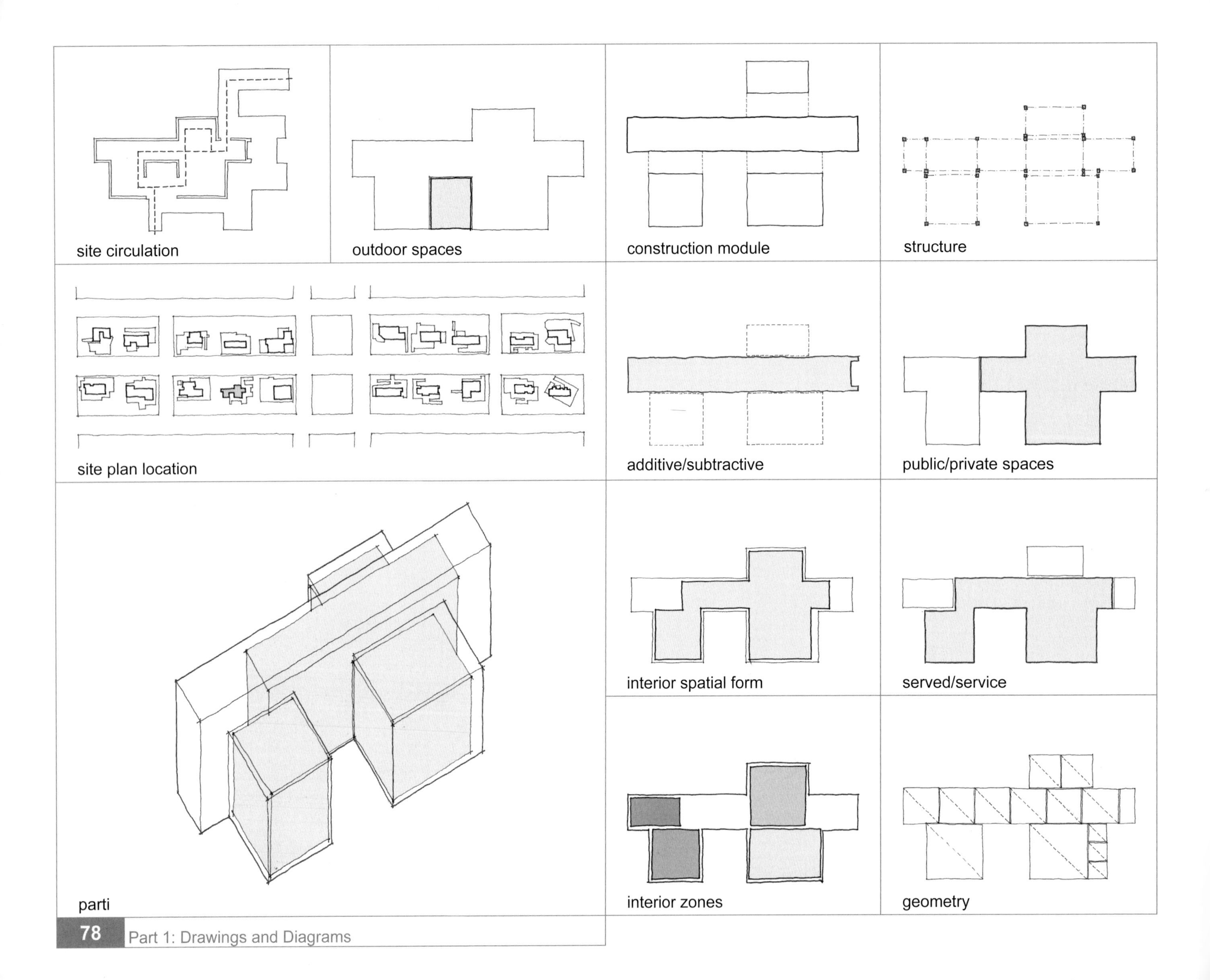
site circulation
outdoor spaces
construction module
structure
site plan location
additive/subtractive
public/private spaces
parti
interior spatial form
served/service
interior zones
geometry

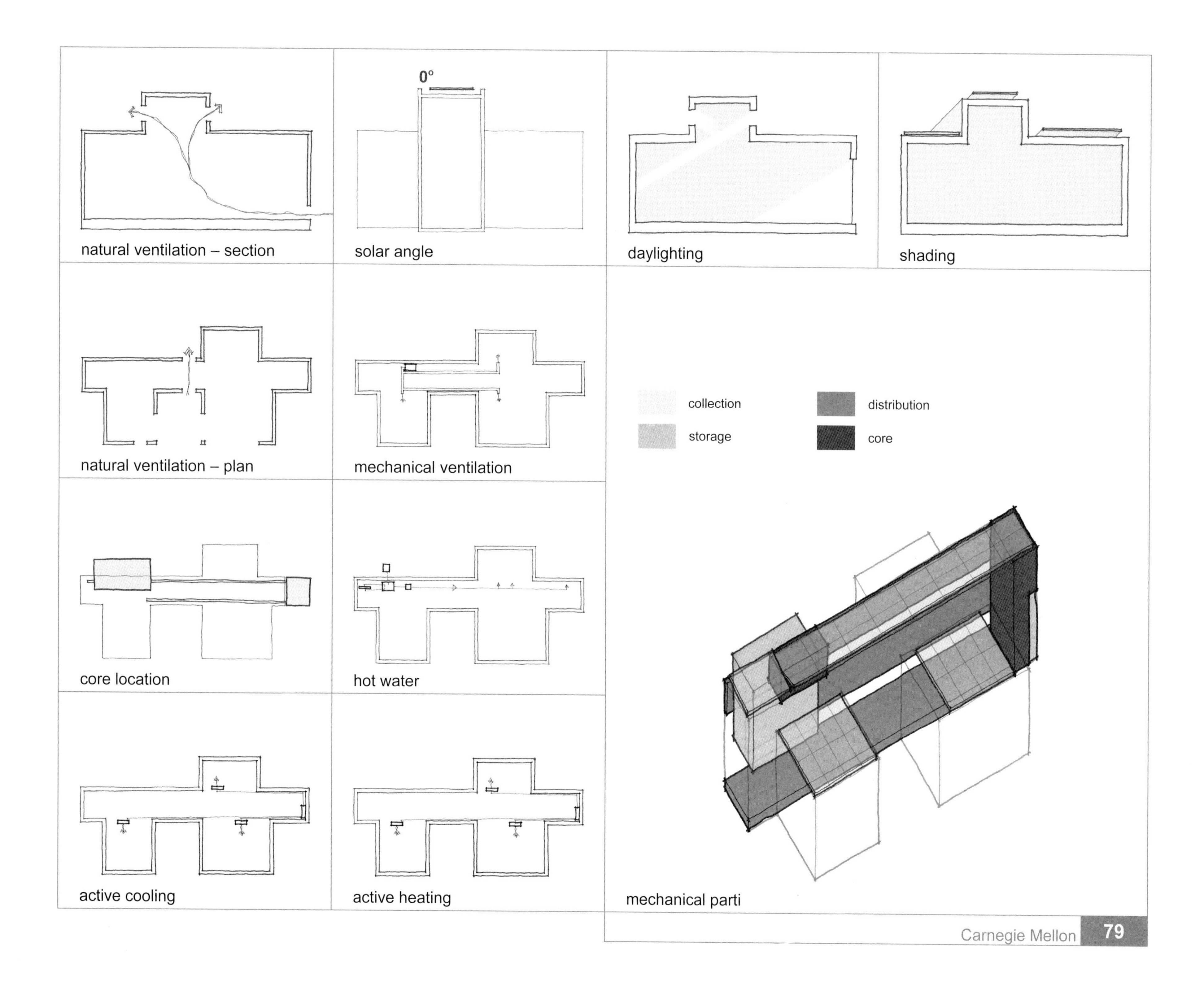
0°
natural ventilation – section
solar angle
daylighting
shading
natural ventilation – plan
mechanical ventilation
collection
storage
distribution
core
core location
hot water
active cooling
active heating
mechanical parti

15

The architectural concept of the Cincinnati house involves the integration of the service systems of the home into the experience of the home. The central spine of the house acts as a collector and distributor of water and energy systems. The transference of the collected energy takes place on the interior of this spine. Incorporated into it are the ventilation systems, plumbing, floor heating and electrical systems. This primary element engages the roof, the walls and the interior partitions of the home, allowing the occupants to experience these systems effectively as part of their dwelling experience.

[re]form house

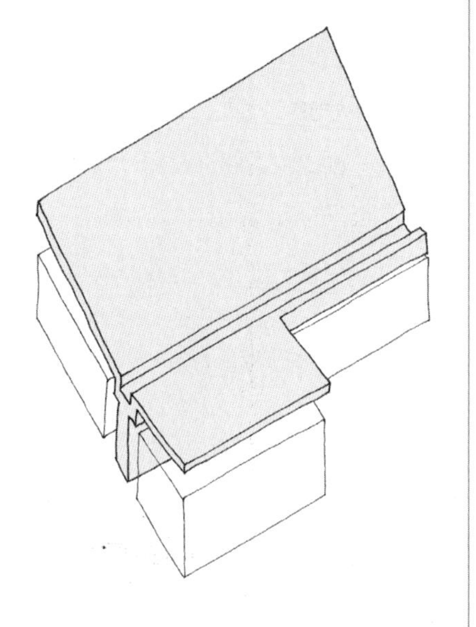

img #29

img #30

University of Cincinnati

Location: Cincinnati, OH, USA
Latitude: 39.14° N Longitude: 84.50° W

Overall Ranking: 15
Architecture: 5
Engineering: 18
Market Viability: 16
Communications: 15
Comfort Zone: 20
Appliances: 19
Hot Water: 9
Lighting: 19
Energy Balance: 1
Getting Around: 17

Total Project Cost: $434,900

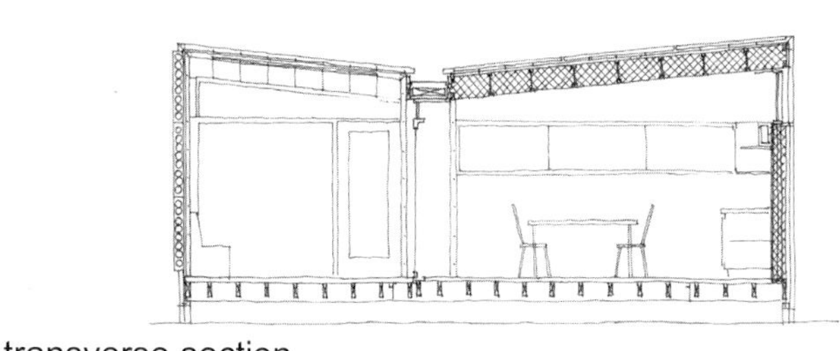

transverse section

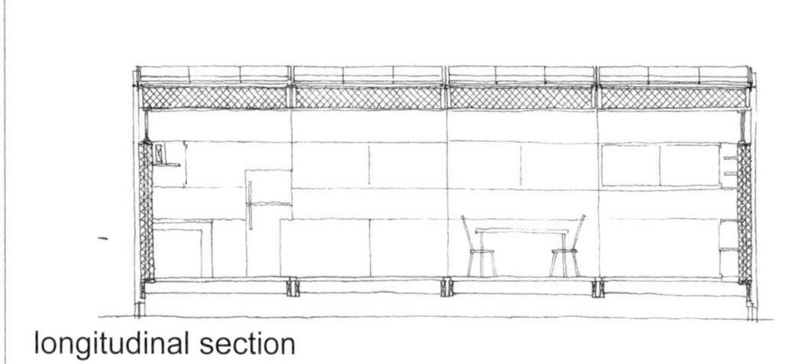

longitudinal section

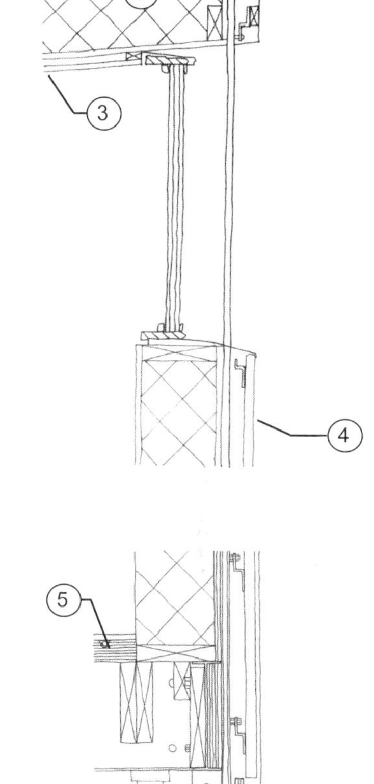

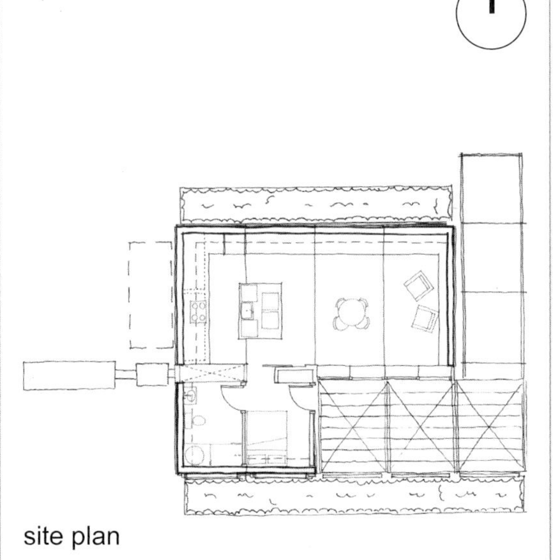

site plan

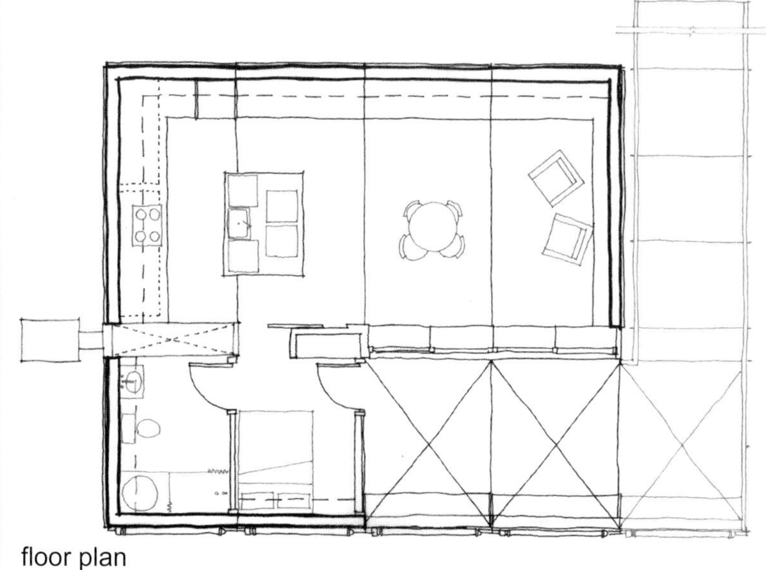

floor plan

wall section

1. metal roofing
2. spray-on foam insulation
3. 1/2" plywood
4. metal-panel rainscreen
5. 1 1/4" plywood radiant subfloor

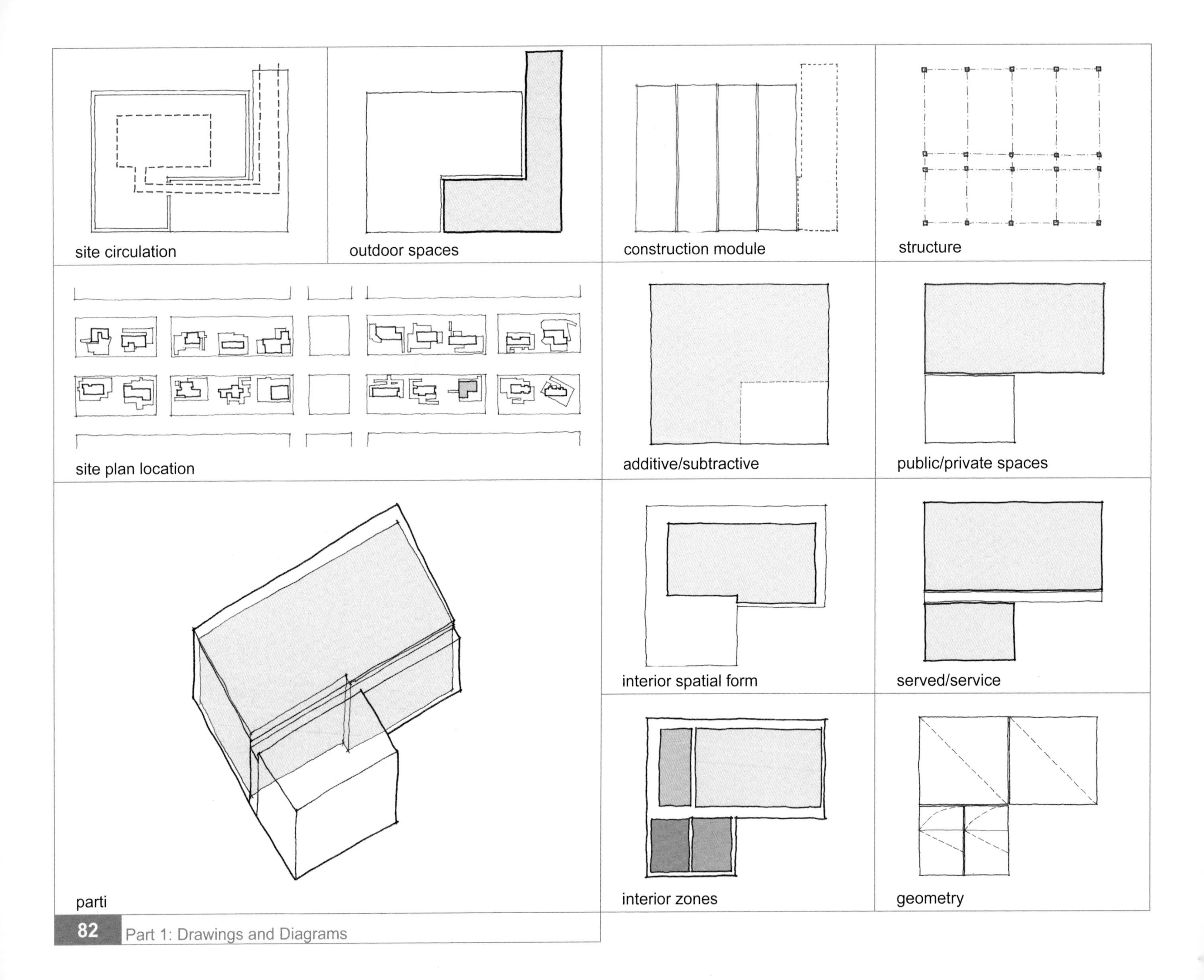
site circulation
outdoor spaces
construction module
structure
site plan location
additive/subtractive
public/private spaces
parti
interior spatial form
served/service
interior zones
geometry

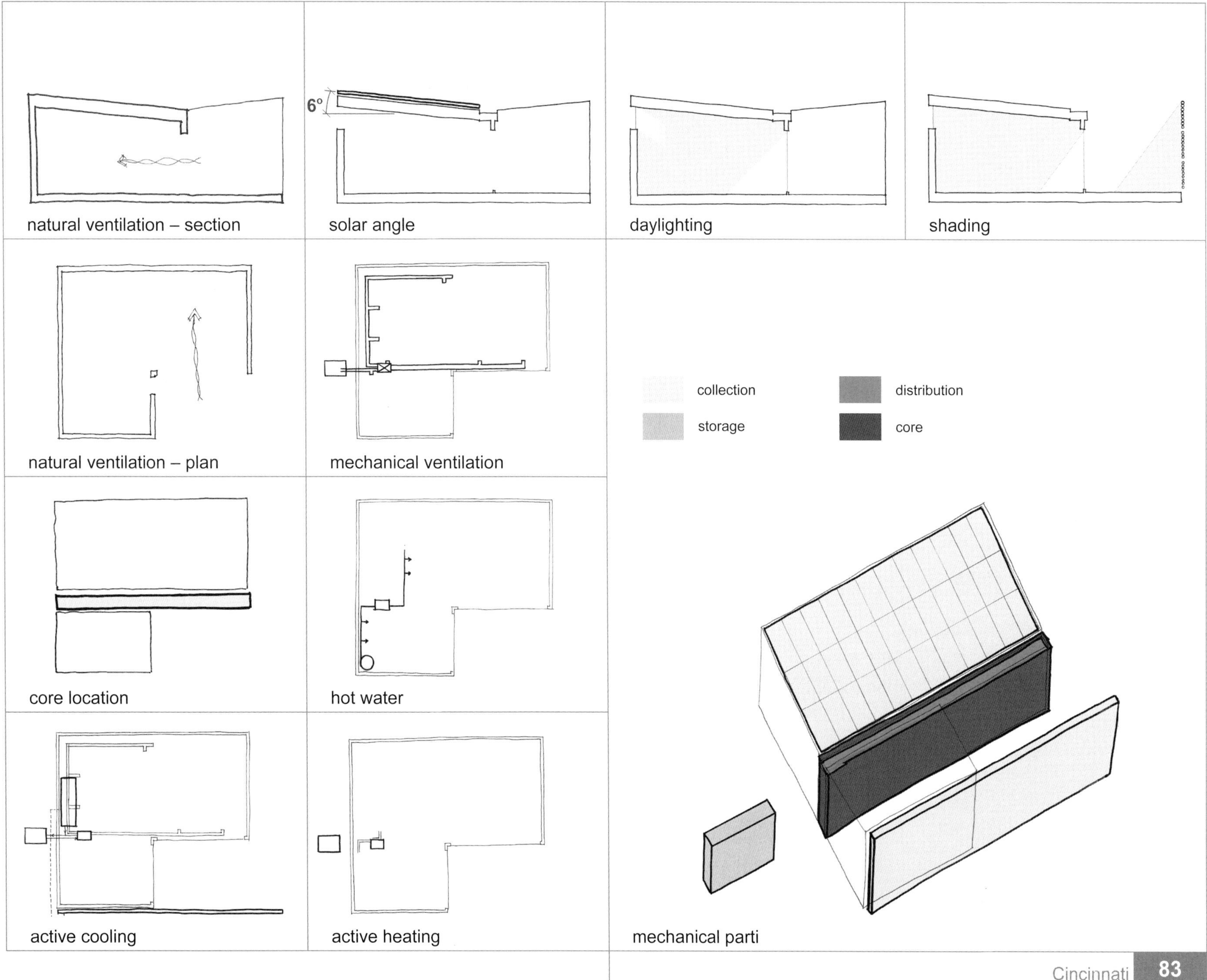
natural ventilation – section
6°
solar angle
daylighting
shading
natural ventilation – plan
mechanical ventilation
collection
distribution
storage
core
core location
hot water
active cooling
active heating
mechanical parti

16

the Cell

The Puerto Rico team was influenced by using "nature as a model for design." They took a metaphorical approach to the spatial organization and architectonic language of their house, deriving them from the "cell." This approach emphasized the plant cell and its central location in order to influence the service core location as well as the shell of the home, which is flexible in its ability to allow light and ventilation to penetrate the "cell membrane." These elements can be controlled by the occupants of the house in a process which ensures comfort and further emphasizes the cellular nature, as the cell membrane can be opened and closed according to their own needs.

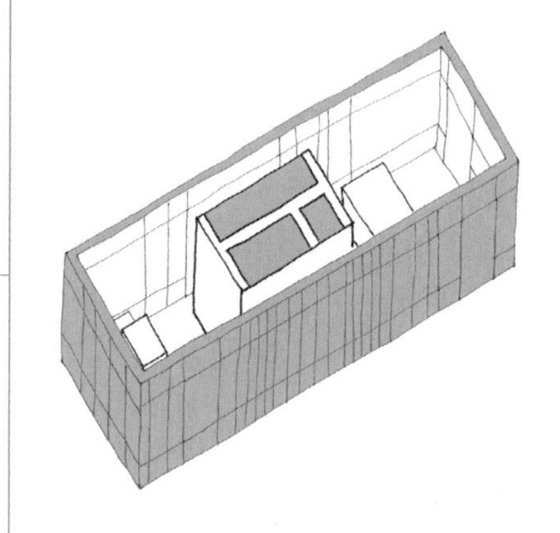

img #31

img #32

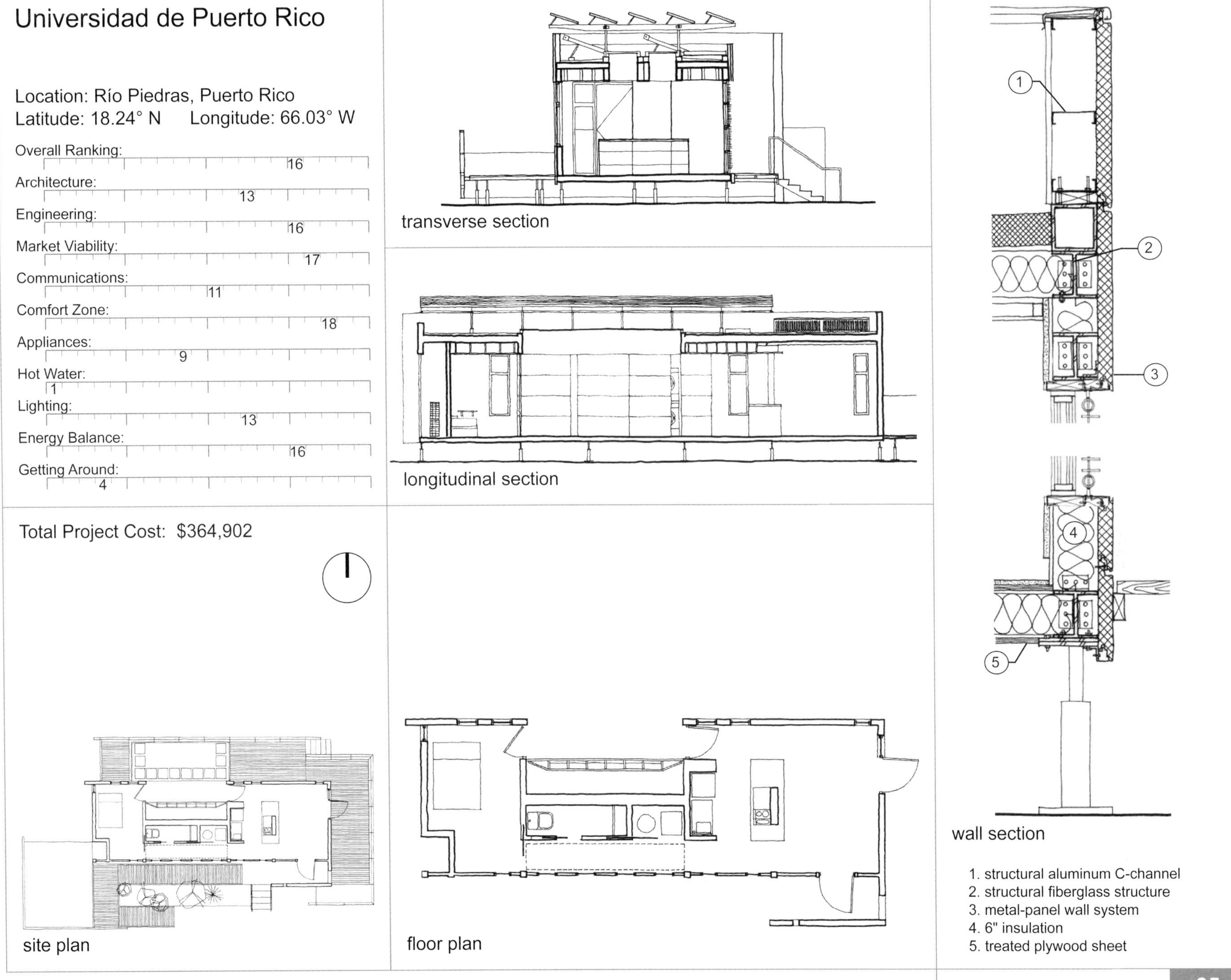

Universidad de Puerto Rico
Location: Río Piedras, Puerto Rico
Latitude: 18.24° N Longitude: 66.03° W
Overall Ranking:
16
Architecture:
13
Engineering:
16
Market Viability:
17
Communications:
11
Comfort Zone:
18
Appliances:
9
Hot Water:
1
Lighting:
13
Energy Balance:
16
Getting Around:
4
Total Project Cost: $364,902
transverse section
longitudinal section
site plan
floor plan
wall section
1. structural aluminum C-channel
2. structural fiberglass structure
3. metal-panel wall system
4. 6" insulation
5. treated plywood sheet

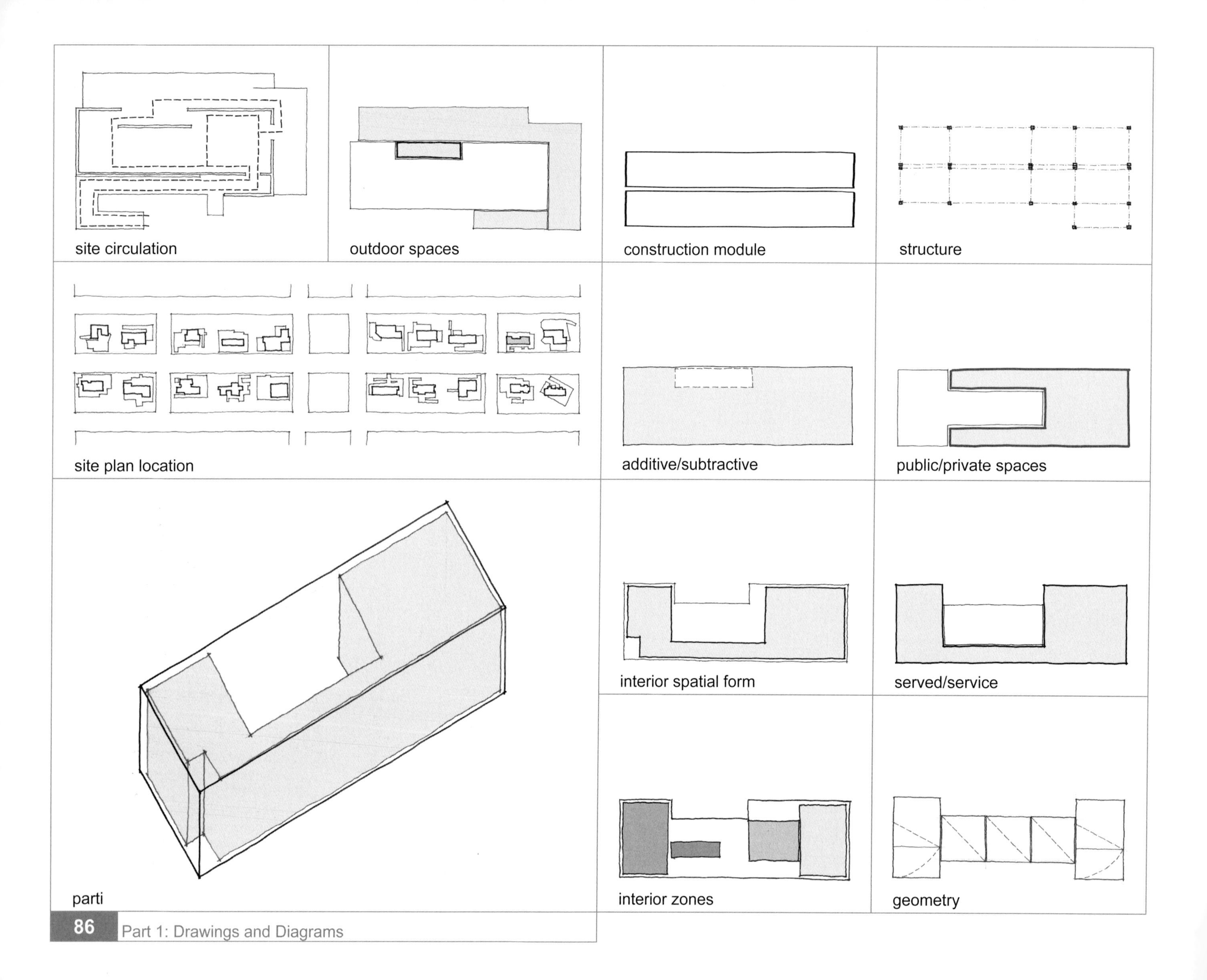
site circulation
outdoor spaces
construction module
structure
site plan location
additive/subtractive
public/private spaces
parti
interior spatial form
served/service
interior zones
geometry

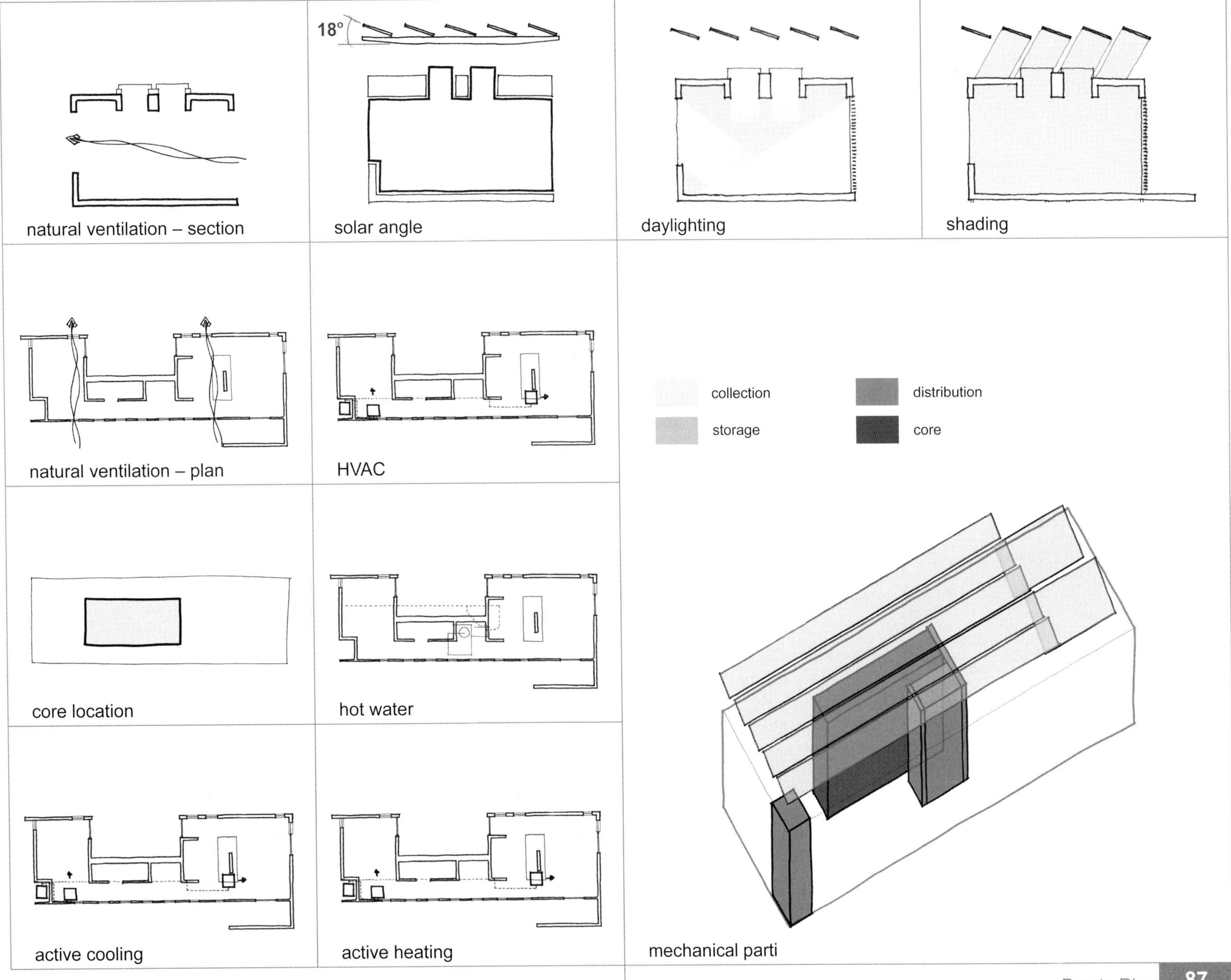
18°
natural ventilation – section
solar angle
daylighting
shading
natural ventilation – plan
HVAC
collection
distribution
storage
core
core location
hot water
active cooling
active heating
mechanical parti

17

The Texas A&M house is a modularized, dimensionally coordinated, open-source kit-of-parts. The larger space units, which attach to each other and to additional entry-space units, create the basis of the house. The structural system is a smaller-scale steel-frame and cable system, which fits between the larger units. Next, "gro walls" are added to the existing units, which contain mostly services and complement the interior space. These units include kitchen equipment, "edutainment" equipment, HVAC equipment, bathroom facilities, etc. This system is intended to be adapted both to single residential units and to much larger-scale community developments.

GroHome

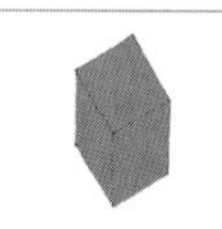

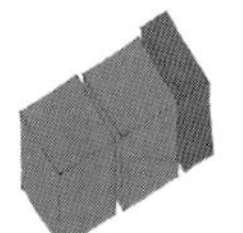

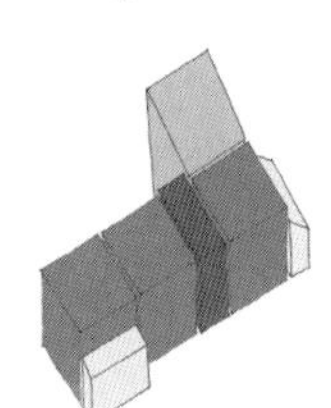

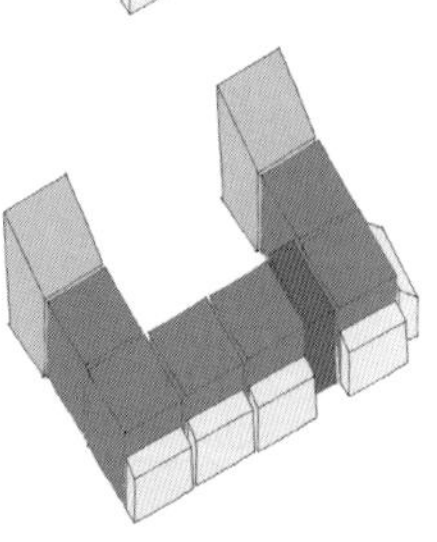

img #33

img #34

Texas A&M University

Location: College Station, TX, USA
Latitude: 30.67° N   Longitude: 96.37° W

Overall Ranking: 17
Architecture: 16
Engineering: 20
Market Viability: 8
Communications: 17
Comfort Zone: 9
Appliances: 1
Hot Water: 9
Lighting: 15
Energy Balance: 16
Getting Around: 9

Total Project Cost: $550,000

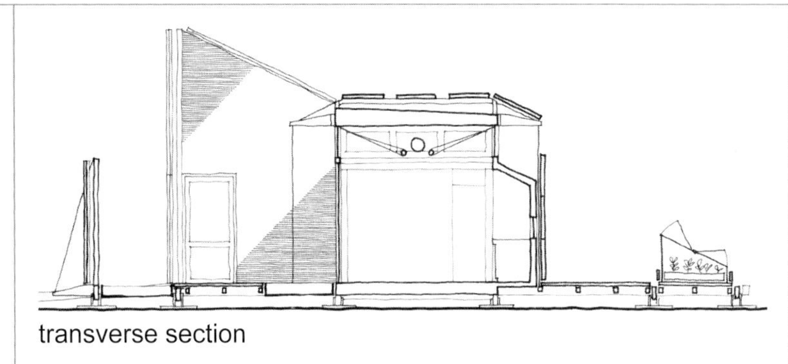
transverse section

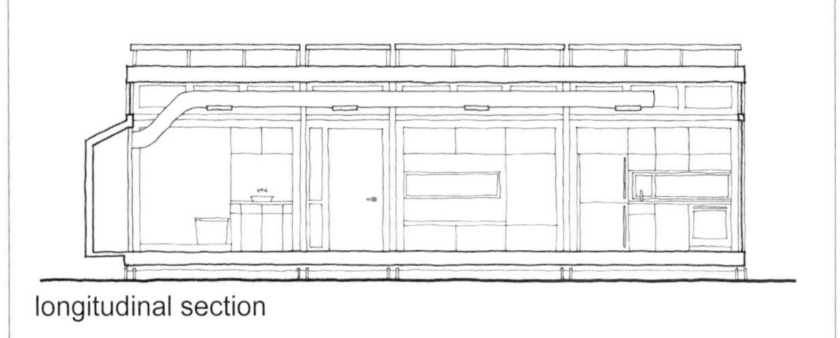
longitudinal section

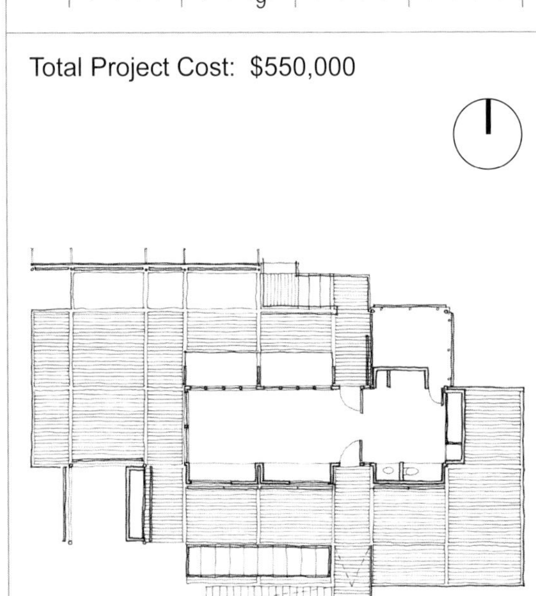
site plan

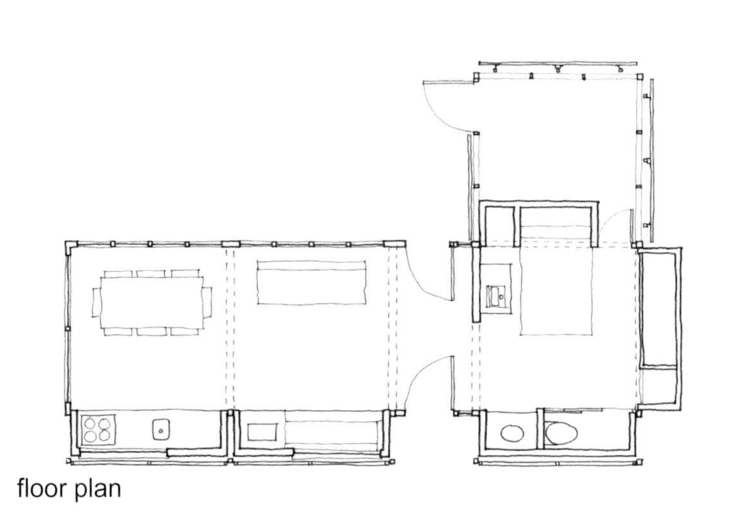
floor plan

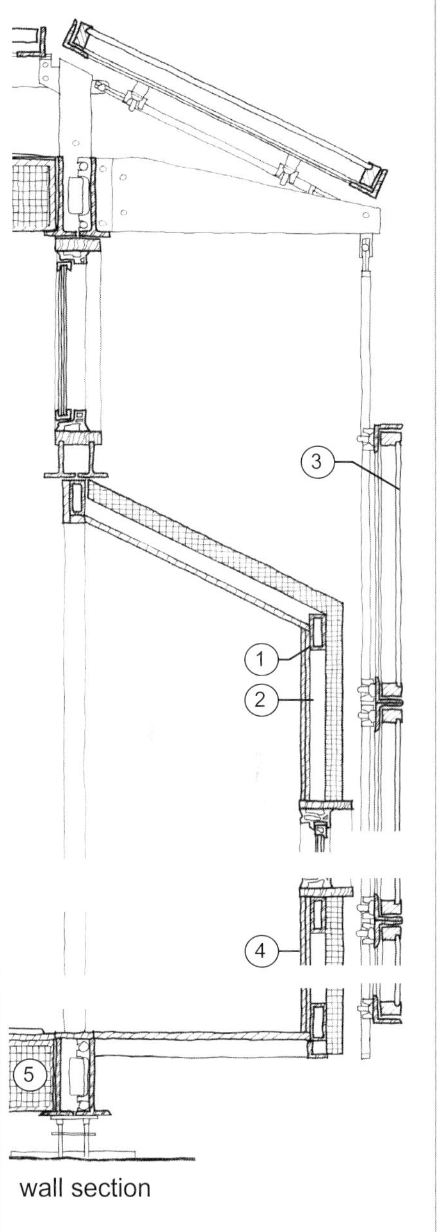

wall section

1. 2" x 2" tube steel
2. bio-based spray-foam infill
3. "light-thru" photovoltaic panel
4. 1/2" plywood
5. 8 1/4" structural insulated panel

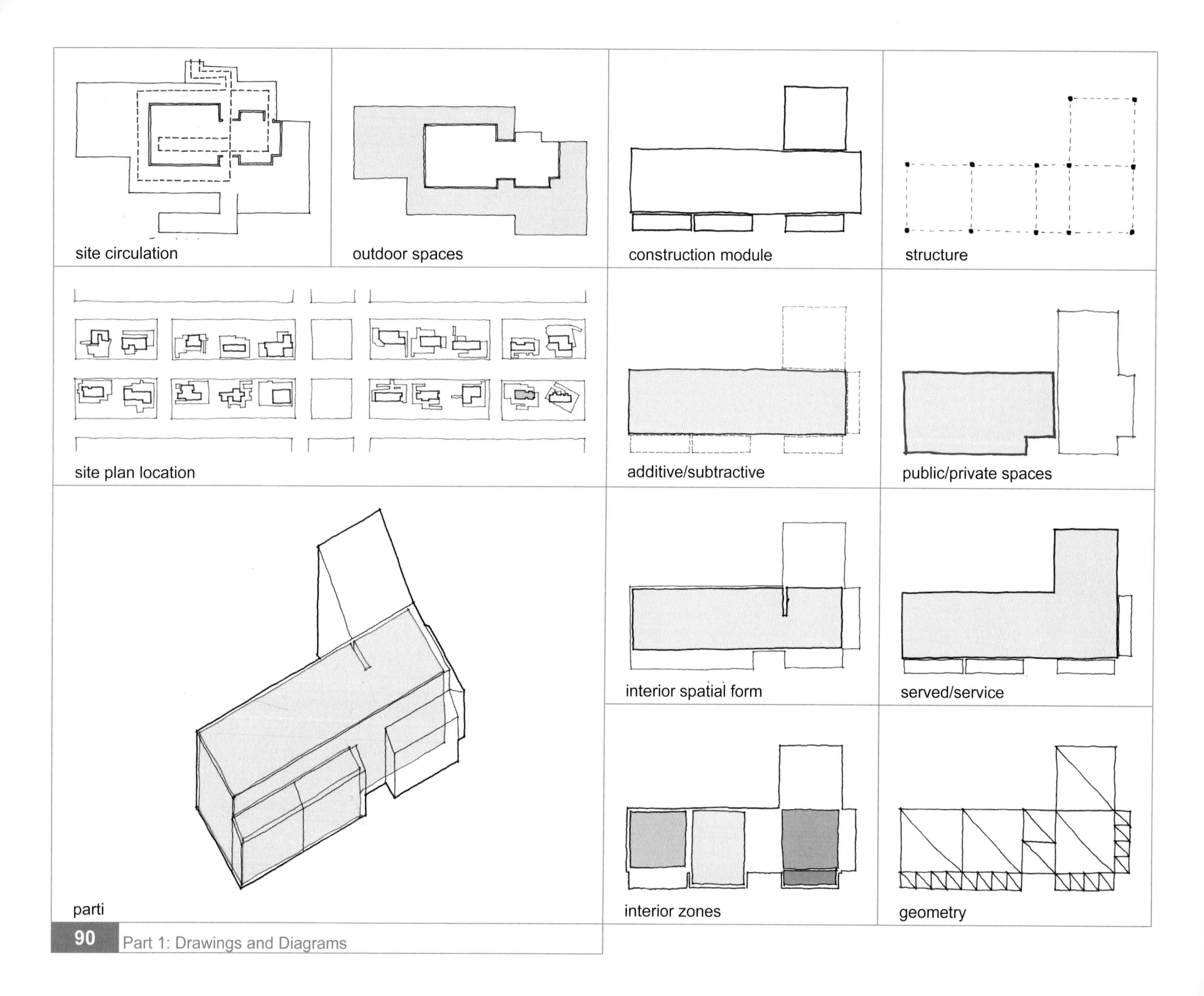
site circulation
outdoor spaces
construction module
structure
site plan location
additive/subtractive
public/private spaces
parti
interior spatial form
served/service
interior zones
geometry

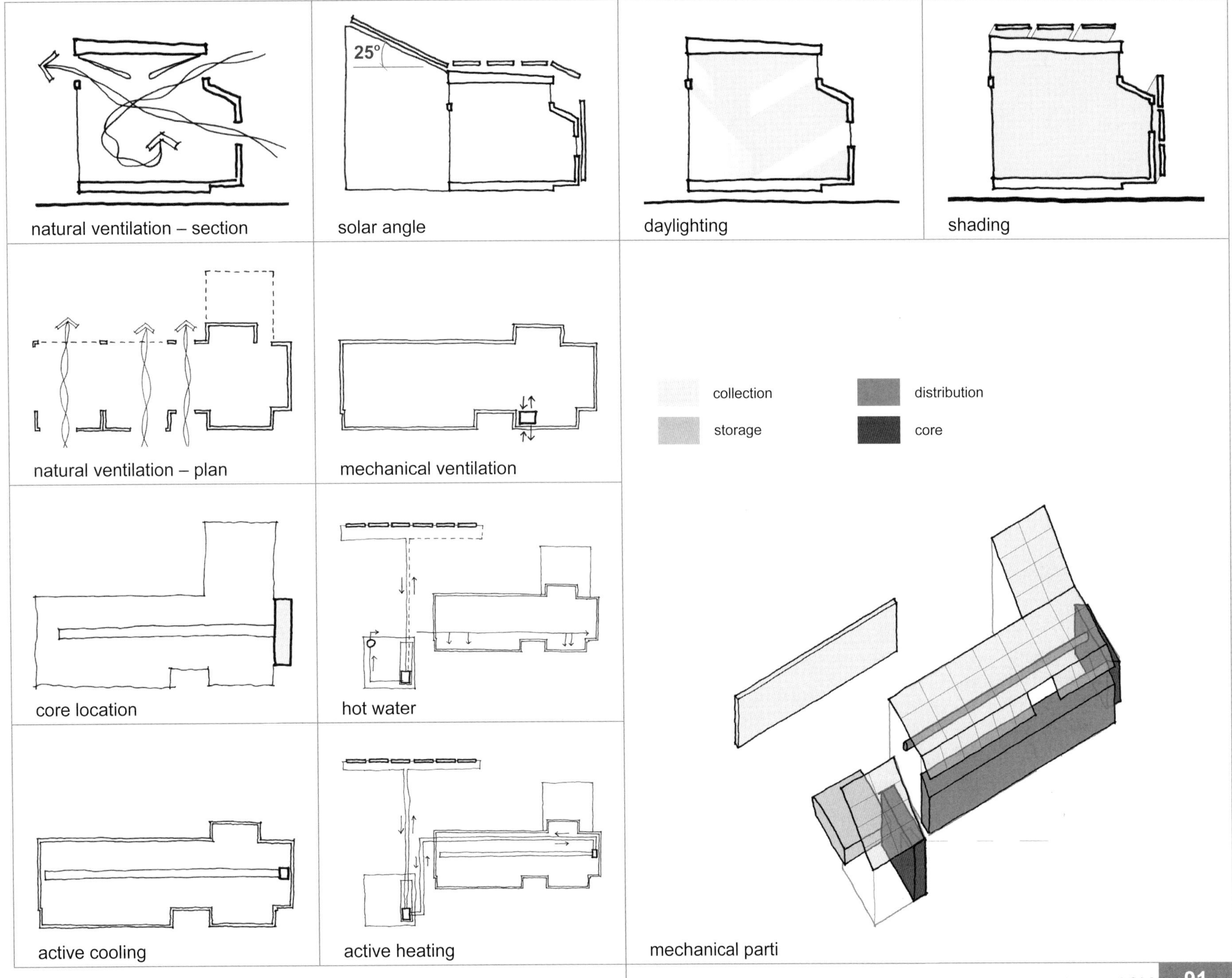
25°
natural ventilation – section
solar angle
daylighting
shading
natural ventilation – plan
mechanical ventilation
collection
storage
distribution
core
core location
hot water
active cooling
active heating
mechanical parti

18

Redirecting Expectations

The primary goal of the Kansas team was to redirect the expectations for a home, and particularly how a potential homeowner views size. The floor plan is an asymmetrical, longitudinally weighted core that is intended to make the space to feel more open. The interior spaces connect to the adjacent outdoor spaces, thereby extending the perceived length of the home. The structural logic reinforces the idea of openness by using a continuous plane to encapsulate the floor, almost the entire south wall and the roof of the home, effectively leaving the north side open to the outdoors.

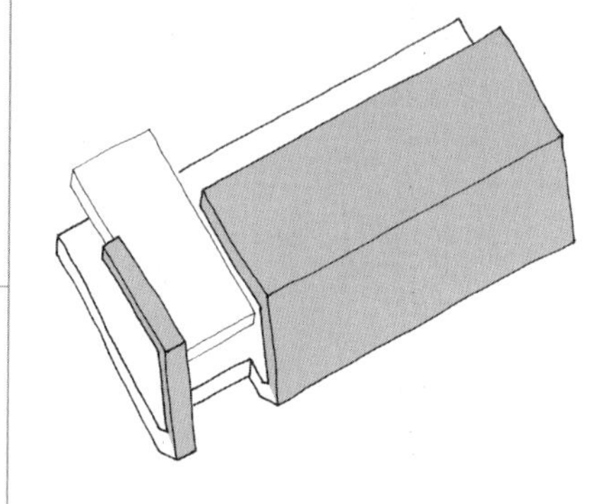

img #35

img #36

Kansas State University and The University of Kansas

Location: Manhattan, KS, USA
Latitude: 39.19° N Longitude: 96.60° W

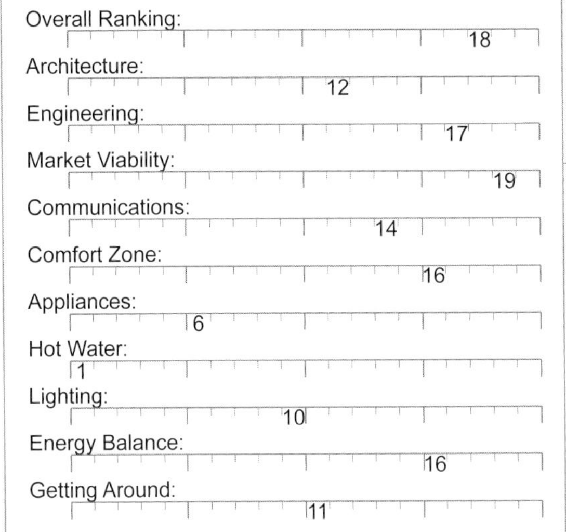

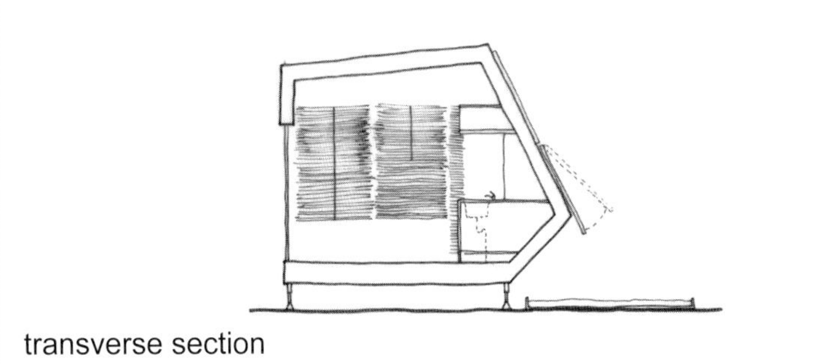

transverse section

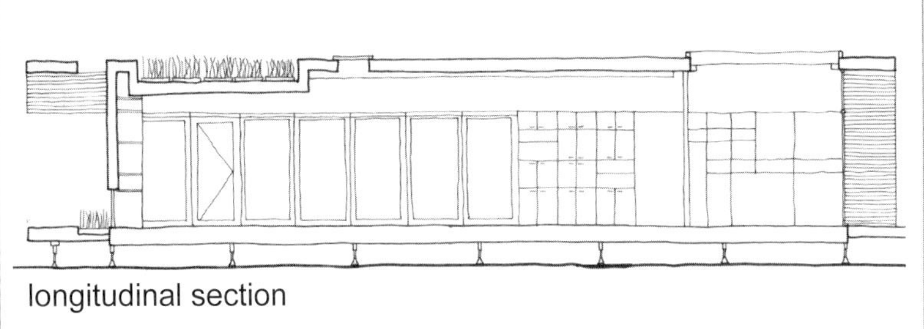

longitudinal section

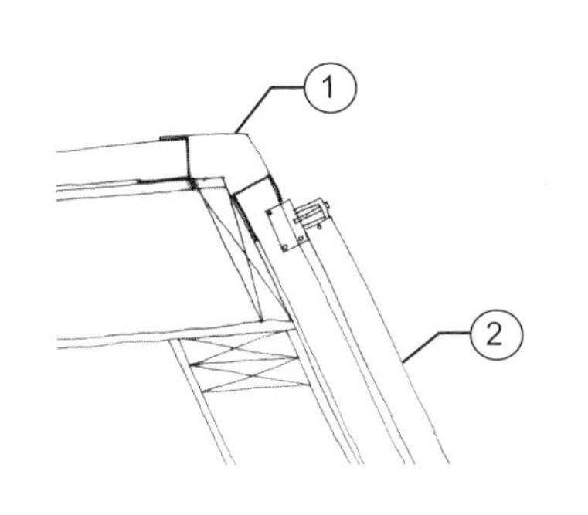

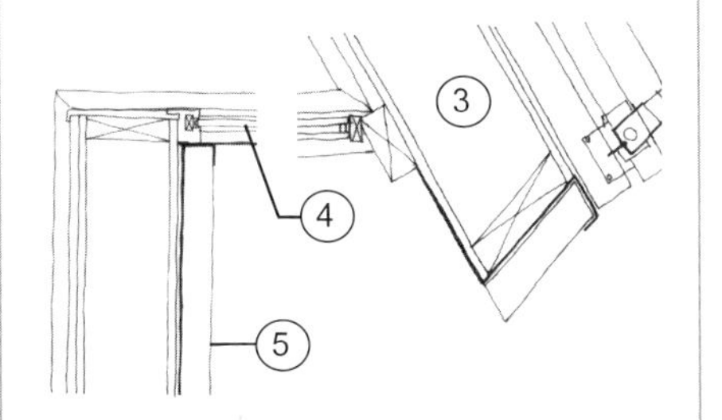

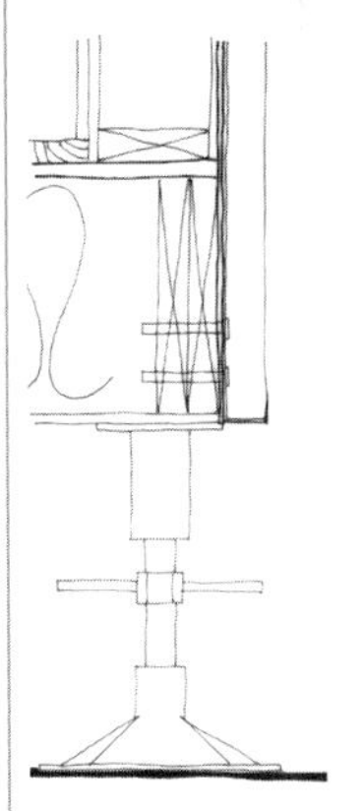

Total Project Cost: $412,808

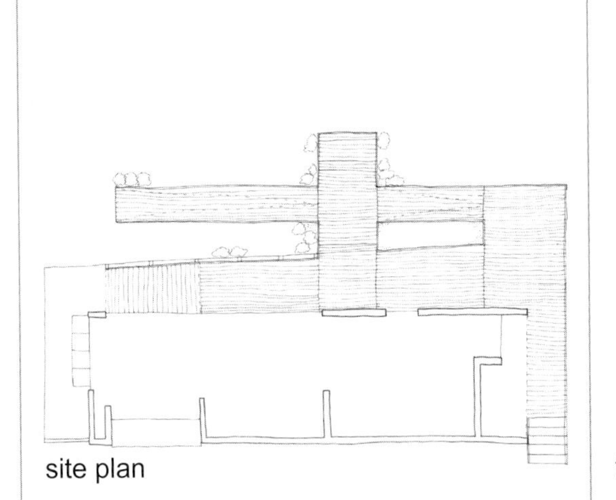

site plan

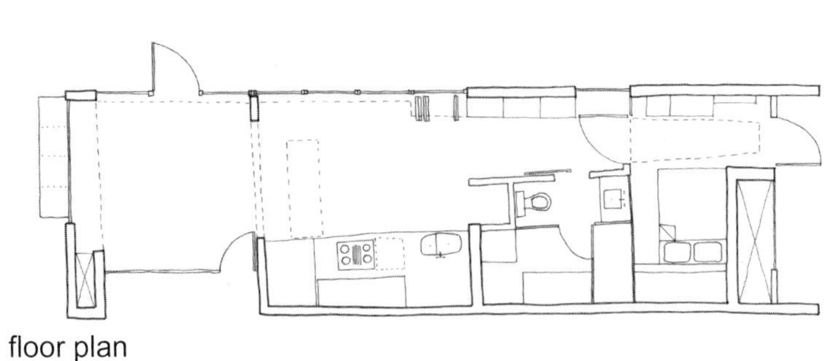

floor plan

wall section

1. 2" standing-seam metal roof
2. photovoltaic panel
3. 6 1/2" structural insulated panel
4. double-glazed, low-e skylight
5. standing-seam metal wall panel

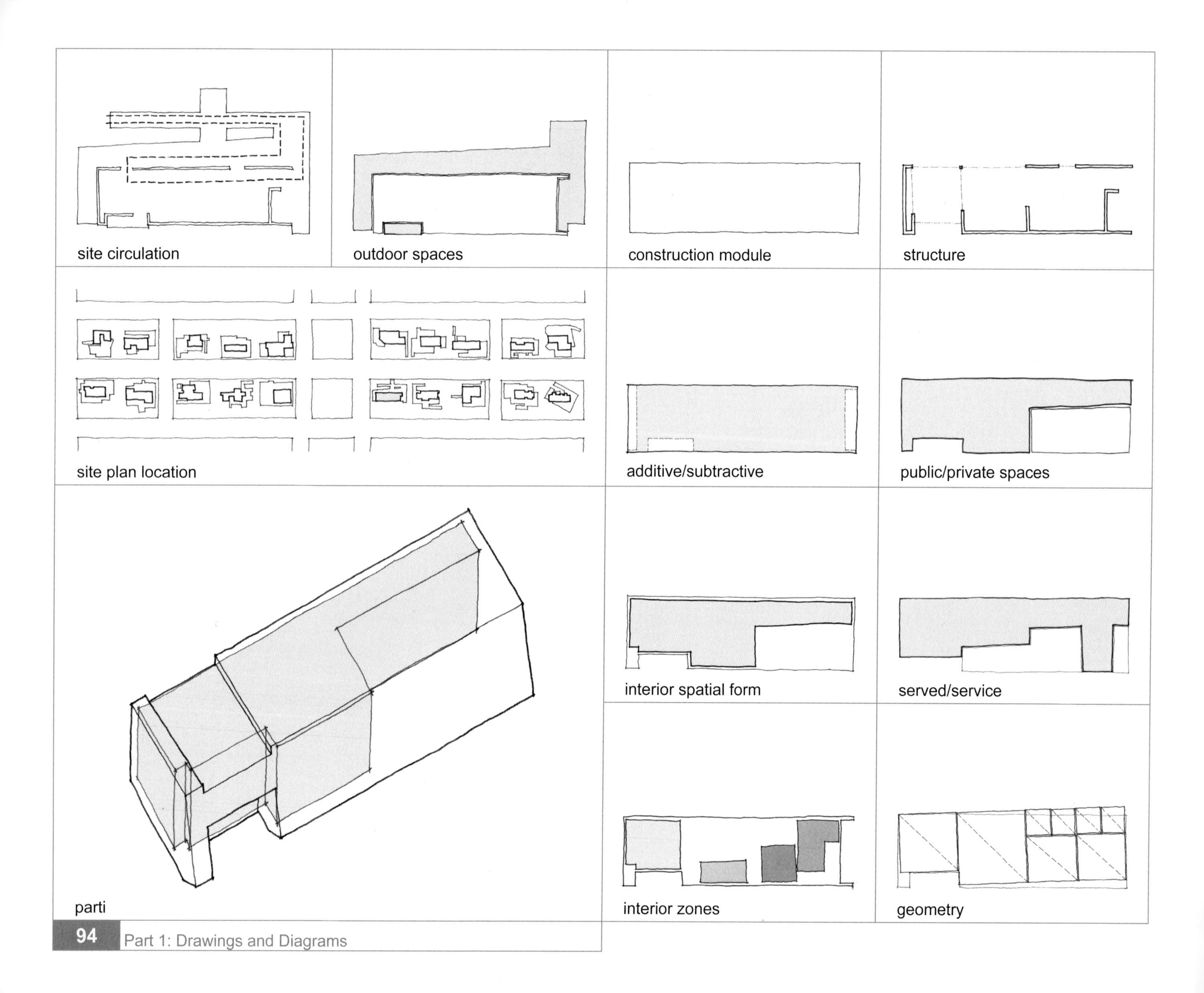
site circulation
outdoor spaces
construction module
structure
site plan location
additive/subtractive
public/private spaces
interior spatial form
served/service
parti
interior zones
geometry

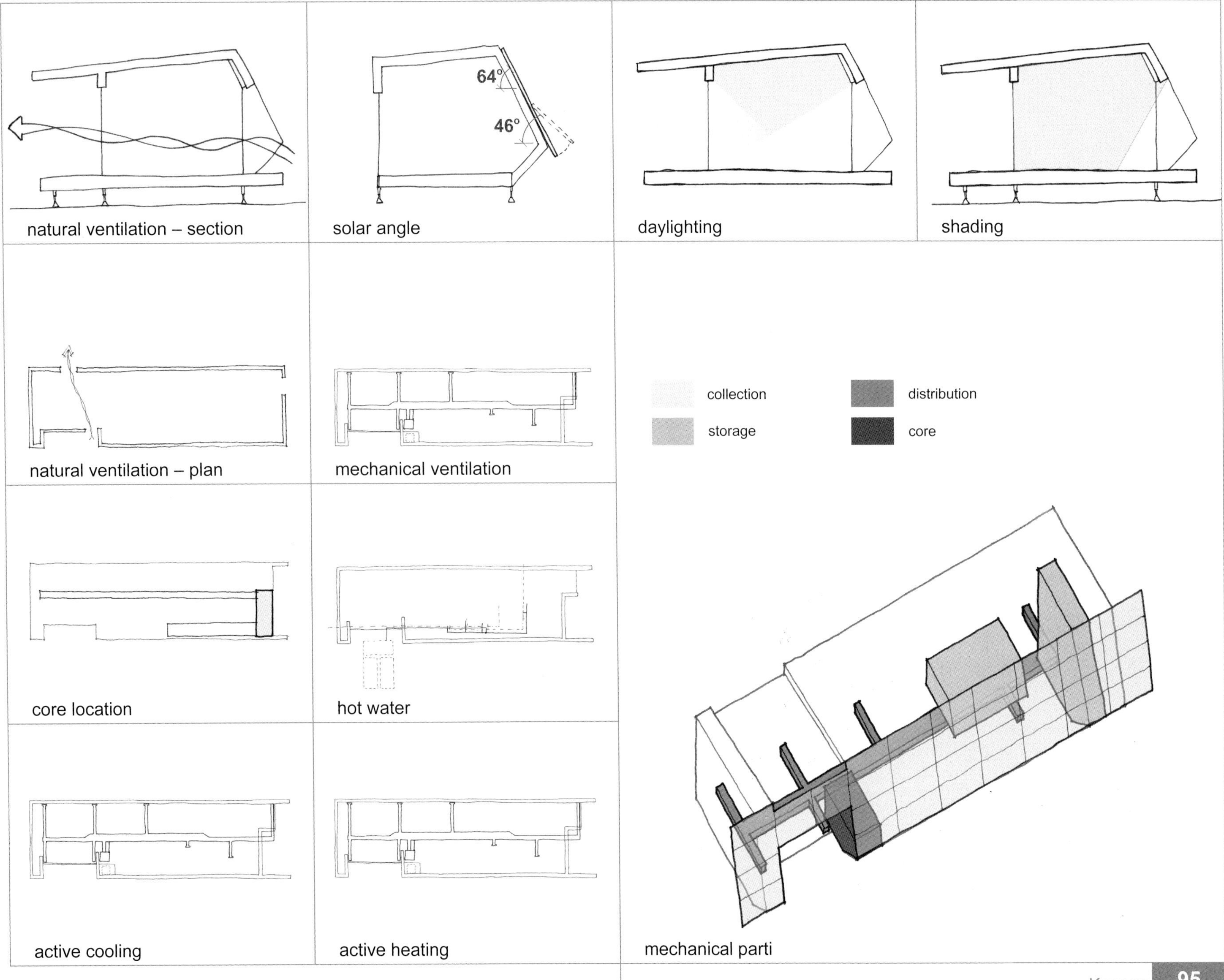
64°
46°
natural ventilation – section
solar angle
daylighting
shading
natural ventilation – plan
mechanical ventilation
collection
distribution
storage
core
core location
hot water
active cooling
active heating
mechanical parti

19

Light Canopy

The Cornell house incorporates a semi-independent canopy structure, which houses most of its fast-changing technological and sustainable components. This "light canopy" concept drove all of the functional and aesthetic decisions of the design. Underneath it sits the house, which interacts with the canopy through its structure and its systems. This framework allows the technological and environmental systems to operate independently of the space below. Additionally, this strategy allows the technological systems to be moved, replaced with newer technologies in the future or simply removed when no longer needed.

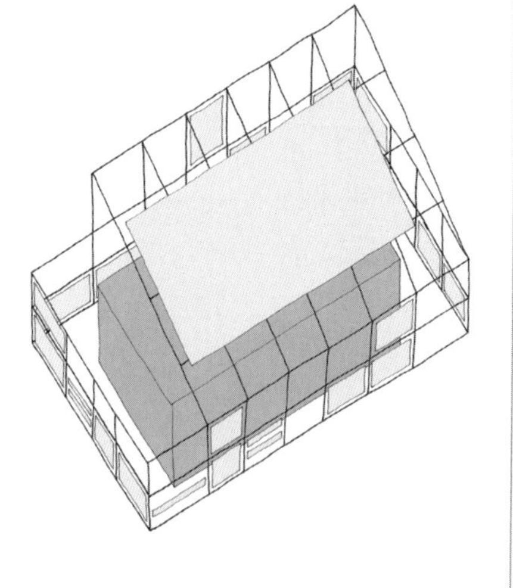

img #37

img #38

Cornell University

Location: Ithaca, NY, USA
Latitude: 42.42° N Longitude: 76.51° W

Overall Ranking: 19

Architecture: 11

Engineering: 9

Market Viability: 14

Communications: 7

Comfort Zone: 12

Appliances: 16

Hot Water: 16

Lighting: 17

Energy Balance: 16

Getting Around: 19

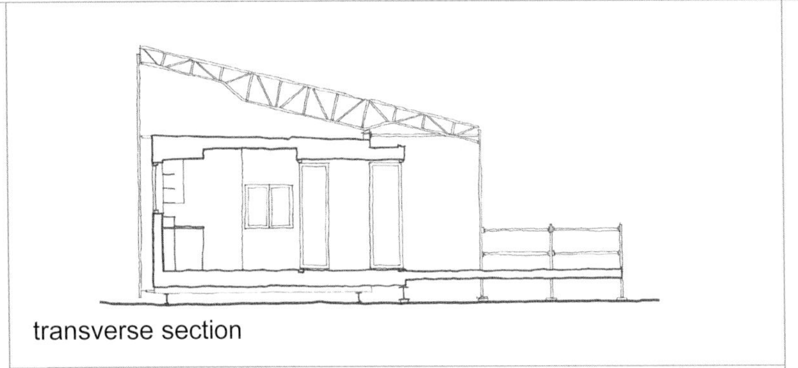

transverse section

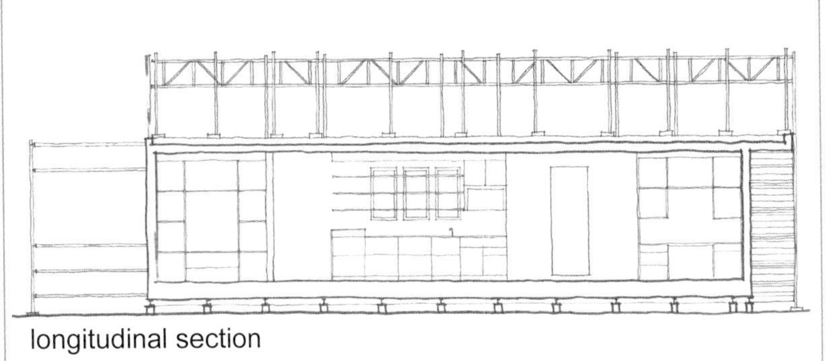

longitudinal section

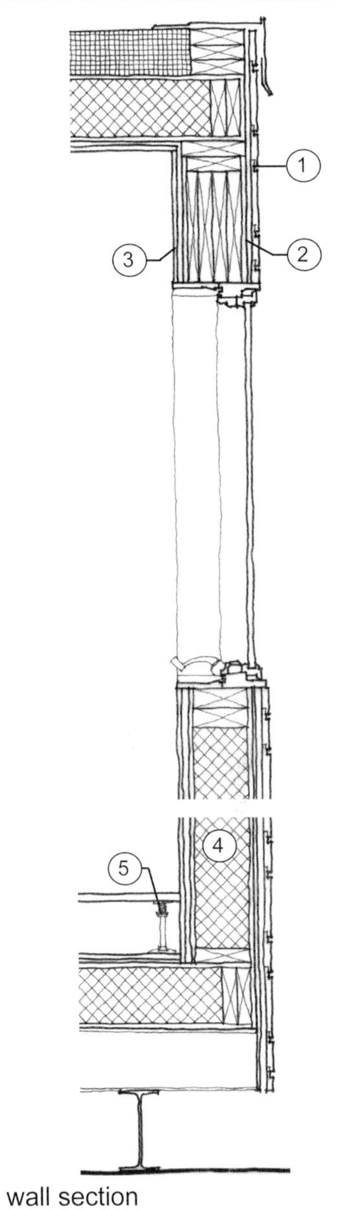

wall section

1. 3/4" cedar ship-lapped siding
2. waterproof membrane
3. 1/2" gypsum wall board
4. structural insulated panel
5. steel subfloor support

Total Project Cost: $675,000

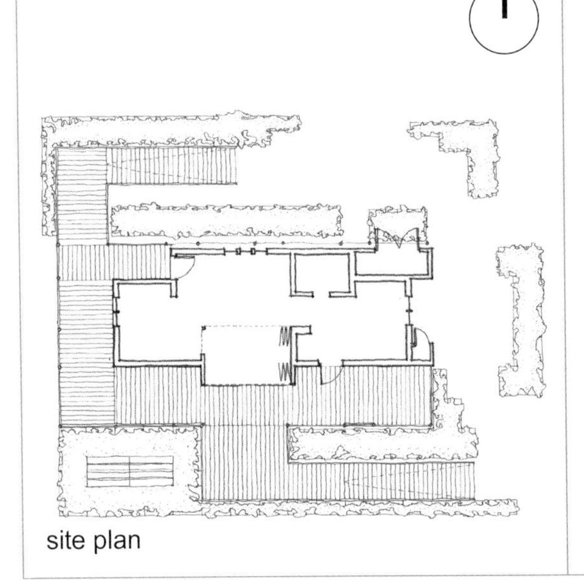

site plan

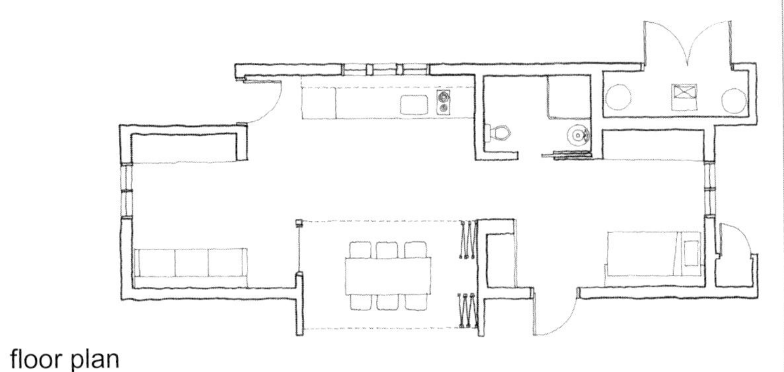

floor plan

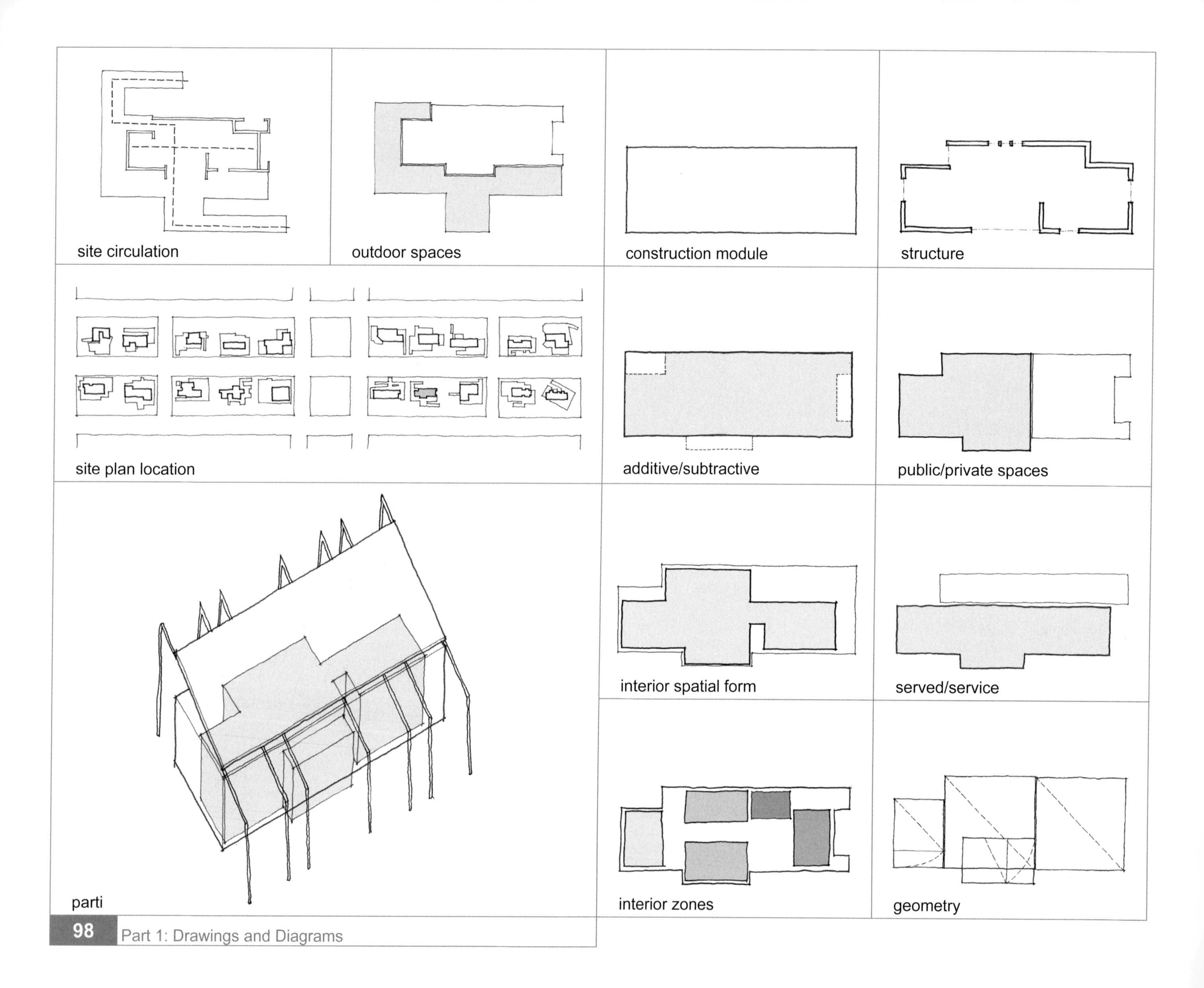
site circulation
outdoor spaces
construction module
structure
site plan location
additive/subtractive
public/private spaces
parti
interior spatial form
served/service
interior zones
geometry

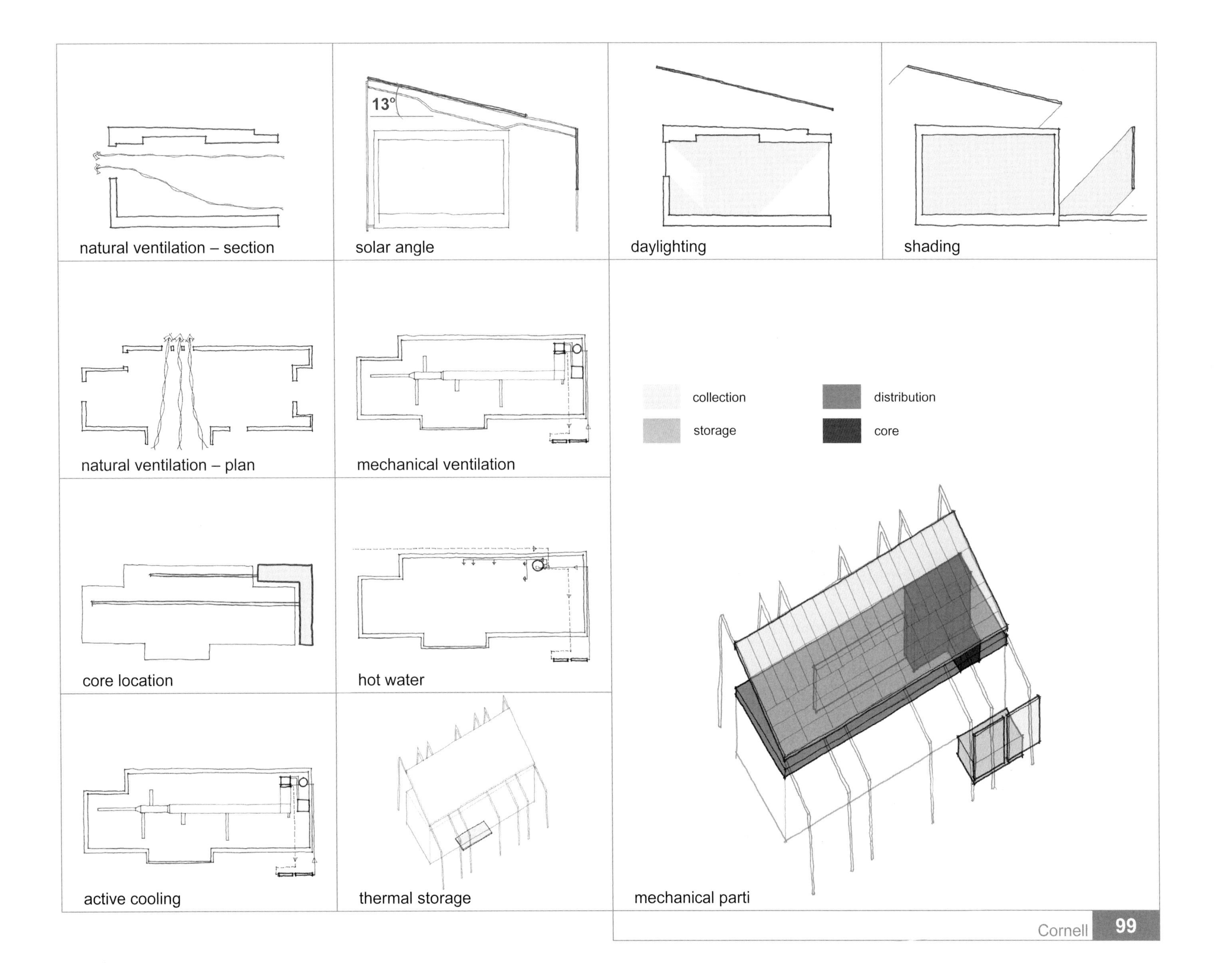
natural ventilation – section
13°
solar angle
daylighting
shading
natural ventilation – plan
mechanical ventilation
collection
distribution
storage
core
core location
hot water
active cooling
thermal storage
mechanical parti

20

The Lawrence Technological University house involves a trellis system that provides structure, shelter, shade and flexibility. While this system acts as the main roofing structure, it also supports the solar panels above. Additionally, this system provides flexibility for the house to self-ventilate through a solar chimney which passes through the trellis as well as via an operable skylight which fits between the trellis members. Finally, outdoor spaces are given a sense of enclosure by the extension of this system beyond the exterior walls of the house and out over the deck, extending the sense of space from the interior to the exterior.

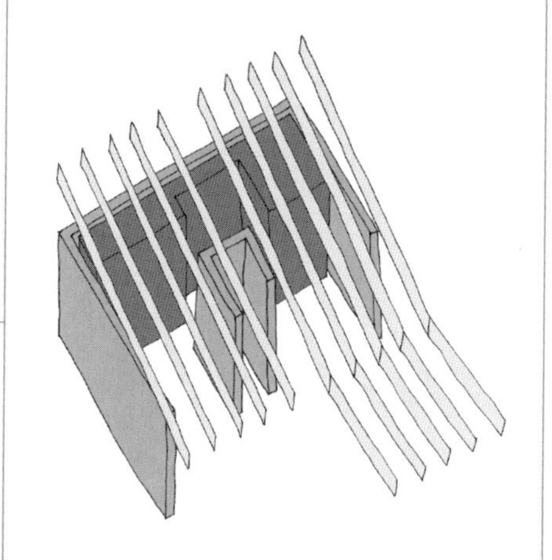

Lawrence Technological University

Location: Southfield, MI, USA
Latitude: 42.49° N Longitude: 83.28° W

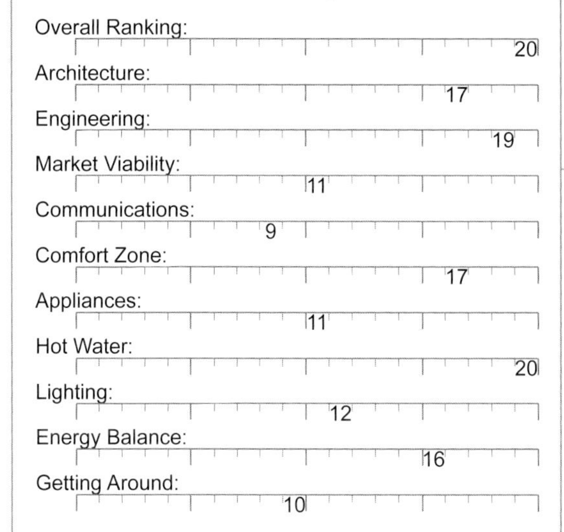

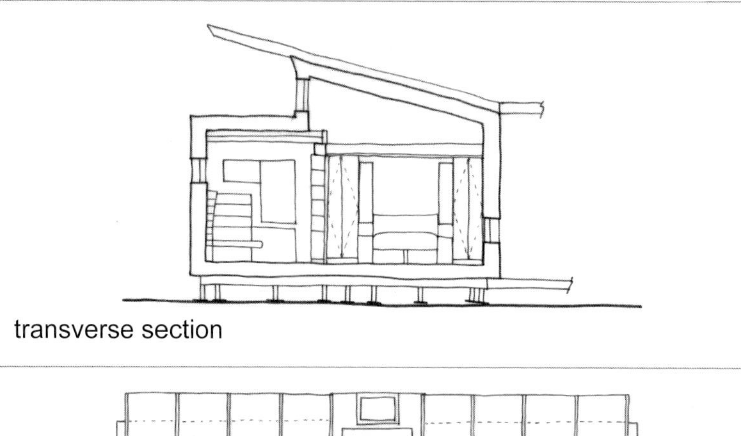

transverse section

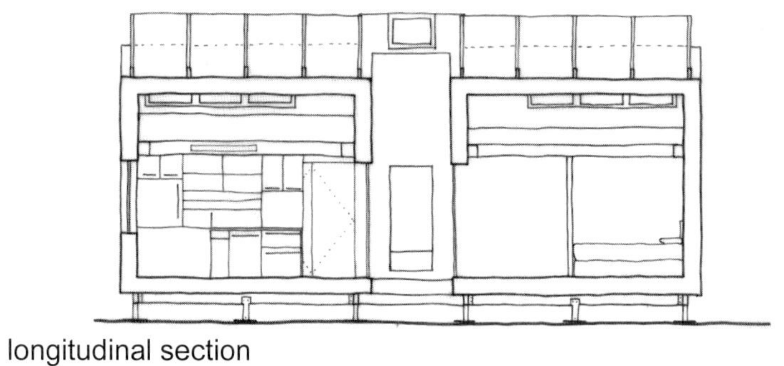

longitudinal section

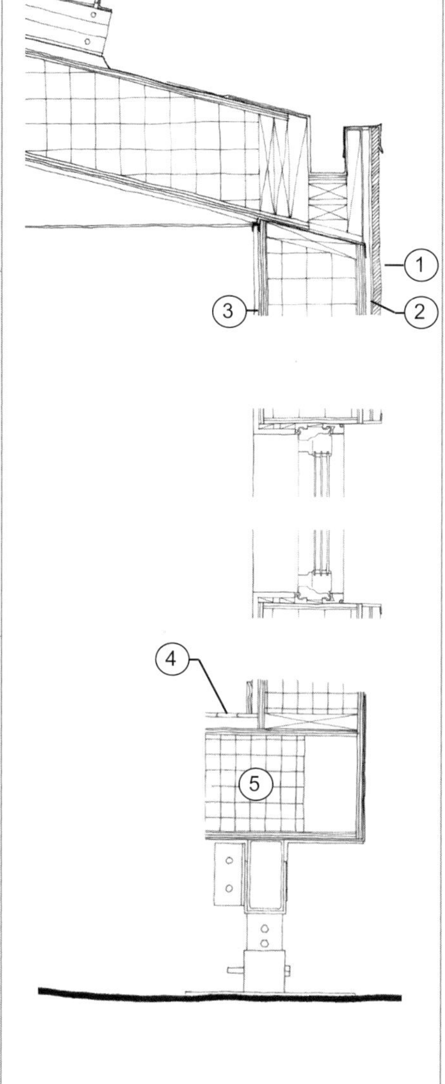

wall section

1. cedar rainscreen
2. 1/4" plywood
3. 1/2" gypsum wall board
4. 1 1/8" radiant floor
5. structural insulated panel

Total Project Cost: $672,000

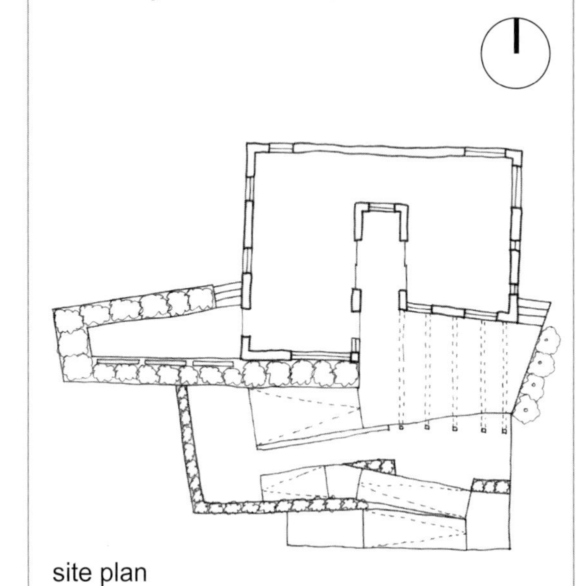

site plan

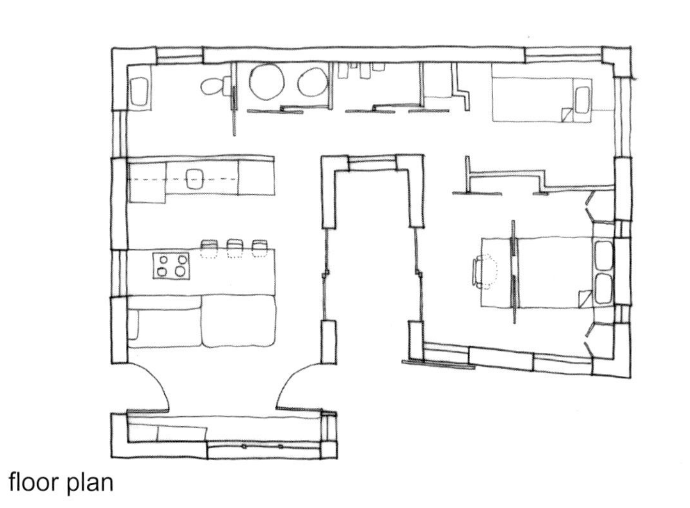

floor plan

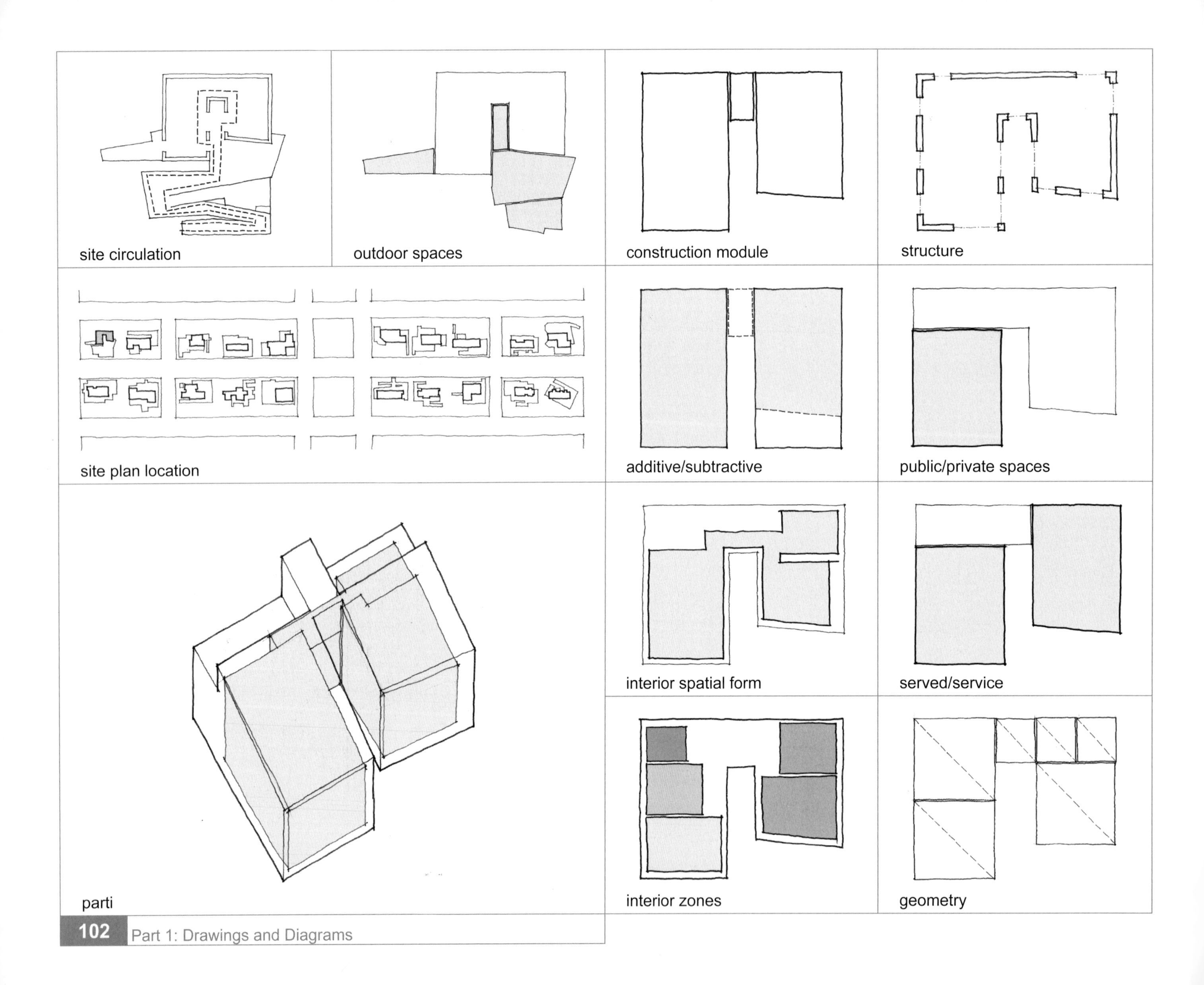
site circulation
outdoor spaces
construction module
structure
site plan location
additive/subtractive
public/private spaces
parti
interior spatial form
served/service
interior zones
geometry

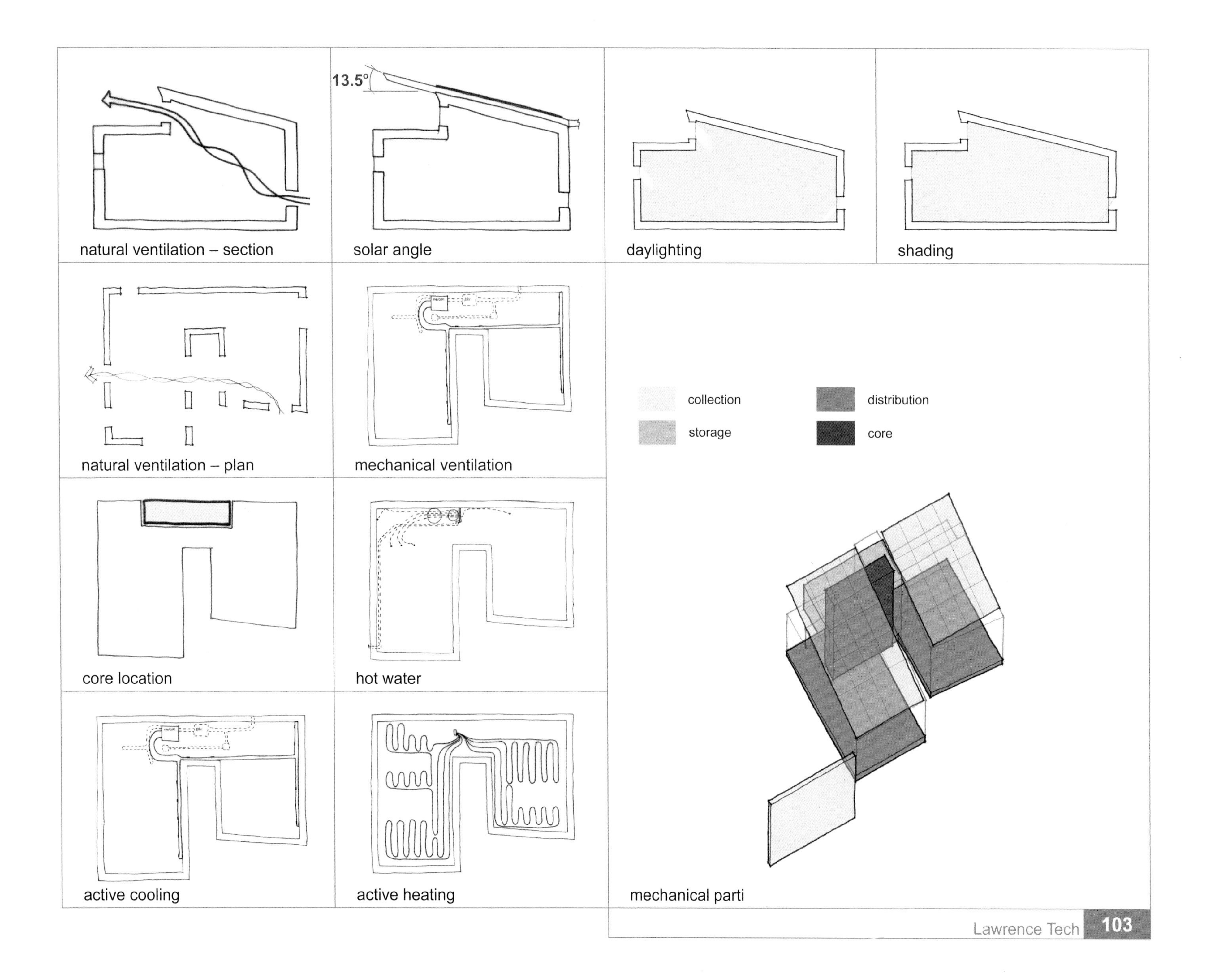
13.5°
natural ventilation – section
solar angle
daylighting
shading
natural ventilation – plan
mechanical ventilation
collection
storage
distribution
core
core location
hot water
active cooling
active heating
mechanical parti

Precedents in Zero-Energy Design:

Part 2: Comparisons

Part 2: Comparisons

Exterior Form Parti

A parti is the basic scheme of an architectural design, which could be sketched in a few lines. This axonometric diagram represents the most distilled expression of the building form in three dimensions. This drawing is an overlay of the exterior form parti and the interior spatial parti.

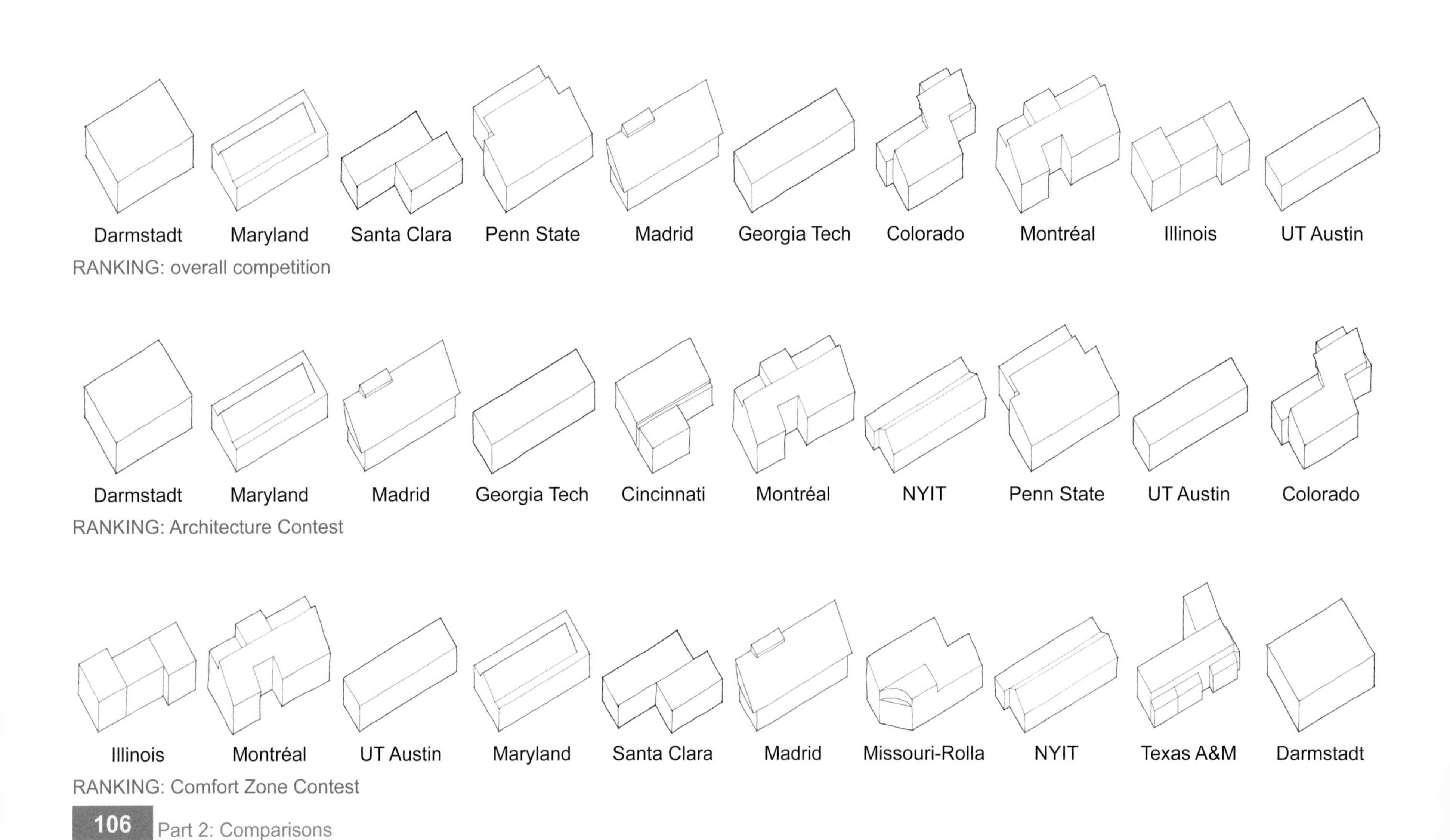

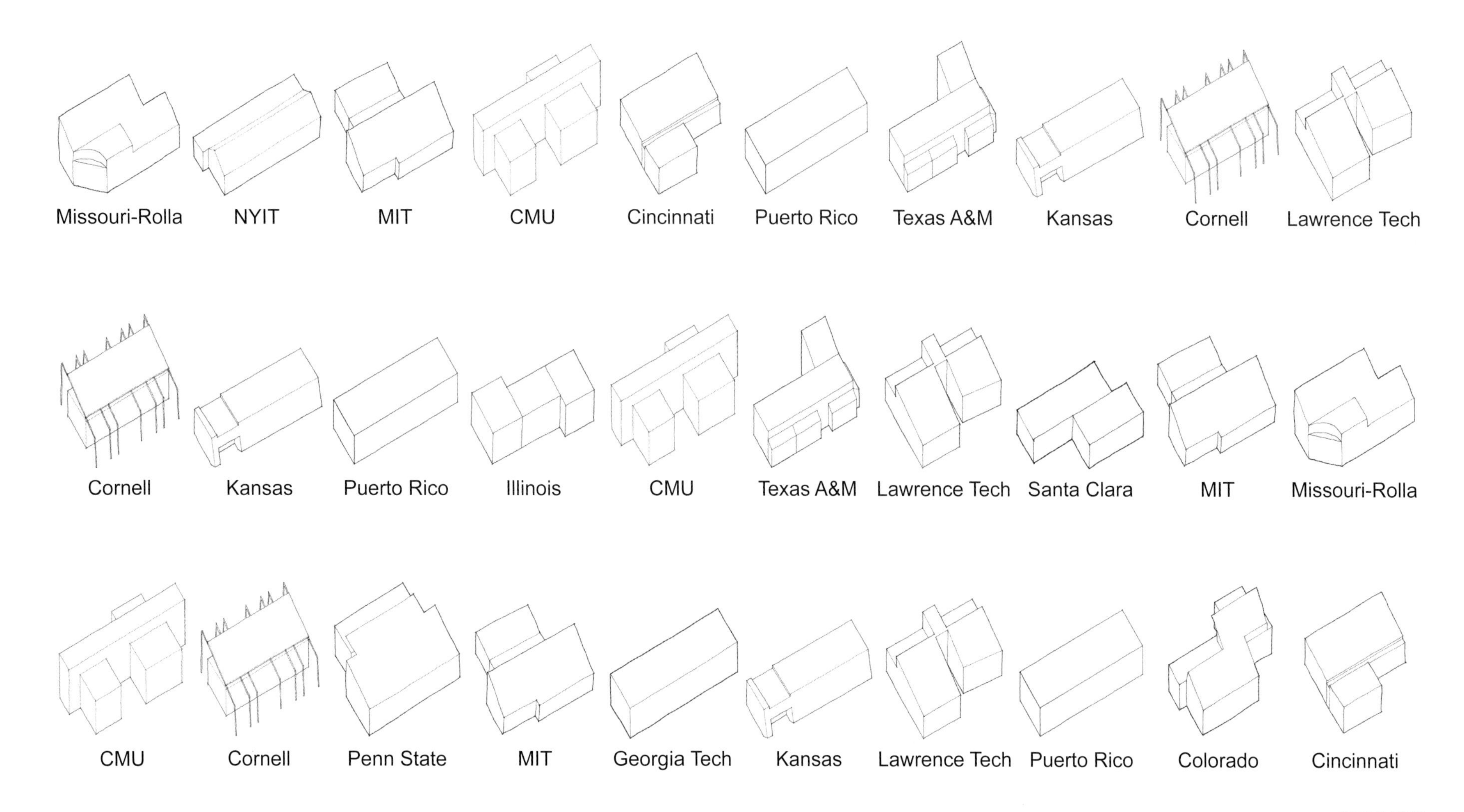
Missouri-Rolla
NYIT
MIT
CMU
Cincinnati
Puerto Rico
Texas A&M
Kansas
Cornell
Lawrence Tech
Cornell
Kansas
Puerto Rico
Illinois
CMU
Texas A&M
Lawrence Tech
Santa Clara
MIT
Missouri-Rolla
CMU
Cornell
Penn State
MIT
Georgia Tech
Kansas
Lawrence Tech
Puerto Rico
Colorado
Cincinnati

Interior Spatial Parti

This axonometric diagram represents the basic shape of the interior usable space in three dimensions.

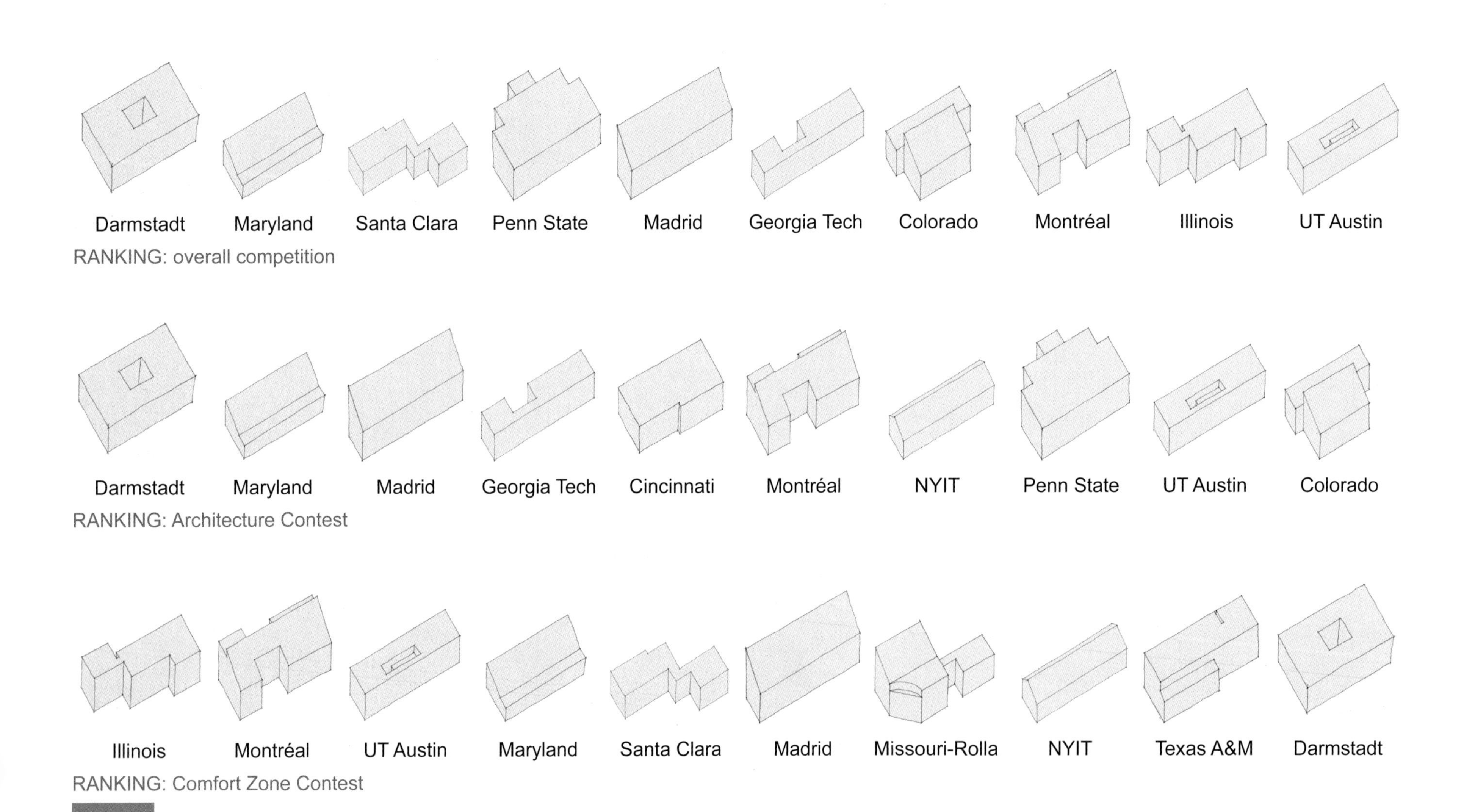

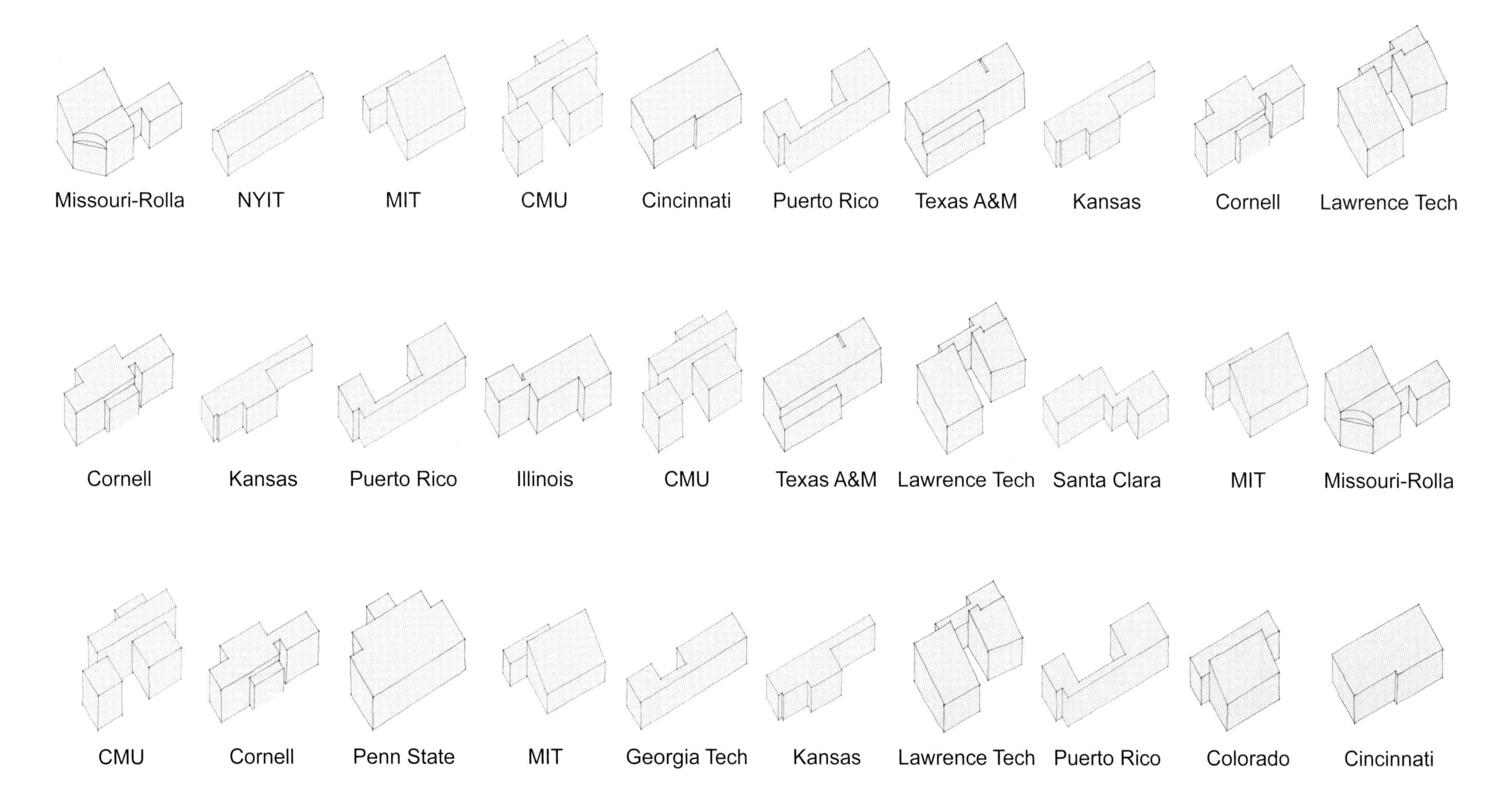
Missouri-Rolla
NYIT
MIT
CMU
Cincinnati
Puerto Rico
Texas A&M
Kansas
Cornell
Lawrence Tech
Cornell
Kansas
Puerto Rico
Illinois
CMU
Texas A&M
Lawrence Tech
Santa Clara
MIT
Missouri-Rolla
CMU
Cornell
Penn State
MIT
Georgia Tech
Kansas
Lawrence Tech
Puerto Rico
Colorado
Cincinnati

Combined Form and Spatial Parti

This axonometric diagram is an overlay of the exterior form parti and the interior spatial parti. This provides an insight into the relationship between the interior spaces and the exterior expression of the house.

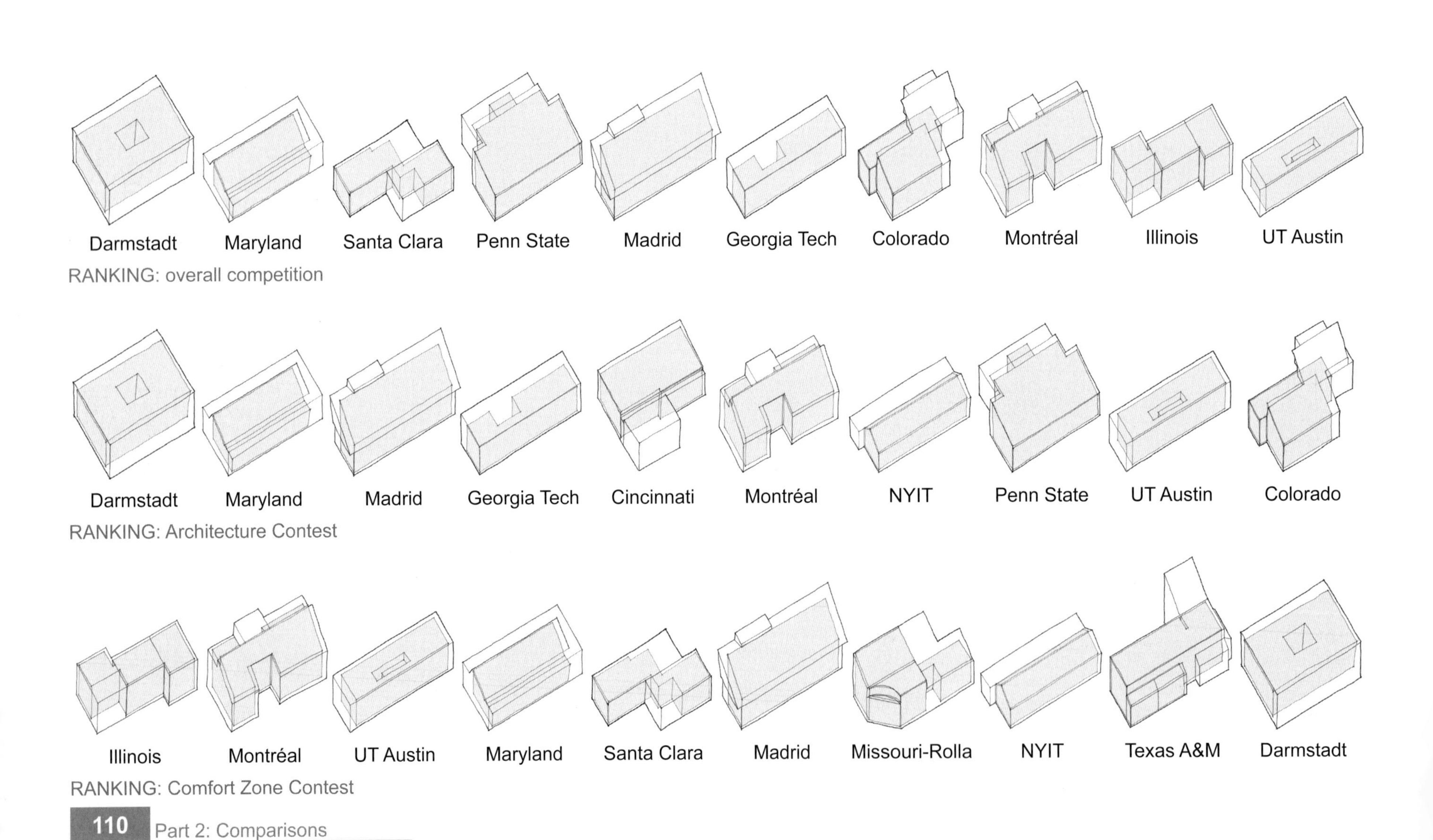

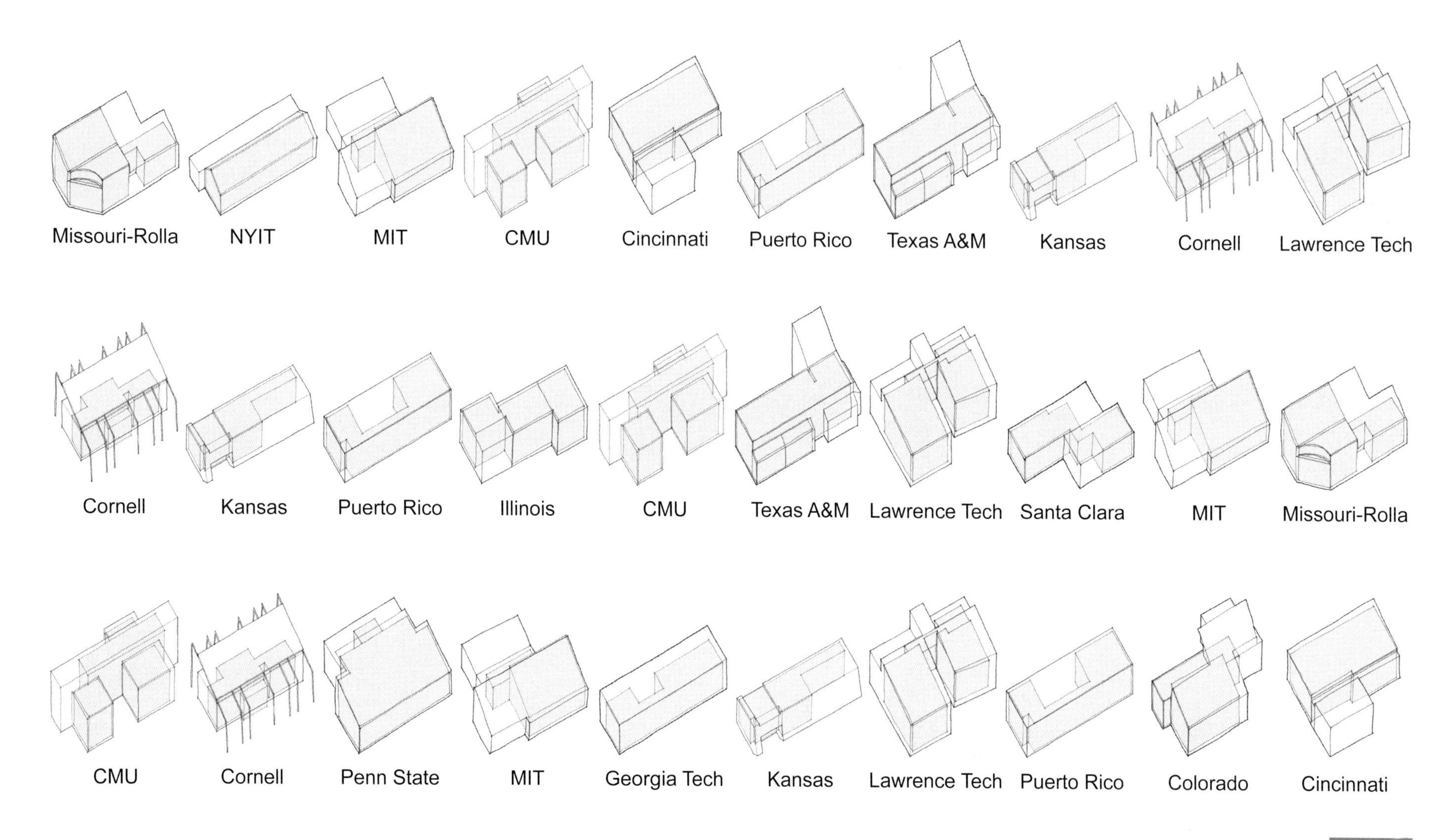
Missouri-Rolla
NYIT
MIT
CMU
Cincinnati
Puerto Rico
Texas A&M
Kansas
Cornell
Lawrence Tech
Cornell
Kansas
Puerto Rico
Illinois
CMU
Texas A&M
Lawrence Tech
Santa Clara
MIT
Missouri-Rolla
CMU
Cornell
Penn State
MIT
Georgia Tech
Kansas
Lawrence Tech
Puerto Rico
Colorado
Cincinnati

Site Circulation

This site plan diagram shows the intended path of a visitor from the main street of the Solar Village into and through the house.

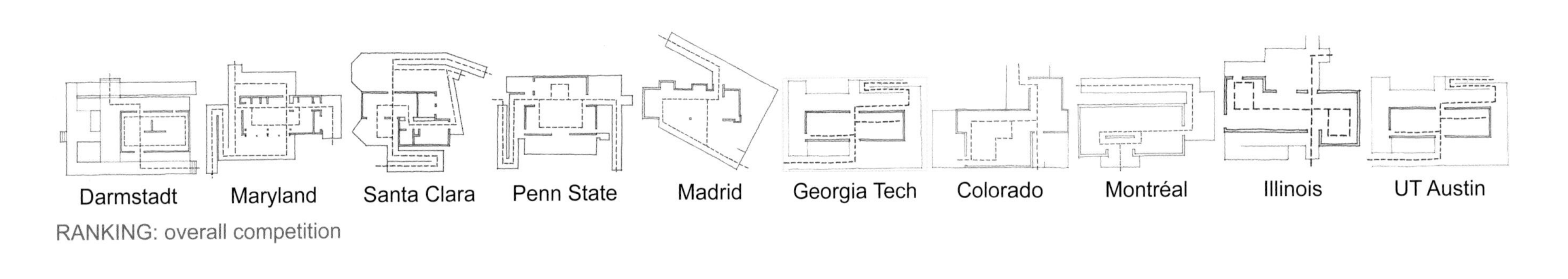

RANKING: overall competition

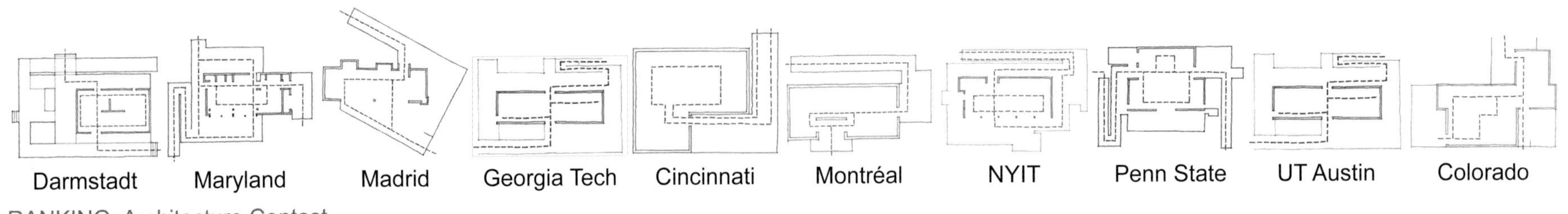

RANKING: Architecture Contest

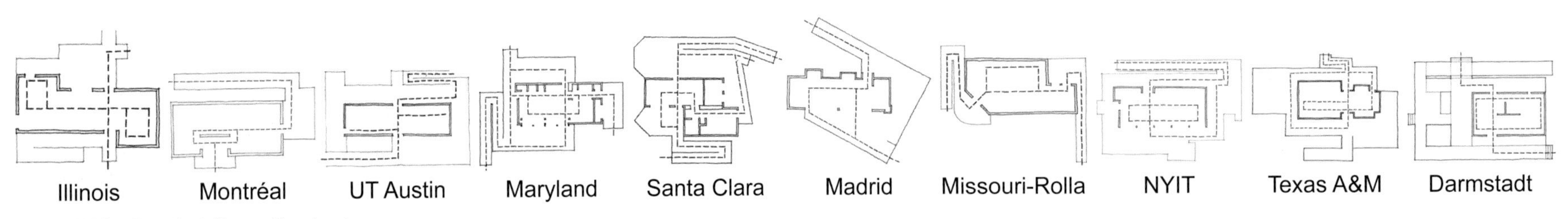

RANKING: Comfort Zone Contest

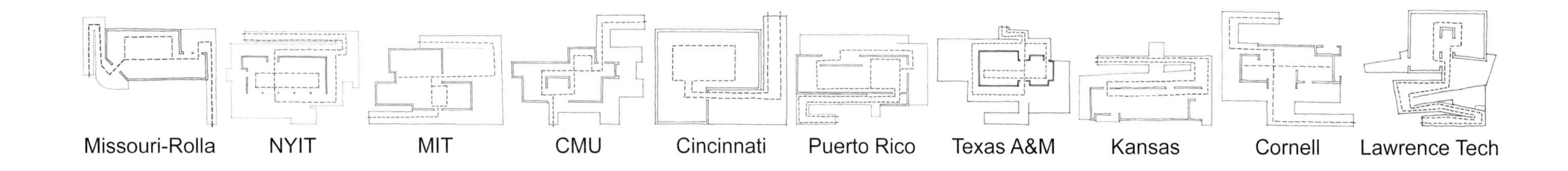

Missouri-Rolla
NYIT
MIT
CMU
Cincinnati
Puerto Rico
Texas A&M
Kansas
Cornell
Lawrence Tech

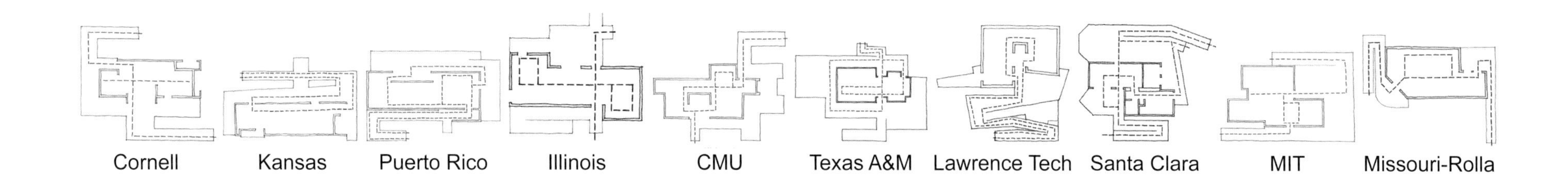

Cornell
Kansas
Puerto Rico
Illinois
CMU
Texas A&M
Lawrence Tech
Santa Clara
MIT
Missouri-Rolla

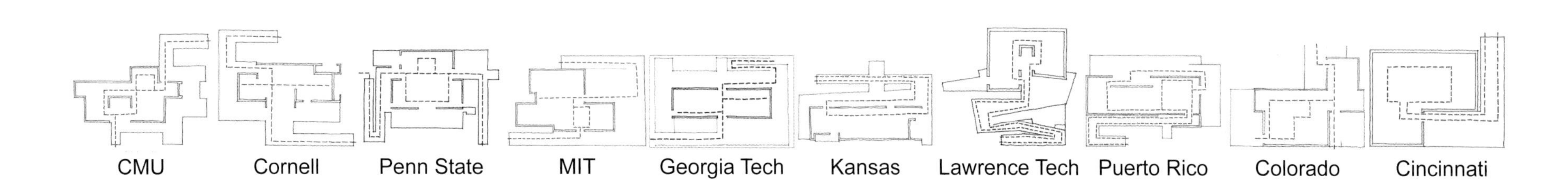

CMU
Cornell
Penn State
MIT
Georgia Tech
Kansas
Lawrence Tech
Puerto Rico
Colorado
Cincinnati

Outdoor Spaces

This site plan diagram shows the outdoor spaces that are created for people to use on the exterior of the house. The house is shown in white. The outdoor spaces are shown in gray.

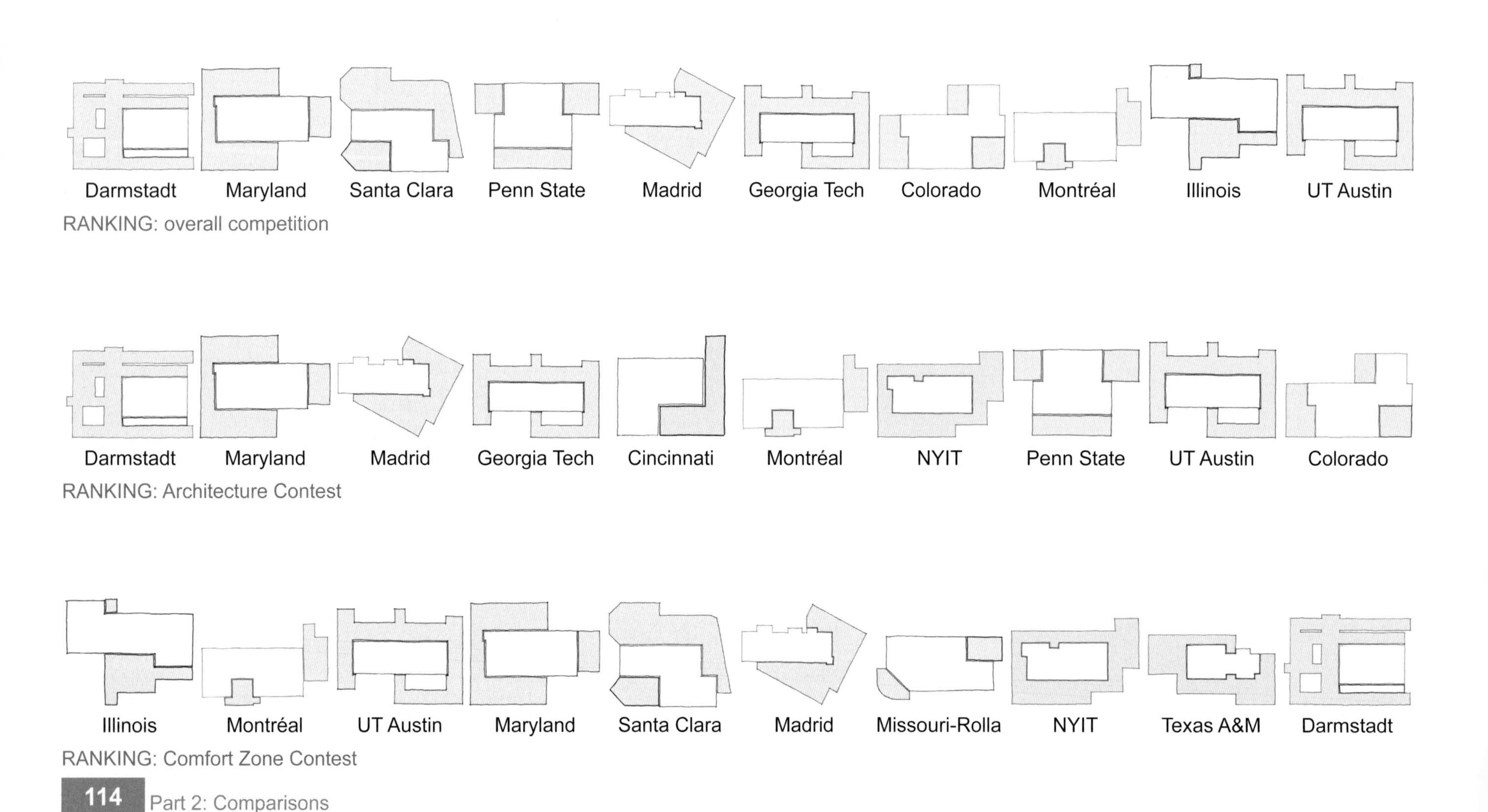

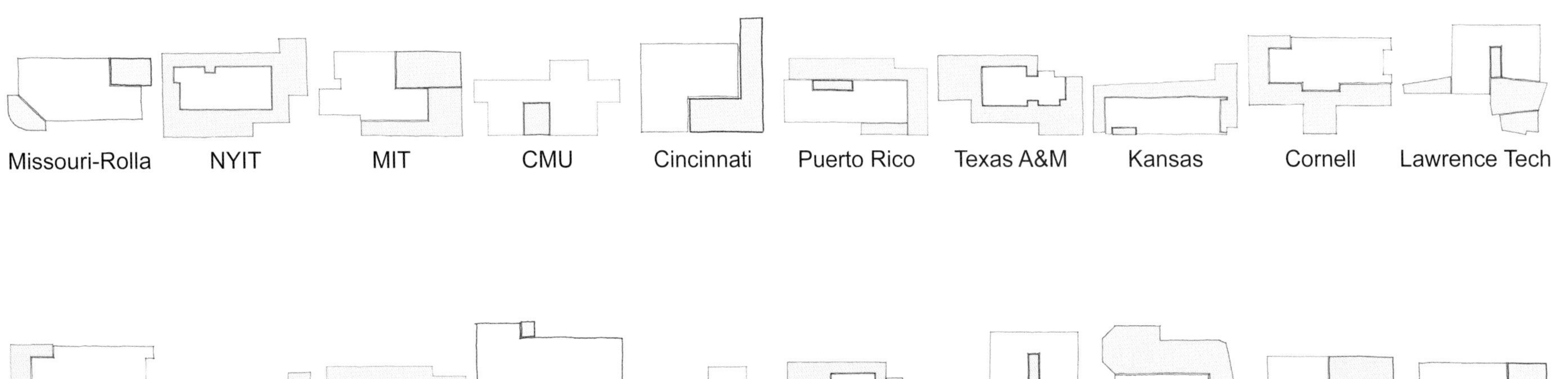
Missouri-Rolla
NYIT
MIT
CMU
Cincinnati
Puerto Rico
Texas A&M
Kansas
Cornell
Lawrence Tech

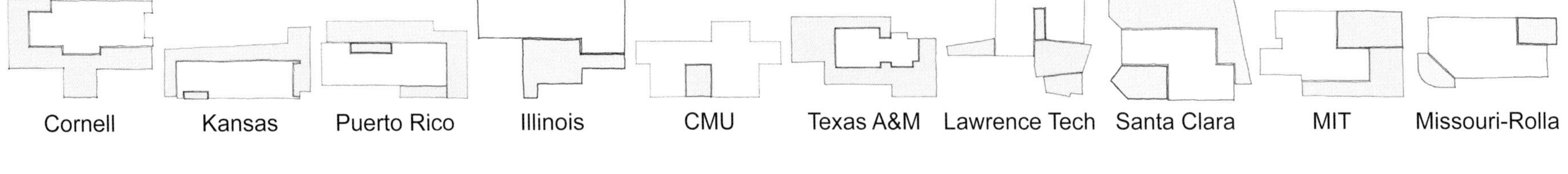
Cornell
Kansas
Puerto Rico
Illinois
CMU
Texas A&M
Lawrence Tech
Santa Clara
MIT
Missouri-Rolla

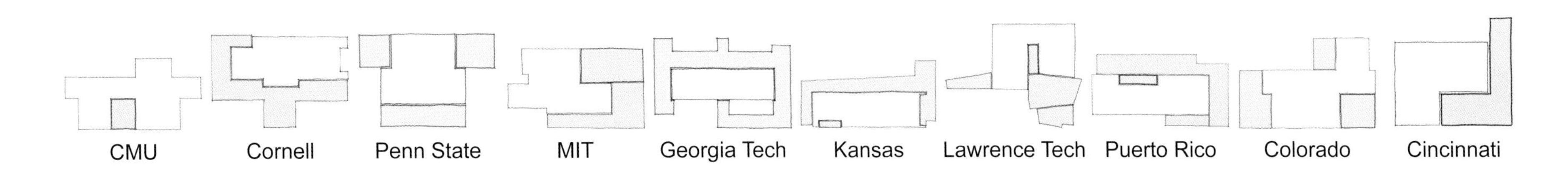
CMU
Cornell
Penn State
MIT
Georgia Tech
Kansas
Lawrence Tech
Puerto Rico
Colorado
Cincinnati

Construction Module

All teams had to transport their house from their university location to the National Mall in Washington, DC. How each team chose to do this varied considerably. Some designed the building as a single volume directly on top of a trailer. These single-volume modules required a crane if they were to be lifted off the trailer. Other teams used smaller modules that could be carried with smaller vehicles and, in some cases, moved without the use of a crane.

This diagram shows the module that was used for transporting the house to the National Mall in Washington, DC. This information was not published, so these are predominantly based on photos of the construction sequence in the construction phase of the competition.

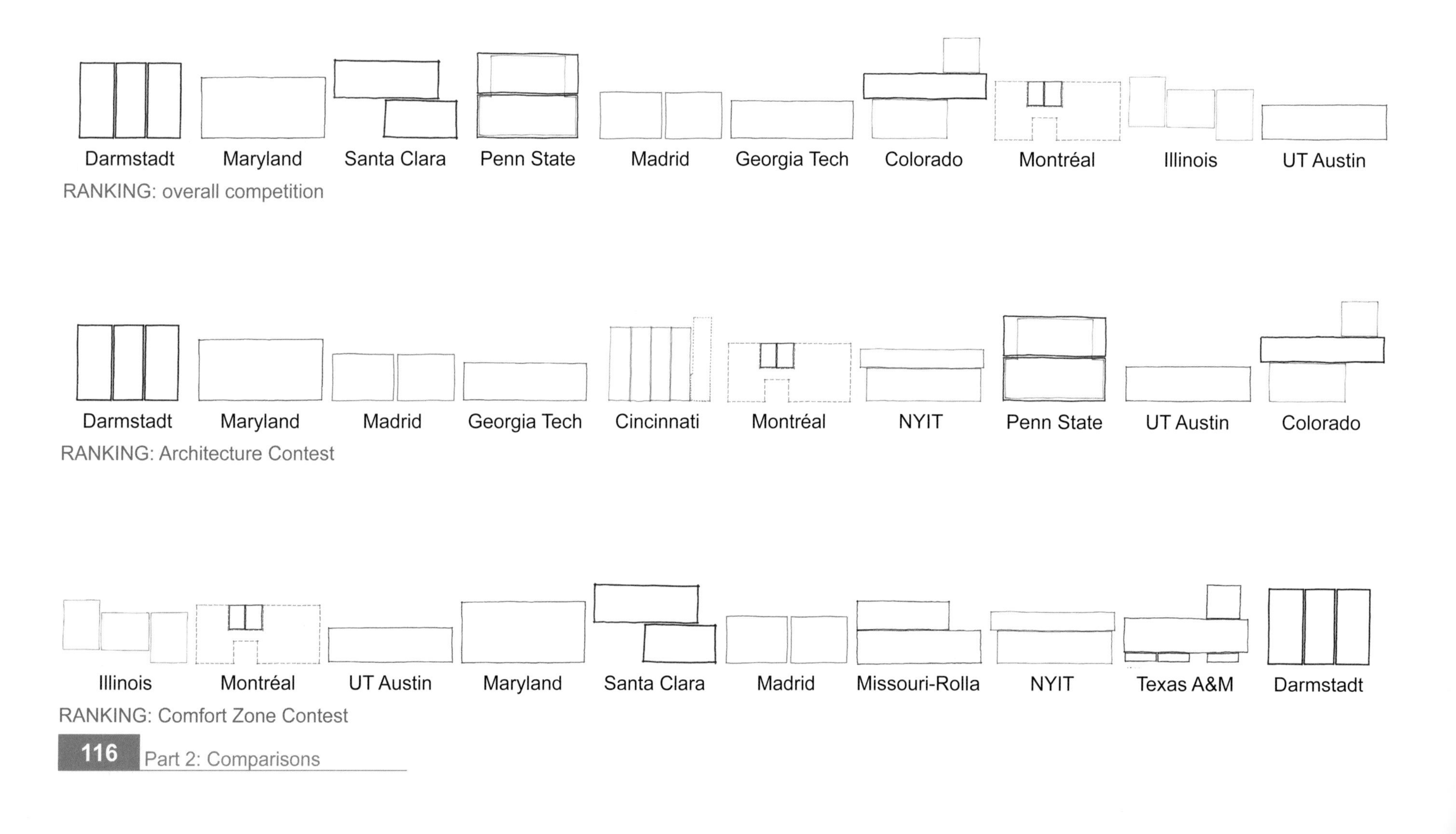

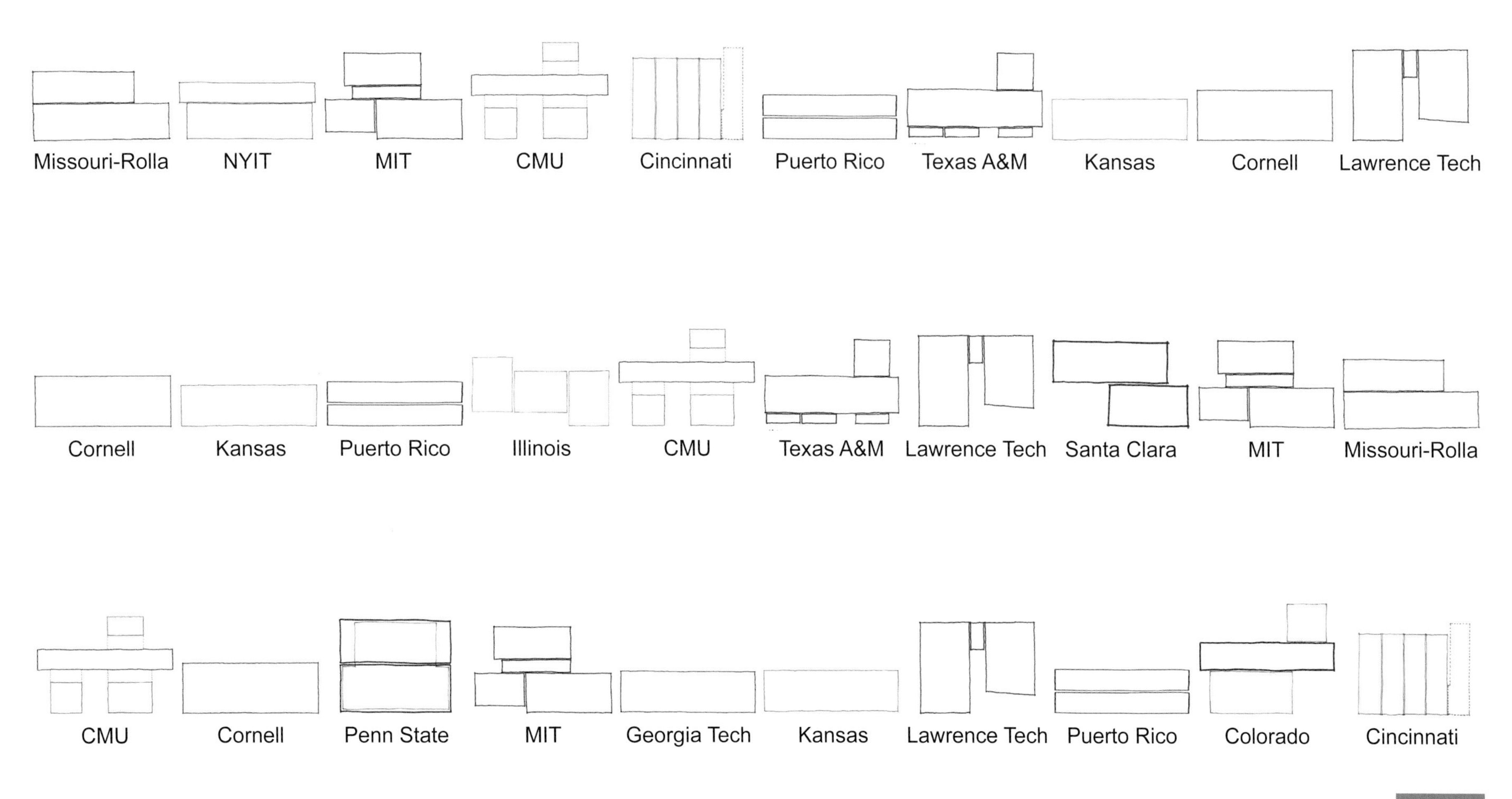
Missouri-Rolla
NYIT
MIT
CMU
Cincinnati
Puerto Rico
Texas A&M
Kansas
Cornell
Lawrence Tech
Cornell
Kansas
Puerto Rico
Illinois
CMU
Texas A&M
Lawrence Tech
Santa Clara
MIT
Missouri-Rolla
CMU
Cornell
Penn State
MIT
Georgia Tech
Kansas
Lawrence Tech
Puerto Rico
Colorado
Cincinnati

Structure

This diagram shows the load-bearing structural elements in solid lines, and the spanning structural elements in dashed lines.

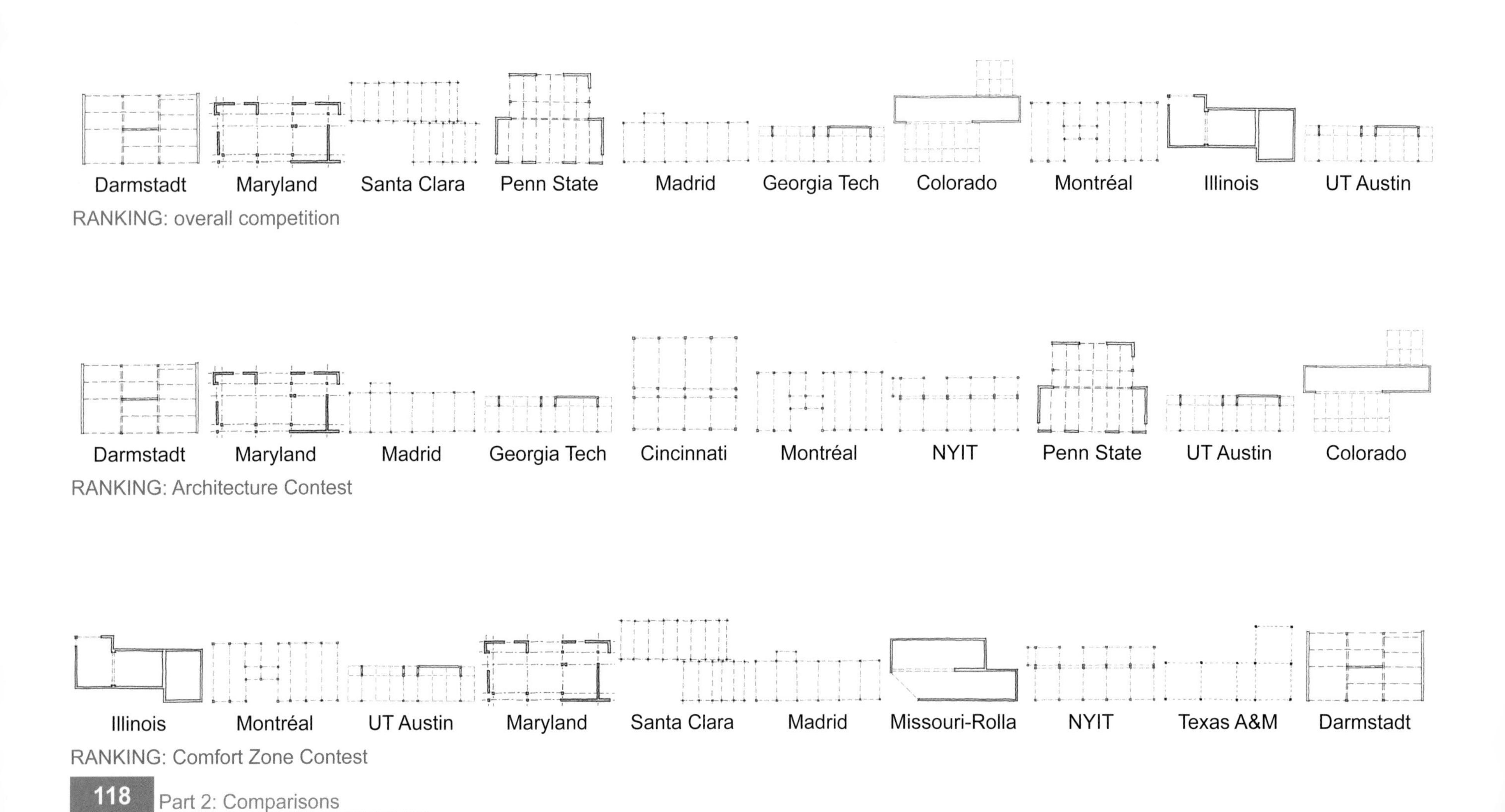

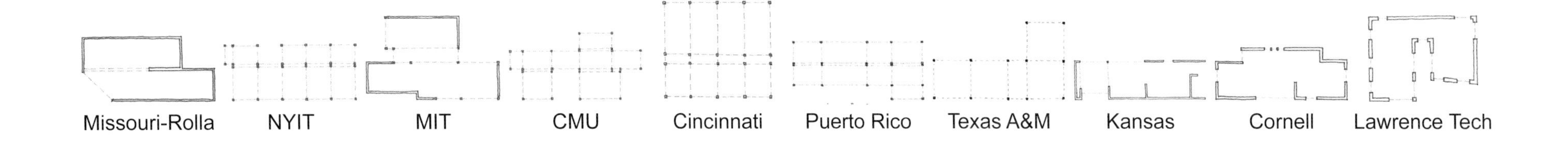
Missouri-Rolla
NYIT
MIT
CMU
Cincinnati
Puerto Rico
Texas A&M
Kansas
Cornell
Lawrence Tech

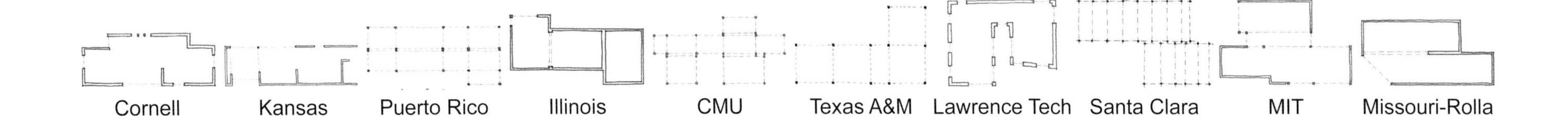
Cornell
Kansas
Puerto Rico
Illinois
CMU
Texas A&M
Lawrence Tech
Santa Clara
MIT
Missouri-Rolla

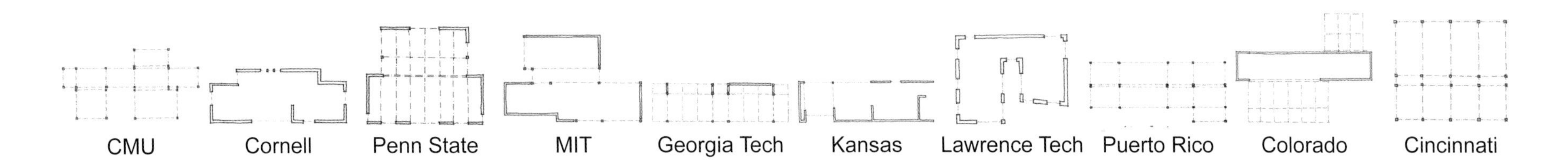
CMU
Cornell
Penn State
MIT
Georgia Tech
Kansas
Lawrence Tech
Puerto Rico
Colorado
Cincinnati

Additive/Subtractive

Beginning with the basic forms, this diagram investigates form additions and subtractions to it. The basic form is shown in gray. Additions and subtractions are shown in white.

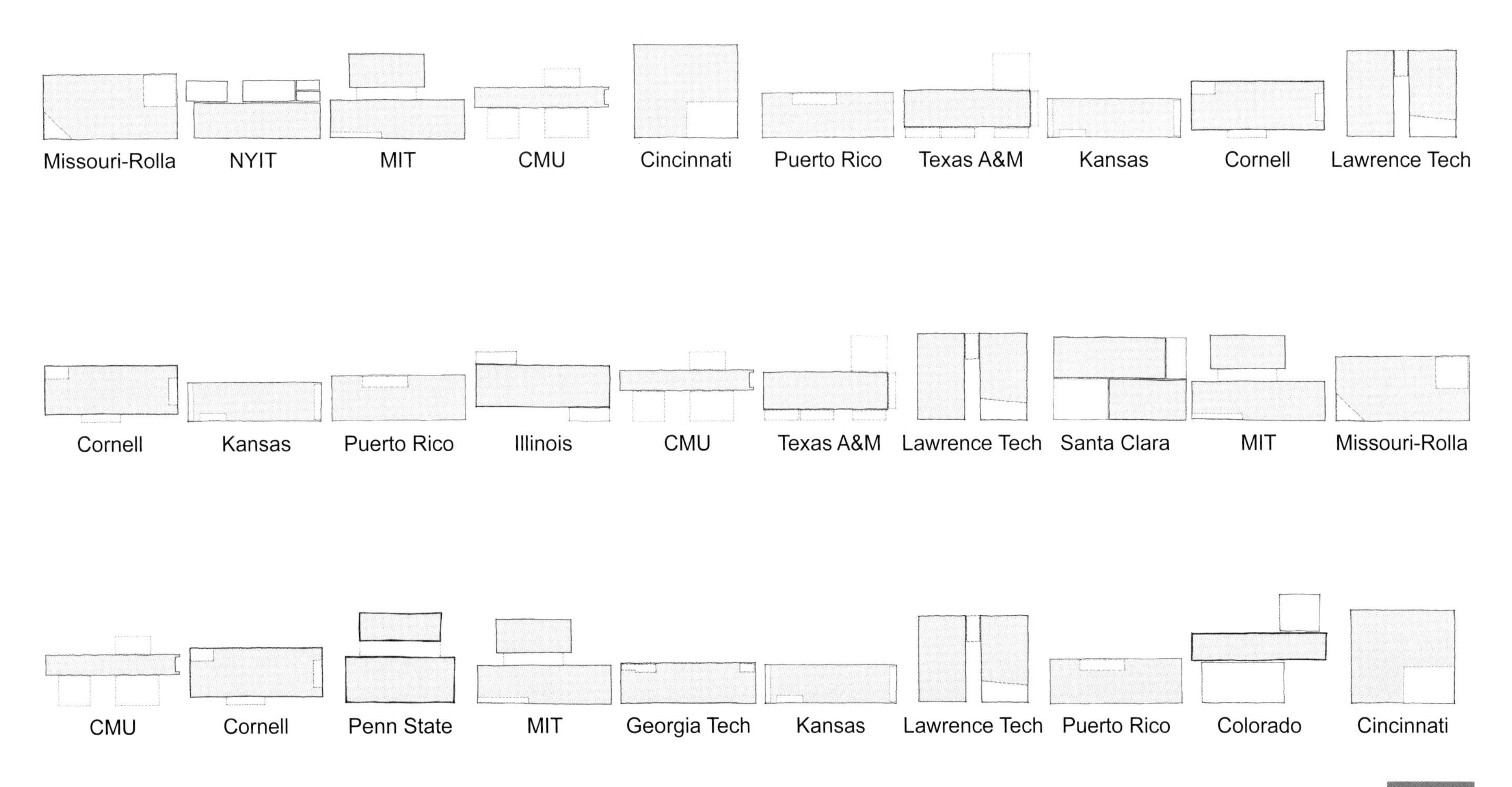
Missouri-Rolla
NYIT
MIT
CMU
Cincinnati
Puerto Rico
Texas A&M
Kansas
Cornell
Lawrence Tech
Cornell
Kansas
Puerto Rico
Illinois
CMU
Texas A&M
Lawrence Tech
Santa Clara
MIT
Missouri-Rolla
CMU
Cornell
Penn State
MIT
Georgia Tech
Kansas
Lawrence Tech
Puerto Rico
Colorado
Cincinnati

Public/Private Spaces

This diagram differentiates between the public and private spaces. The “public” spaces (shown in gray) include the spaces where a interaction might occur with a guest: living/dining space, study and kitchen. The “private” space (shown in white) includes the mechanical services, bathroom and bedroom.

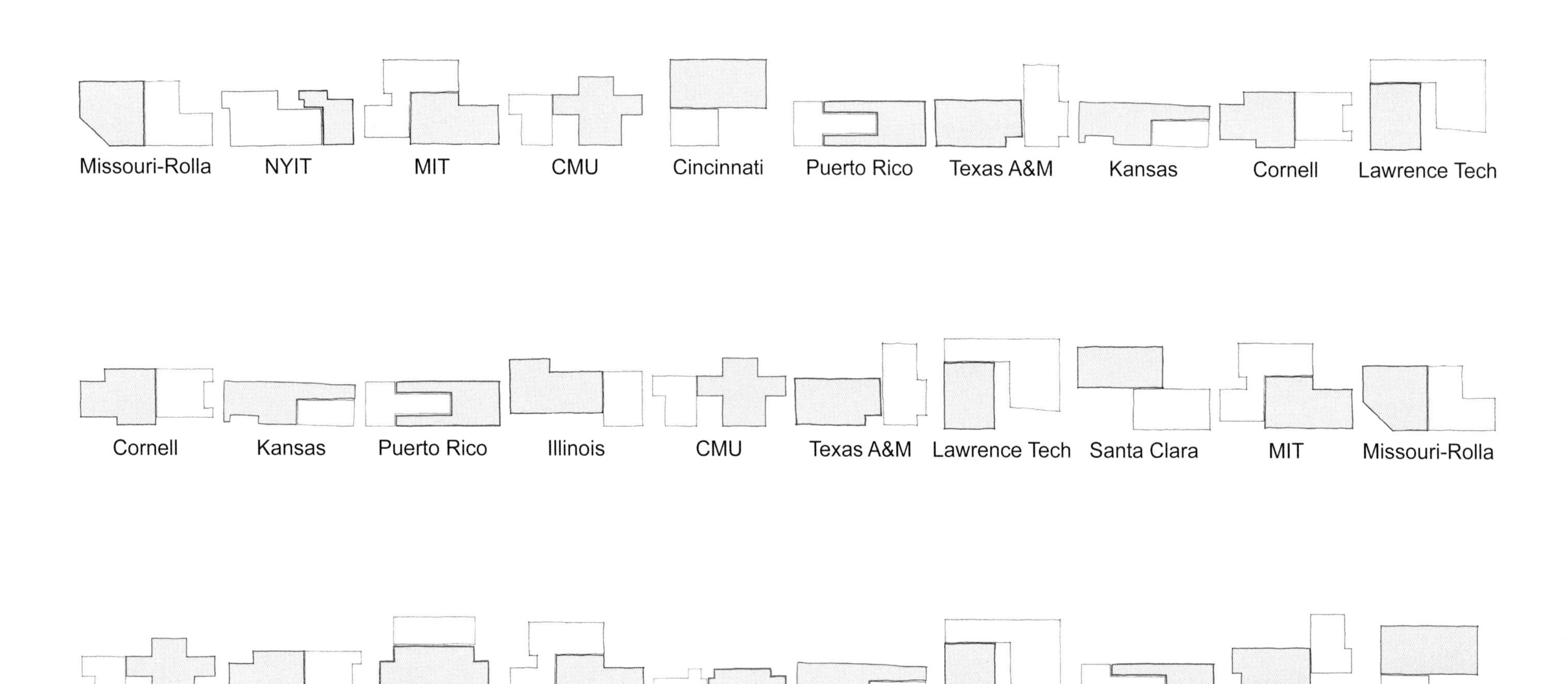
Missouri-Rolla
NYIT
MIT
CMU
Cincinnati
Puerto Rico
Texas A&M
Kansas
Cornell
Lawrence Tech
Cornell
Kansas
Puerto Rico
Illinois
CMU
Texas A&M
Lawrence Tech
Santa Clara
MIT
Missouri-Rolla
CMU
Cornell
Penn State
MIT
Georgia Tech
Kansas
Lawrence Tech
Puerto Rico
Colorado
Cincinnati

Interior Spatial Forms

Beginning with the basic interior forms, this diagram investigates form additions and subtractions to it. The basic interior form is shown in gray within the perimeter of the exterior form.

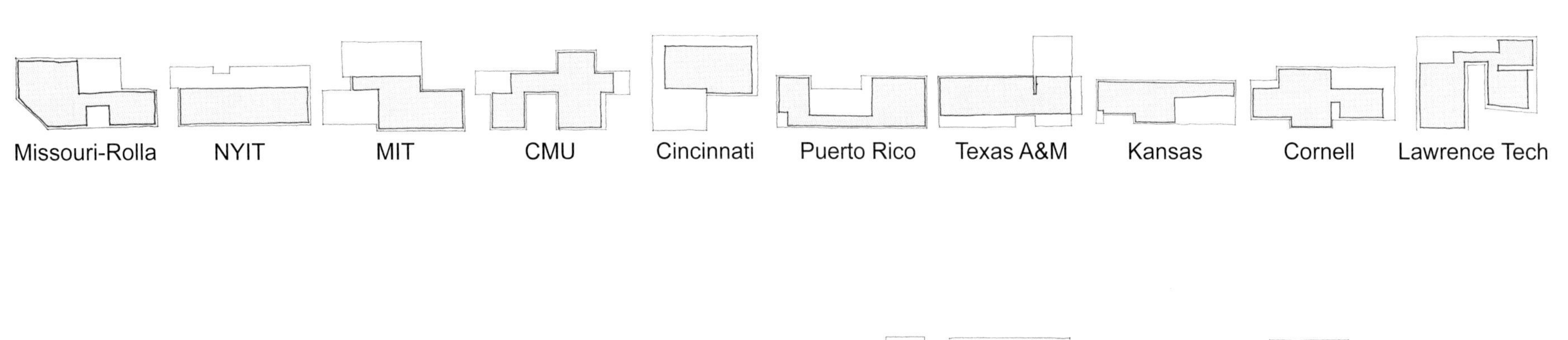
Missouri-Rolla
NYIT
MIT
CMU
Cincinnati
Puerto Rico
Texas A&M
Kansas
Cornell
Lawrence Tech

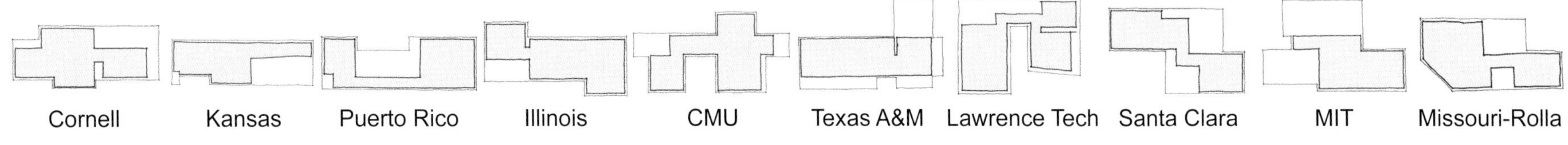
Cornell
Kansas
Puerto Rico
Illinois
CMU
Texas A&M
Lawrence Tech
Santa Clara
MIT
Missouri-Rolla

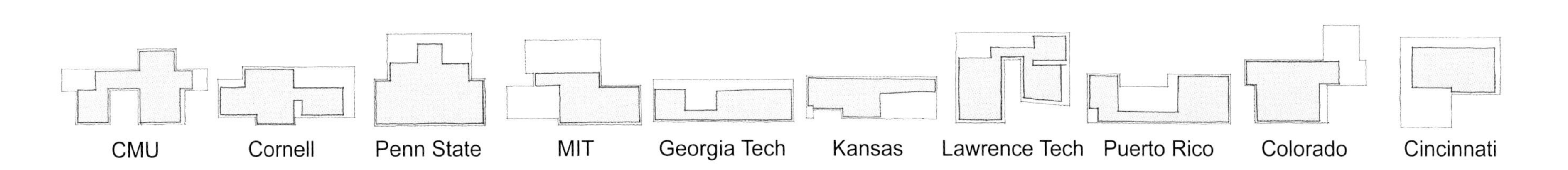
CMU
Cornell
Penn State
MIT
Georgia Tech
Kansas
Lawrence Tech
Puerto Rico
Colorado
Cincinnati

Served/Service

Louis Kahn introduced the concept of "served" and "service" spaces. "Service" space is that which is dedicated to supporting the "served" spaces. The served space is shown in gray with the service spaces in white.

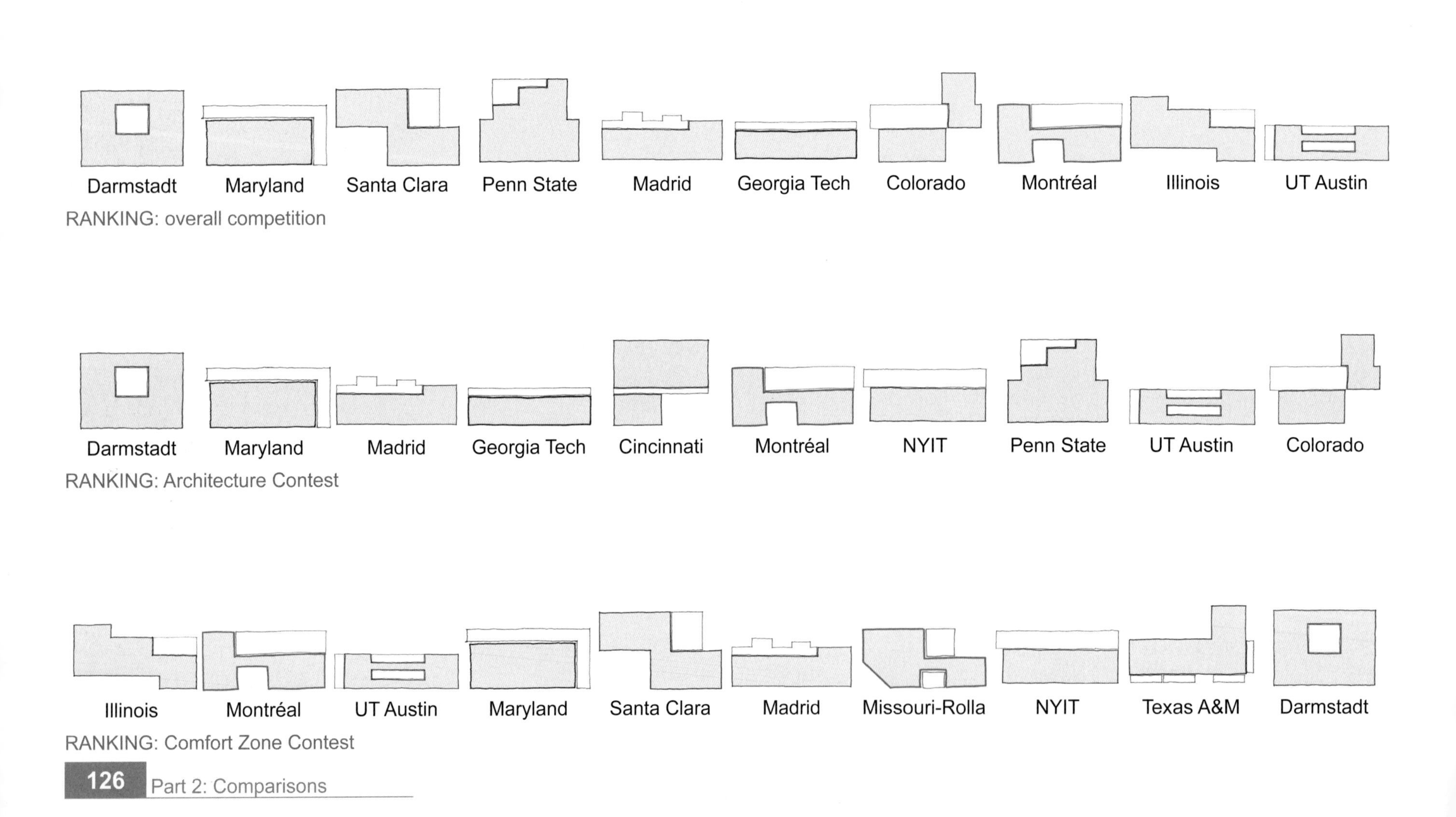

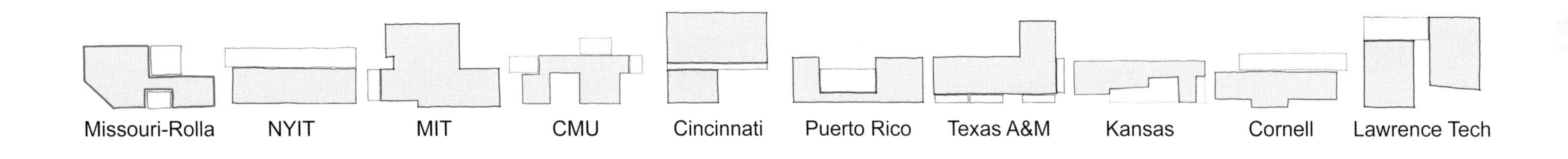
Missouri-Rolla
NYIT
MIT
CMU
Cincinnati
Puerto Rico
Texas A&M
Kansas
Cornell
Lawrence Tech

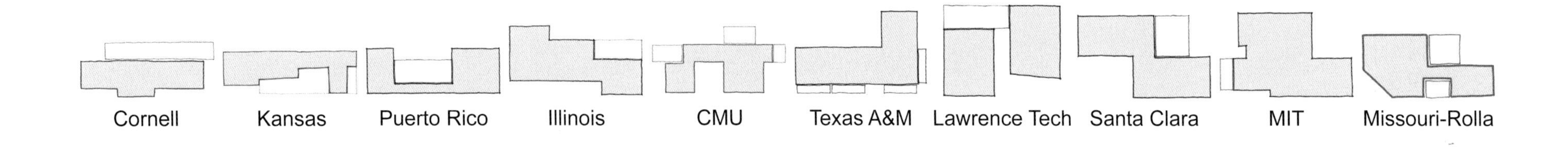
Cornell
Kansas
Puerto Rico
Illinois
CMU
Texas A&M
Lawrence Tech
Santa Clara
MIT
Missouri-Rolla

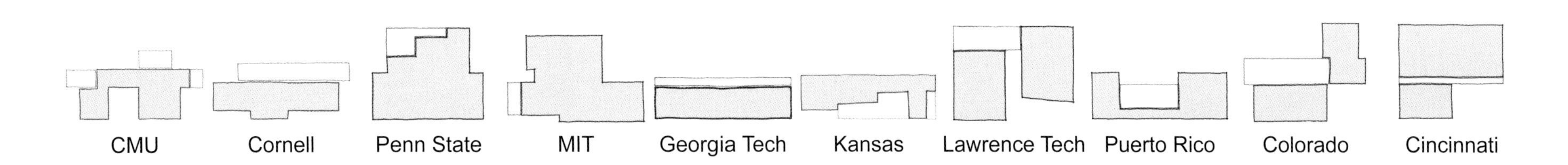
CMU
Cornell
Penn State
MIT
Georgia Tech
Kansas
Lawrence Tech
Puerto Rico
Colorado
Cincinnati

Interior Zones

This diagram uses four tones to differentiate the programmatic zones in each house.

- living space
- kitchen
- bedroom
- bathroom

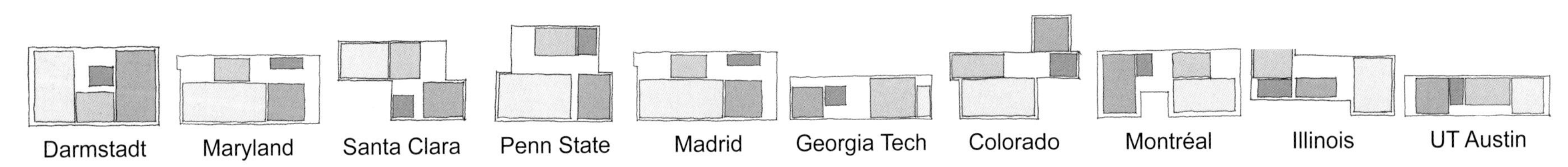

RANKING: overall competition

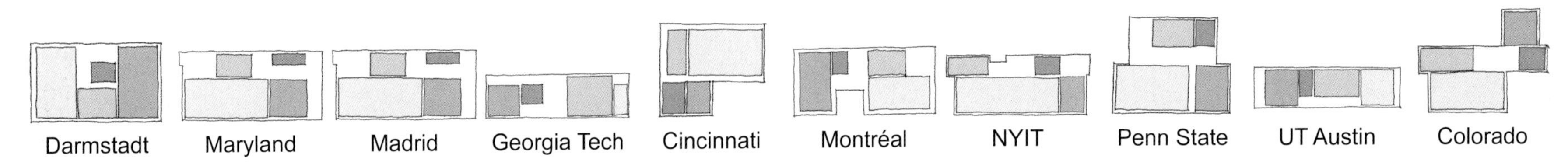

RANKING: Architecture Contest

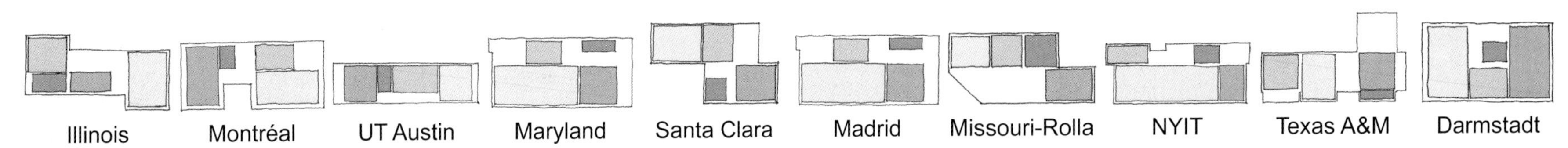

RANKING: Comfort Zone Contest

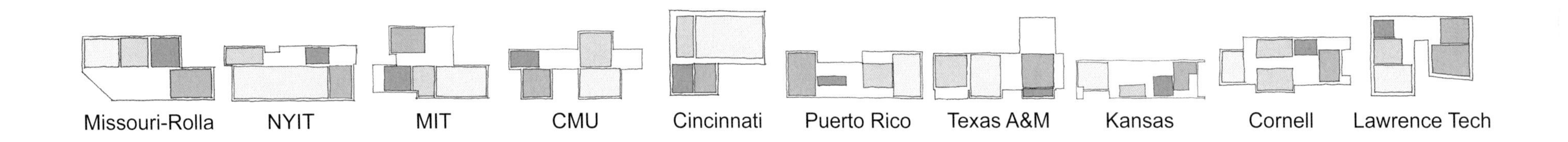
Missouri-Rolla
NYIT
MIT
CMU
Cincinnati
Puerto Rico
Texas A&M
Kansas
Cornell
Lawrence Tech

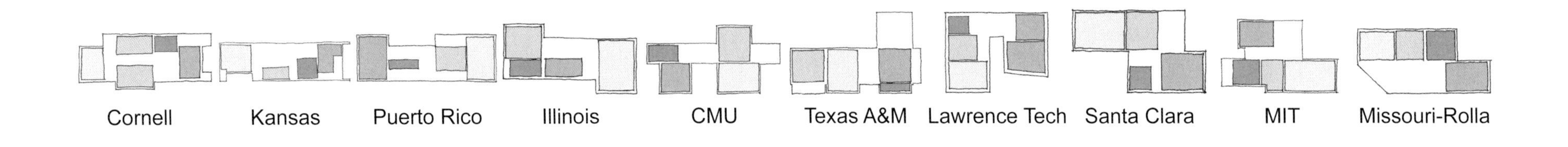
Cornell
Kansas
Puerto Rico
Illinois
CMU
Texas A&M
Lawrence Tech
Santa Clara
MIT
Missouri-Rolla

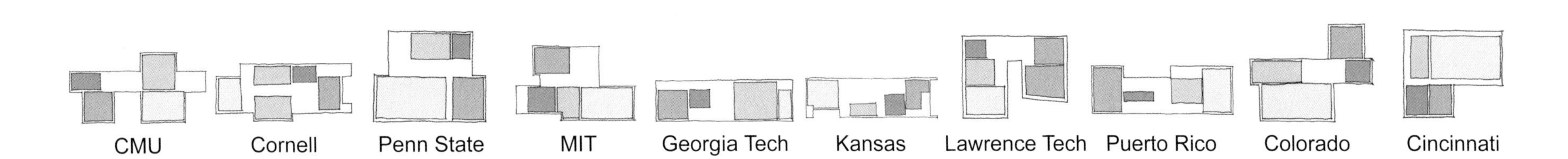
CMU
Cornell
Penn State
MIT
Georgia Tech
Kansas
Lawrence Tech
Puerto Rico
Colorado
Cincinnati

Geometry

This diagram shows basic geometric shapes evident in the plan. These include the square, "√2 rectangle" (1:1.414) and "golden rectangle" (1:1.618).

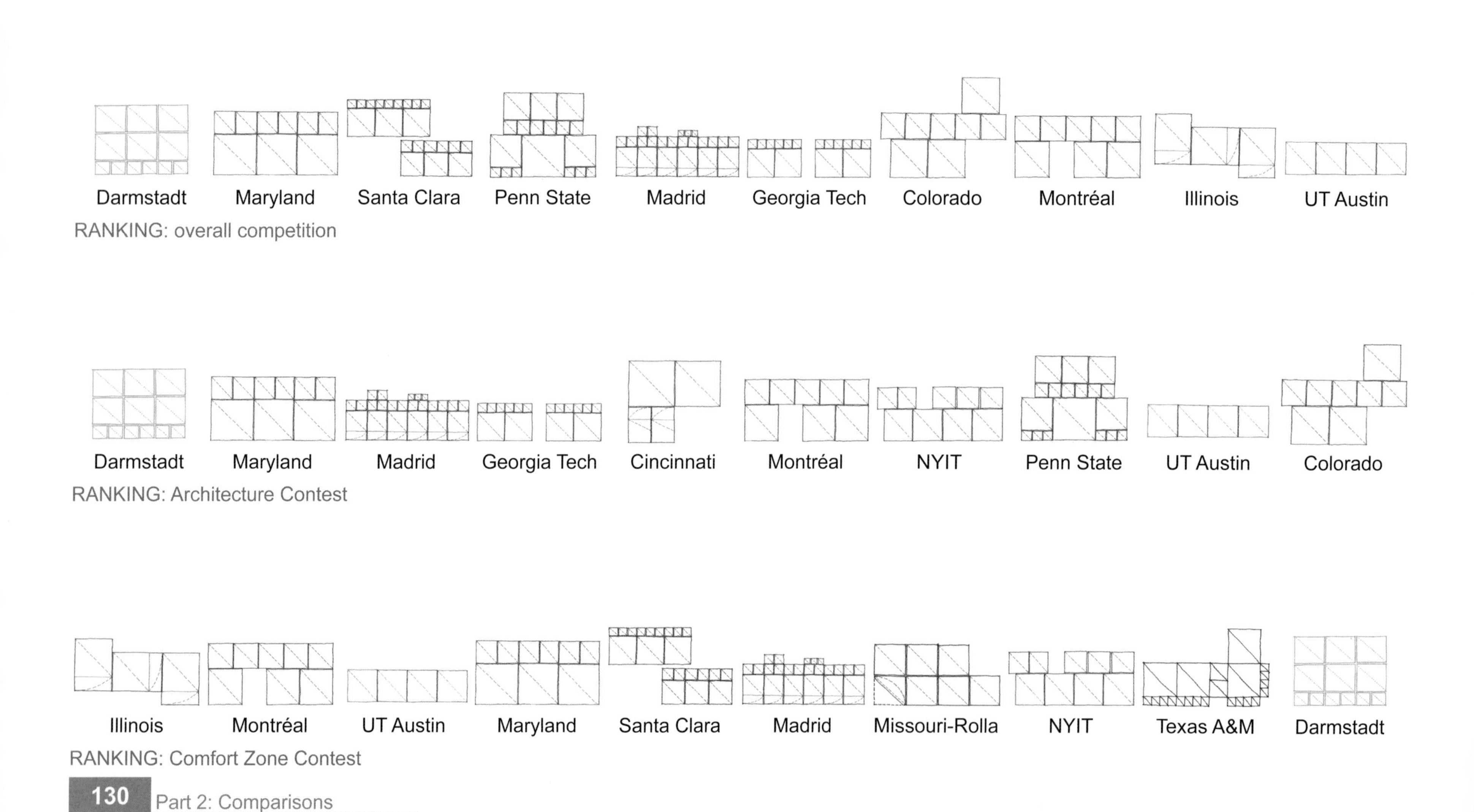

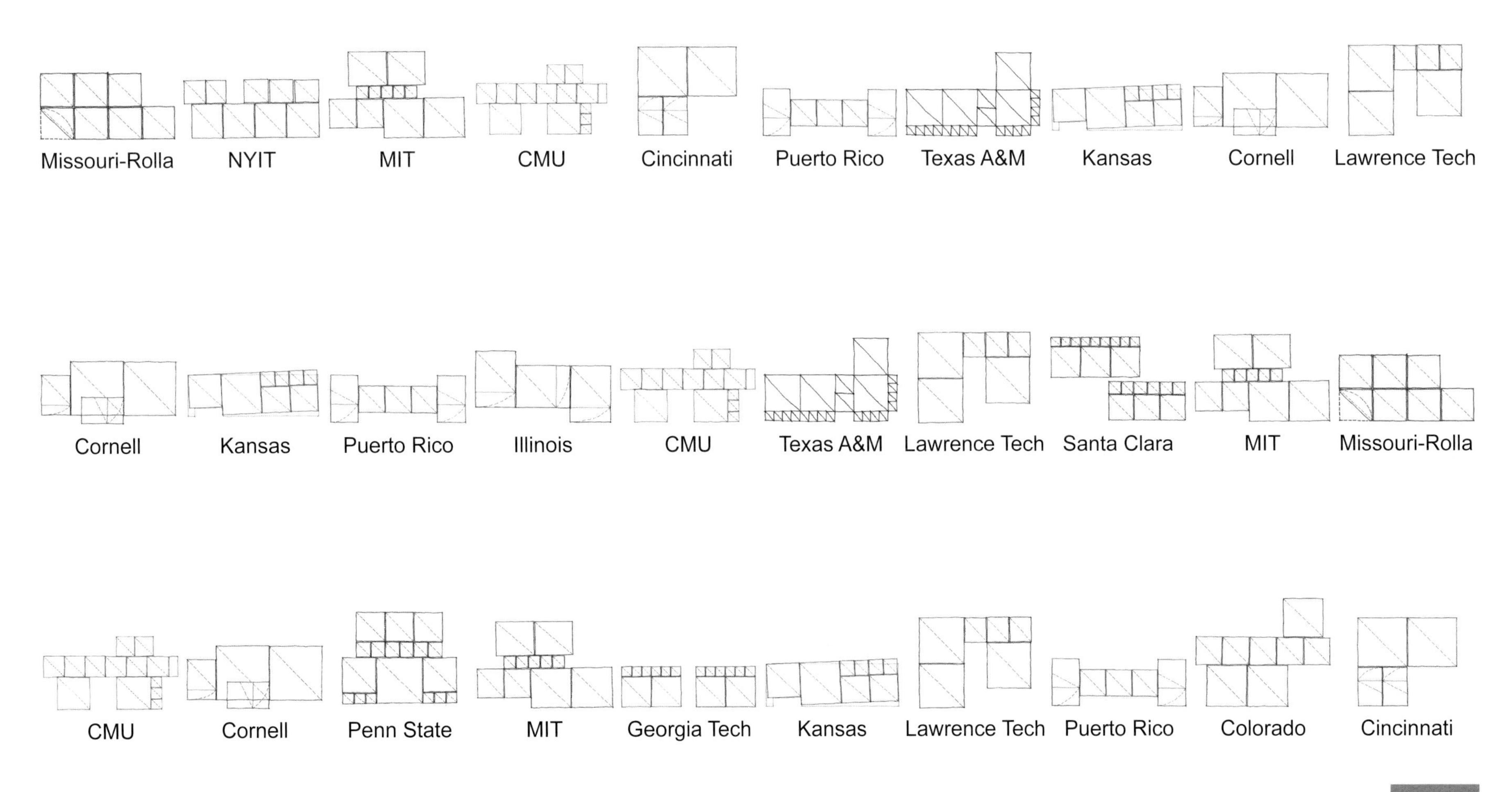
Missouri-Rolla
NYIT
MIT
CMU
Cincinnati
Puerto Rico
Texas A&M
Kansas
Cornell
Lawrence Tech
Cornell
Kansas
Puerto Rico
Illinois
CMU
Texas A&M
Lawrence Tech
Santa Clara
MIT
Missouri-Rolla
CMU
Cornell
Penn State
MIT
Georgia Tech
Kansas
Lawrence Tech
Puerto Rico
Colorado
Cincinnati

Natural Ventilation – Plan

This set of ventilation diagrams allows us to compare how and where natural ventilation is being incorporated in and through the houses. In most cases it is across the short width of the house, although in some cases it is along the length of the house.

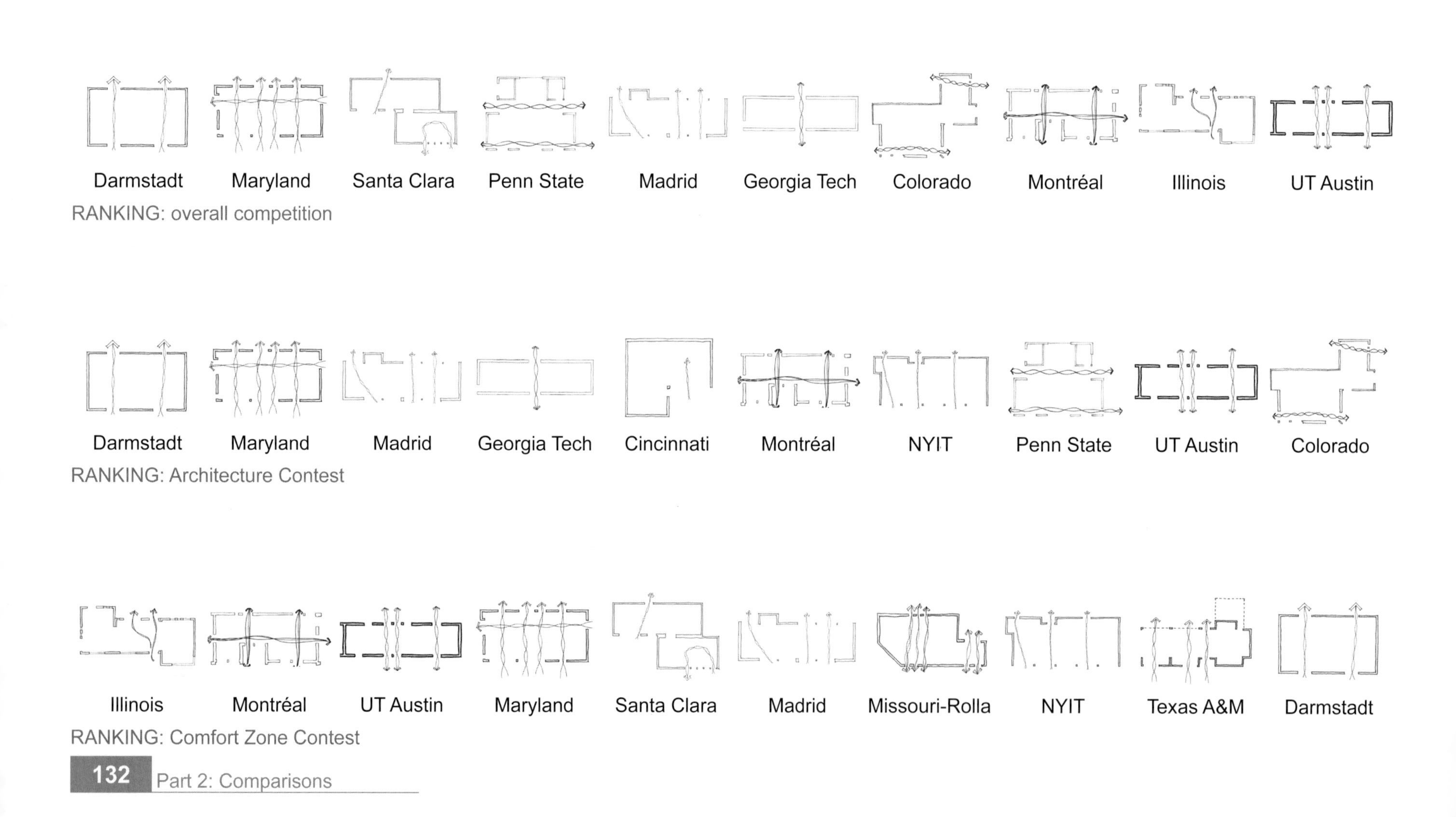

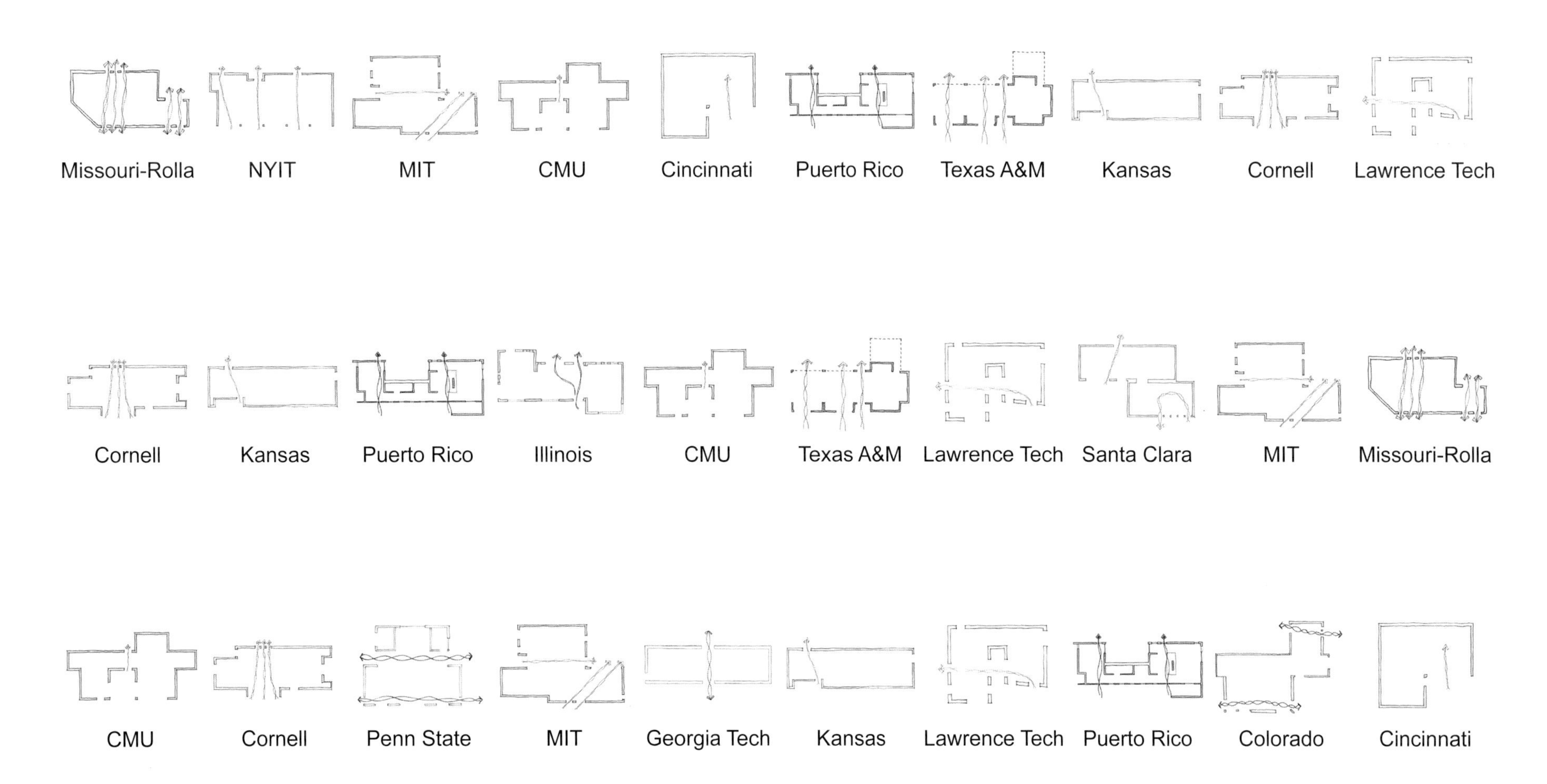
Missouri-Rolla
NYIT
MIT
CMU
Cincinnati
Puerto Rico
Texas A&M
Kansas
Cornell
Lawrence Tech
Cornell
Kansas
Puerto Rico
Illinois
CMU
Texas A&M
Lawrence Tech
Santa Clara
MIT
Missouri-Rolla
CMU
Cornell
Penn State
MIT
Georgia Tech
Kansas
Lawrence Tech
Puerto Rico
Colorado
Cincinnati

Natural Ventilation – Section

This set of ventilation diagrams allows us to compare how and where natural ventilation is being incorporated in and through the houses. In most cases it is across the short width of the house, although in some cases it is along the length of the house. In section, we can see where teams attempted to increase the velocity of air movement by placing air inlets low and outlets high, thereby creating opportunities for stack effect.

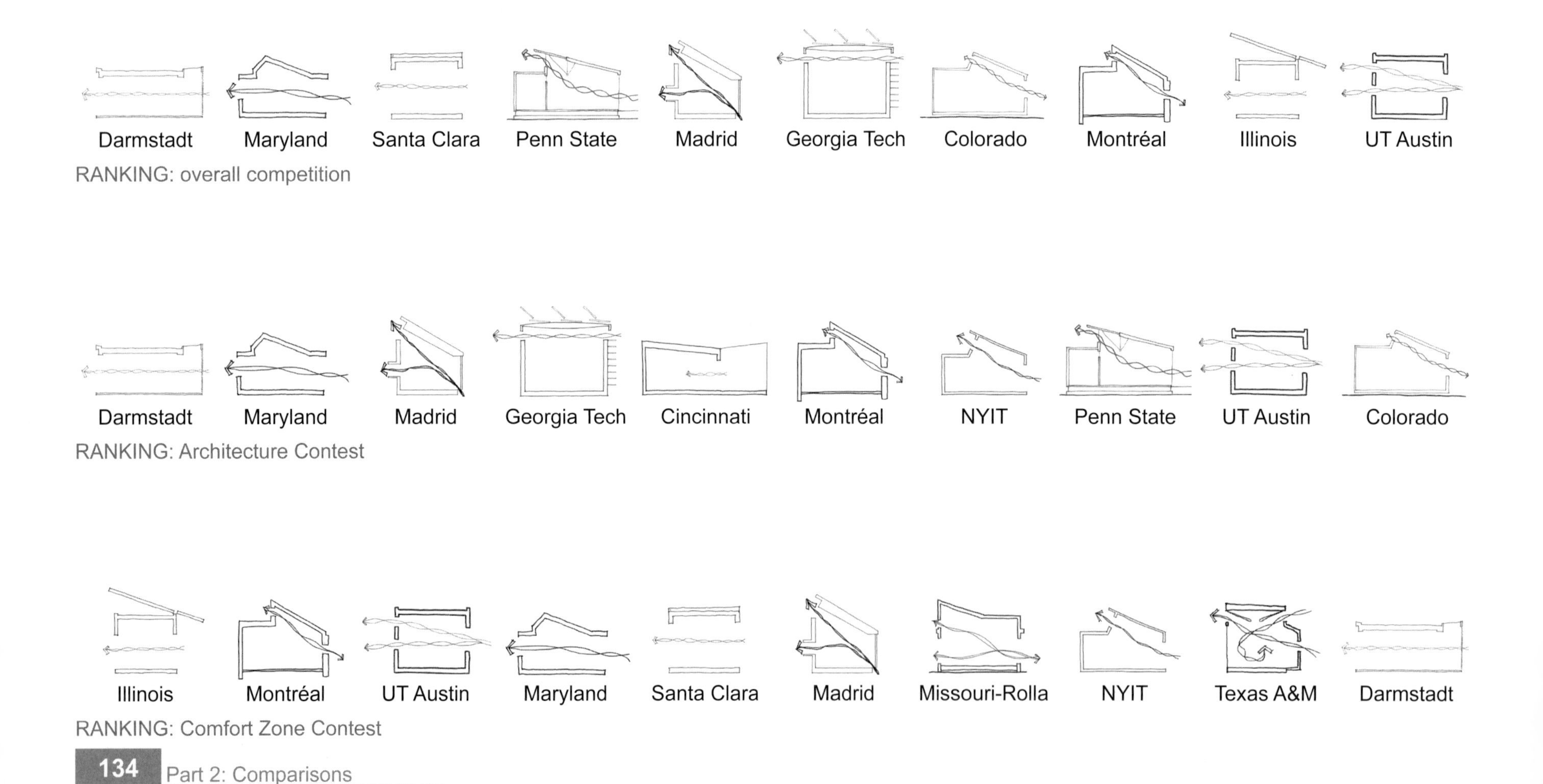

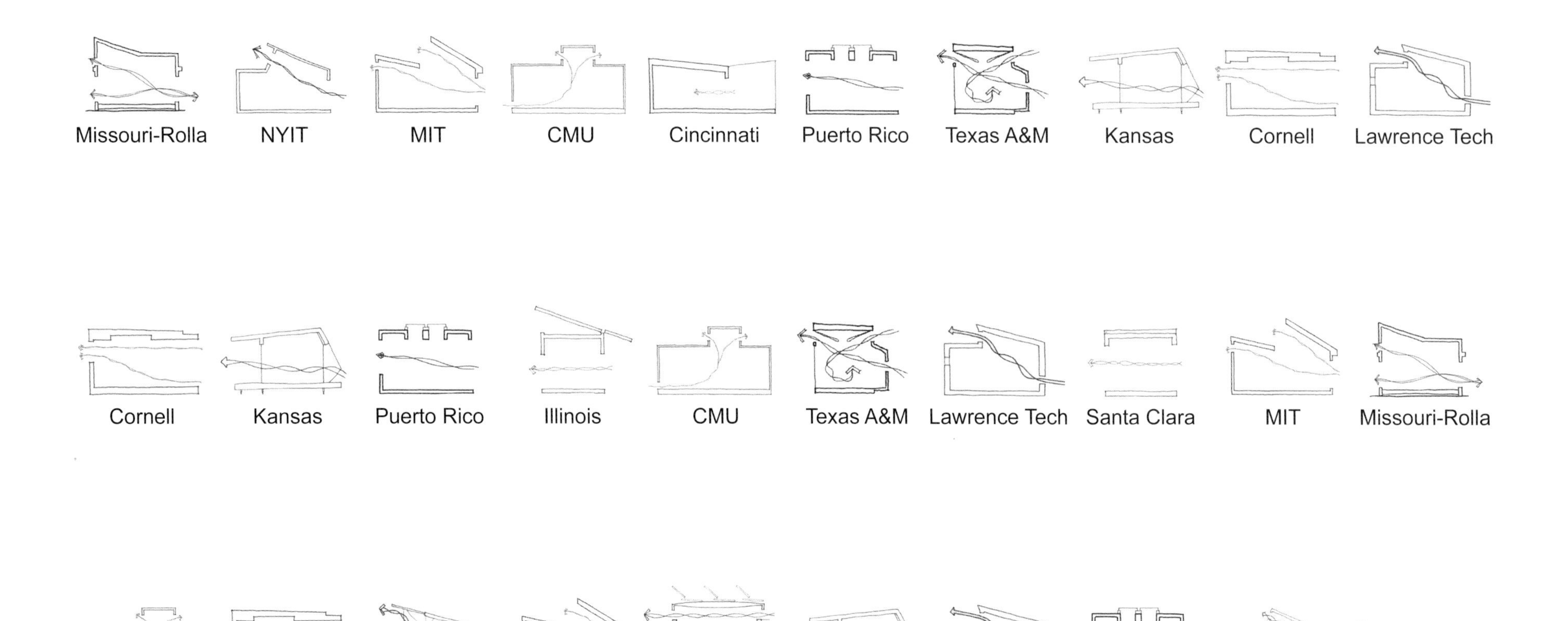
Missouri-Rolla
NYIT
MIT
CMU
Cincinnati
Puerto Rico
Texas A&M
Kansas
Cornell
Lawrence Tech
Cornell
Kansas
Puerto Rico
Illinois
CMU
Texas A&M
Lawrence Tech
Santa Clara
MIT
Missouri-Rolla
CMU
Cornell
Penn State
MIT
Georgia Tech
Kansas
Lawrence Tech
Puerto Rico
Colorado
Cincinnati

Daylighting

This diagram shows where direct and ambient daylight are entering the interior of the houses. Direct light is shown with a clearly bound white area while ambient daylight is shown with a light, gradient hatch that decreases as distance from the daylight source (glazing) increases.

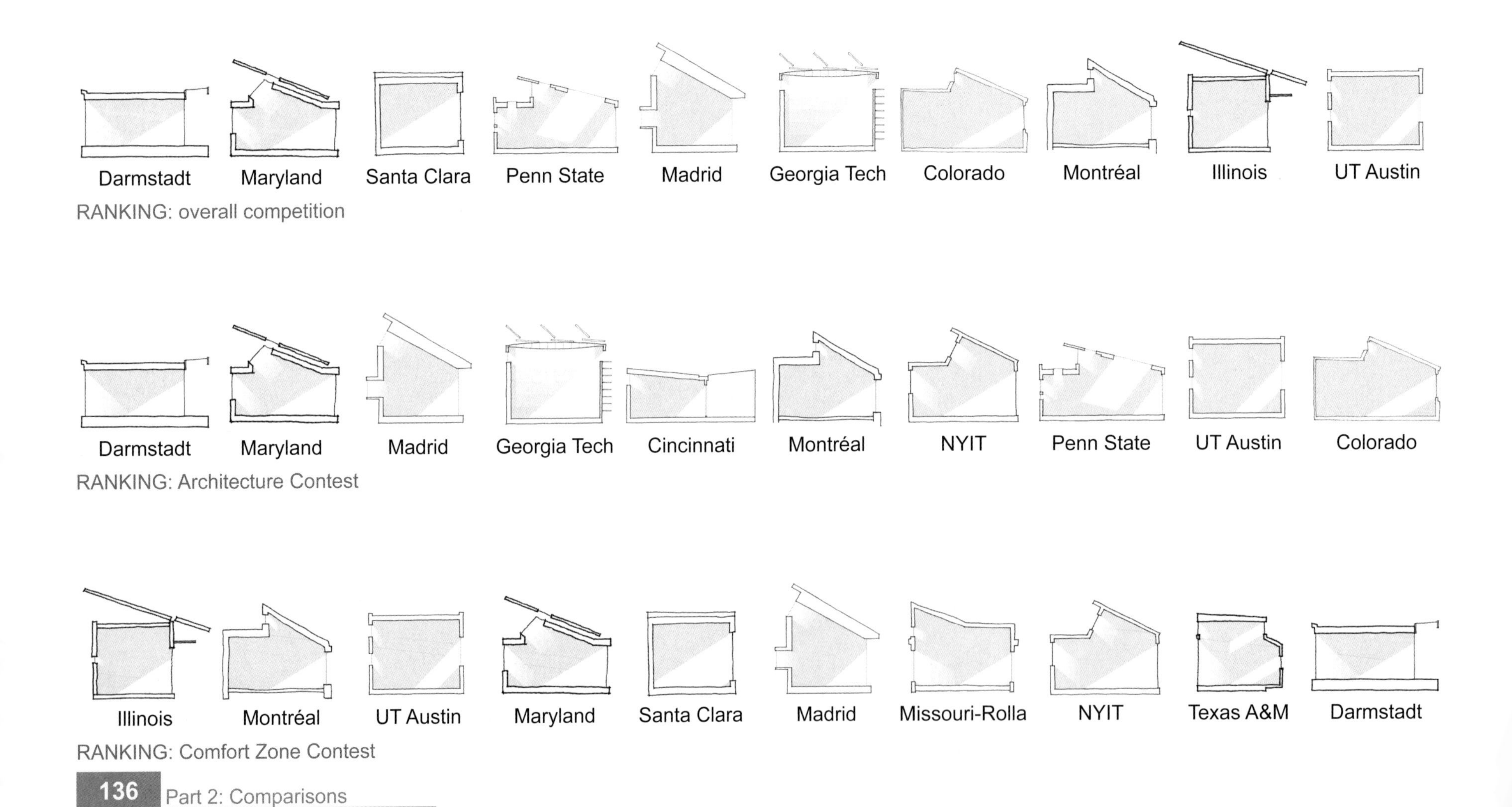

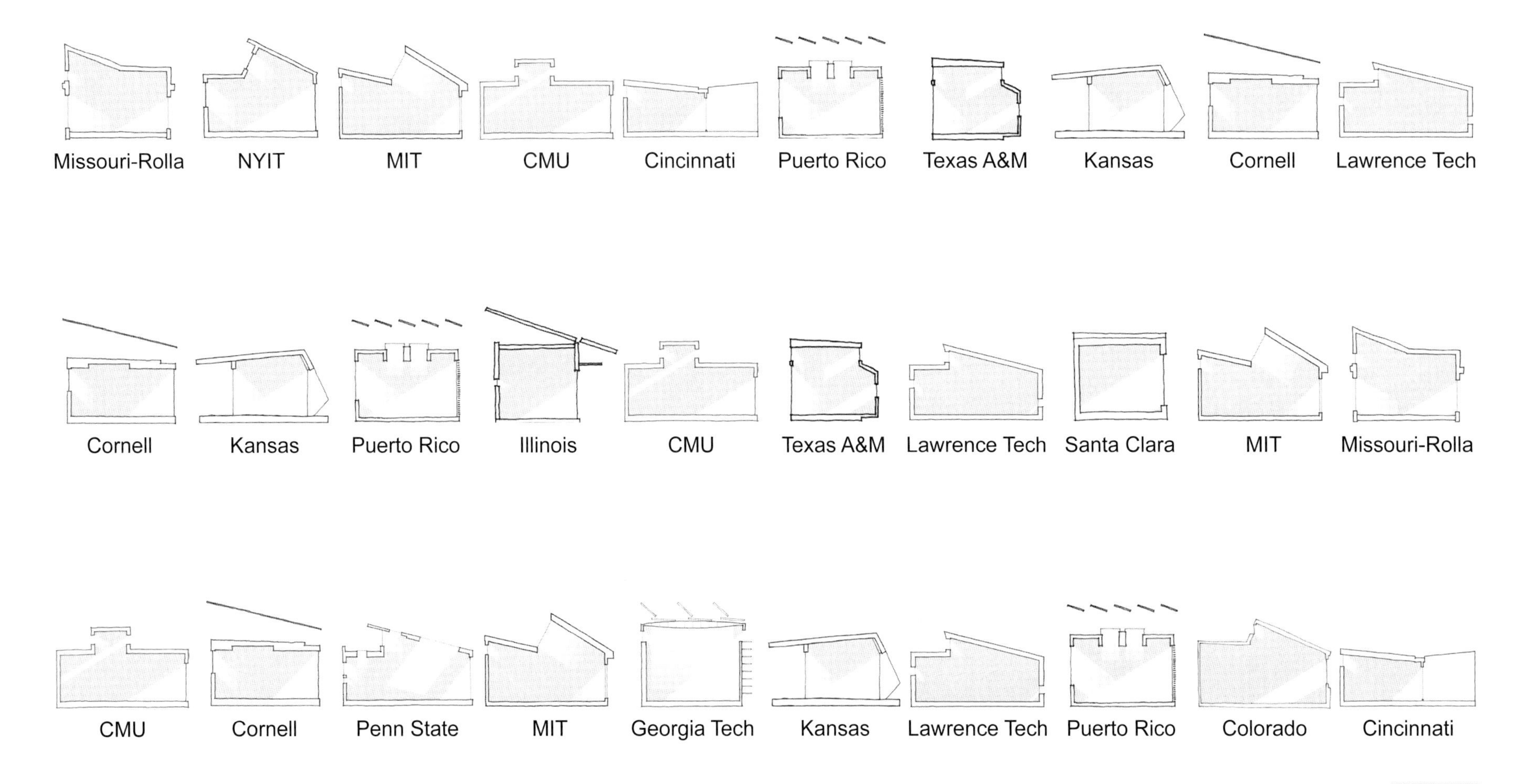
Missouri-Rolla
NYIT
MIT
CMU
Cincinnati
Puerto Rico
Texas A&M
Kansas
Cornell
Lawrence Tech
Cornell
Kansas
Puerto Rico
Illinois
CMU
Texas A&M
Lawrence Tech
Santa Clara
MIT
Missouri-Rolla
CMU
Cornell
Penn State
MIT
Georgia Tech
Kansas
Lawrence Tech
Puerto Rico
Colorado
Cincinnati

Shading

The shading diagram shows where direct light can enter a given space and where direct light is kept out as a result of shading. It also shows the general shape of any physical shade element (overhang, sunshade, etc.).

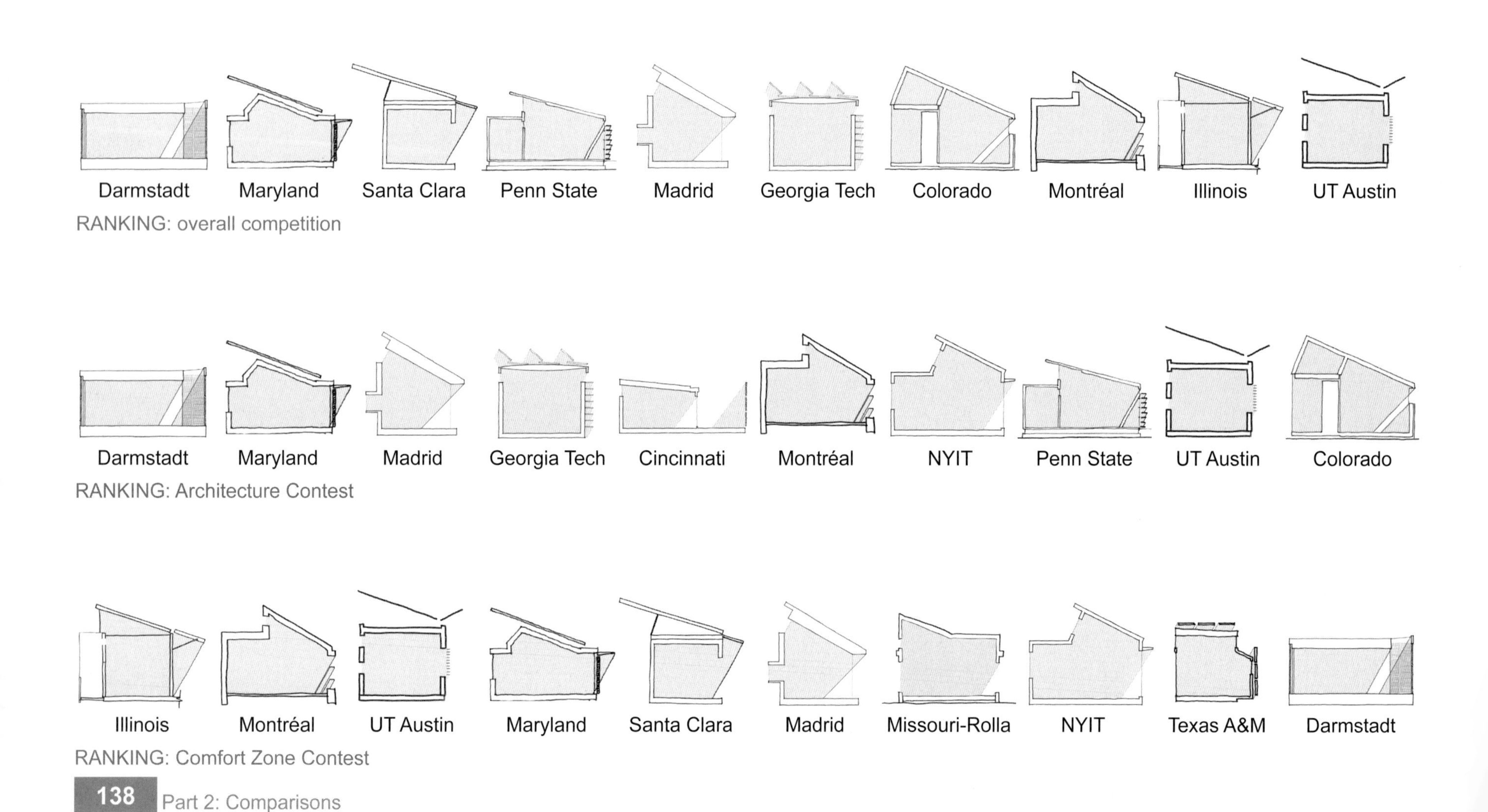

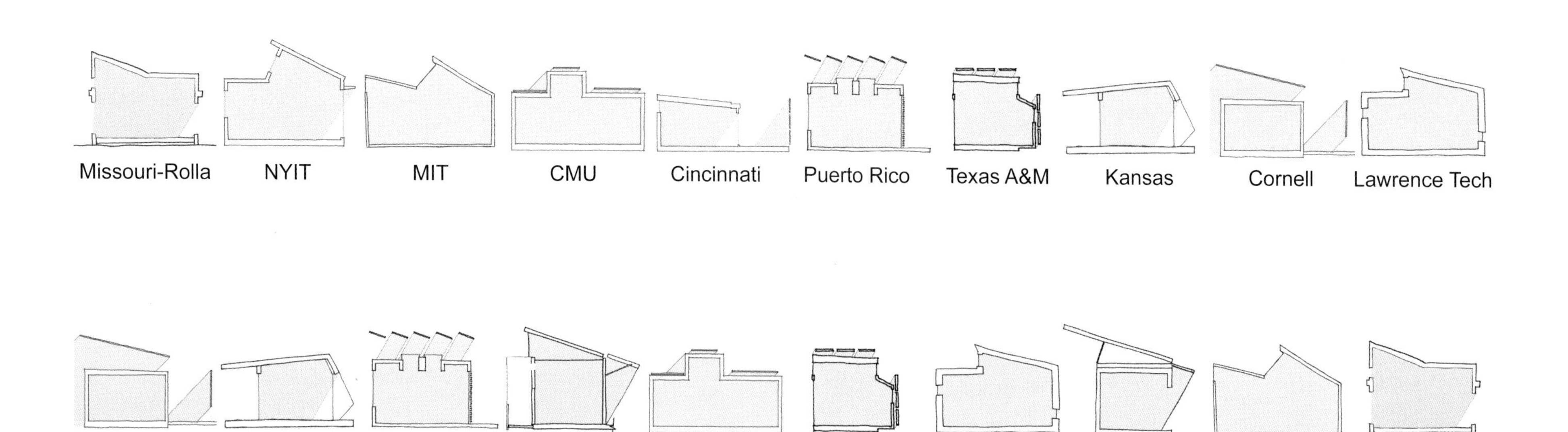

CMU Cornell Penn State MIT Georgia Tech Kansas Lawrence Tech Puerto Rico Colorado Cincinnati

Solar Angle – Angle of PV Array

Every team had to face a critical decision about solar angle in the competition. Unless the house was designed with a photovoltaic system that could be adjusted, a specific angle had to be chosen. Every team was fully aware of the ideal angle in Washington, DC during the competition, but each team was also designing a house that would exist in another location following the event. One example, the Kansas team, chose to optimize its solar angle (64°) for winter in their location. As a result, it was less efficient during the competition than it would have been had its solar angle been shallower. The other factor was total coverage of photovoltaics. Darmstadt had its roof completely covered in a nearly flat array. The low solar angle decreased efficiency, but the production of the array was extremely high as a result of the total amount of photovoltaic panel exposure. (See PV Area Analyses in Part 3.)

Note: Secondary Solar Angle is shown in parentheses

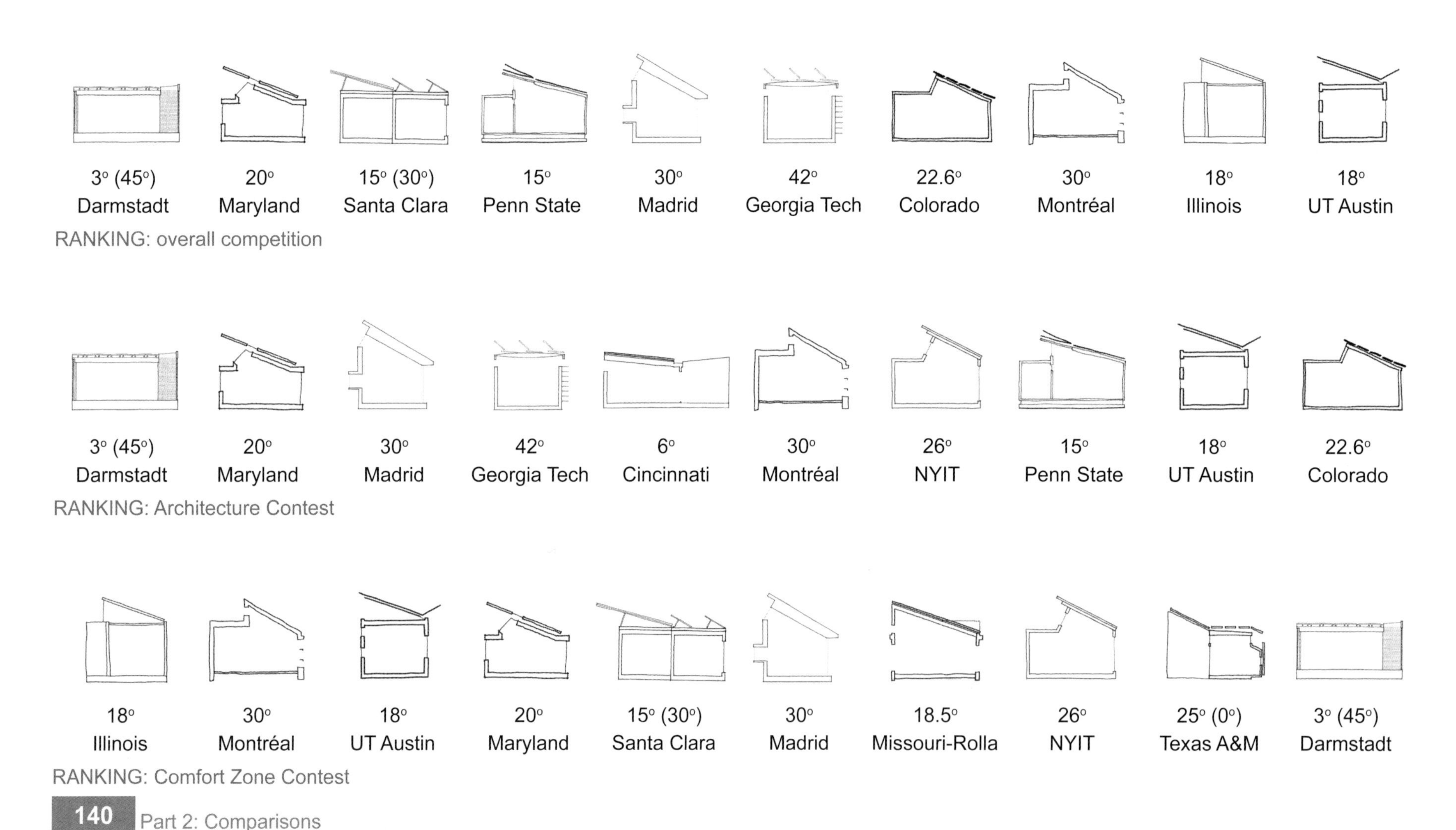

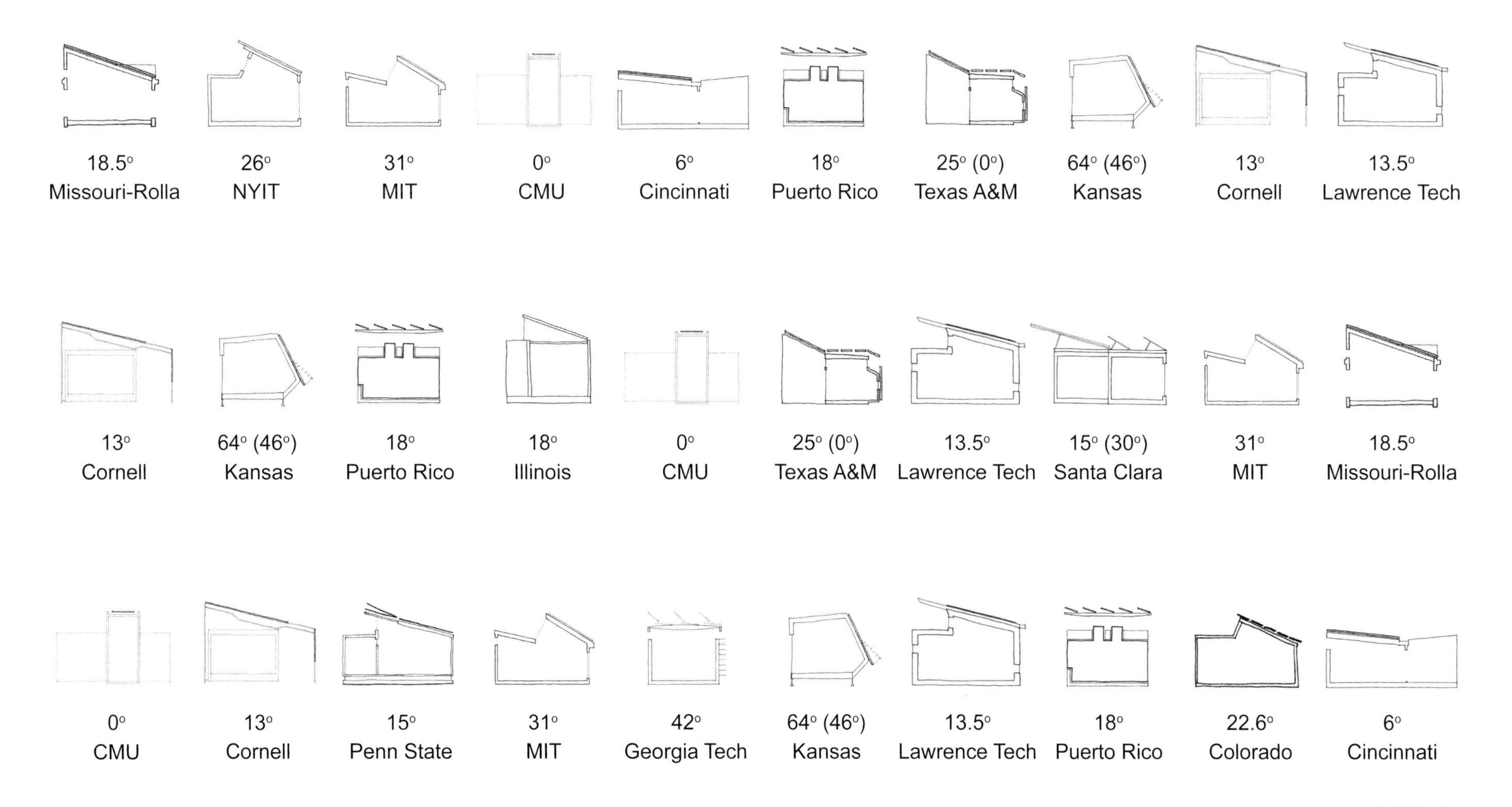
18.5°
Missouri-Rolla
26°
NYIT
31°
MIT
0°
CMU
6°
Cincinnati
18°
Puerto Rico
25° (0°)
Texas A&M
64° (46°)
Kansas
13°
Cornell
13.5°
Lawrence Tech
13°
Cornell
64° (46°)
Kansas
18°
Puerto Rico
18°
Illinois
0°
CMU
25° (0°)
Texas A&M
13.5°
Lawrence Tech
15° (30°)
Santa Clara
31°
MIT
18.5°
Missouri-Rolla
0°
CMU
13°
Cornell
15°
Penn State
31°
MIT
42°
Georgia Tech
64° (46°)
Kansas
13.5°
Lawrence Tech
18°
Puerto Rico
22.6°
Colorado
6°
Cincinnati

Photovoltaic Array

This set of comparisons illustrates the photovoltaic array layout of each house.

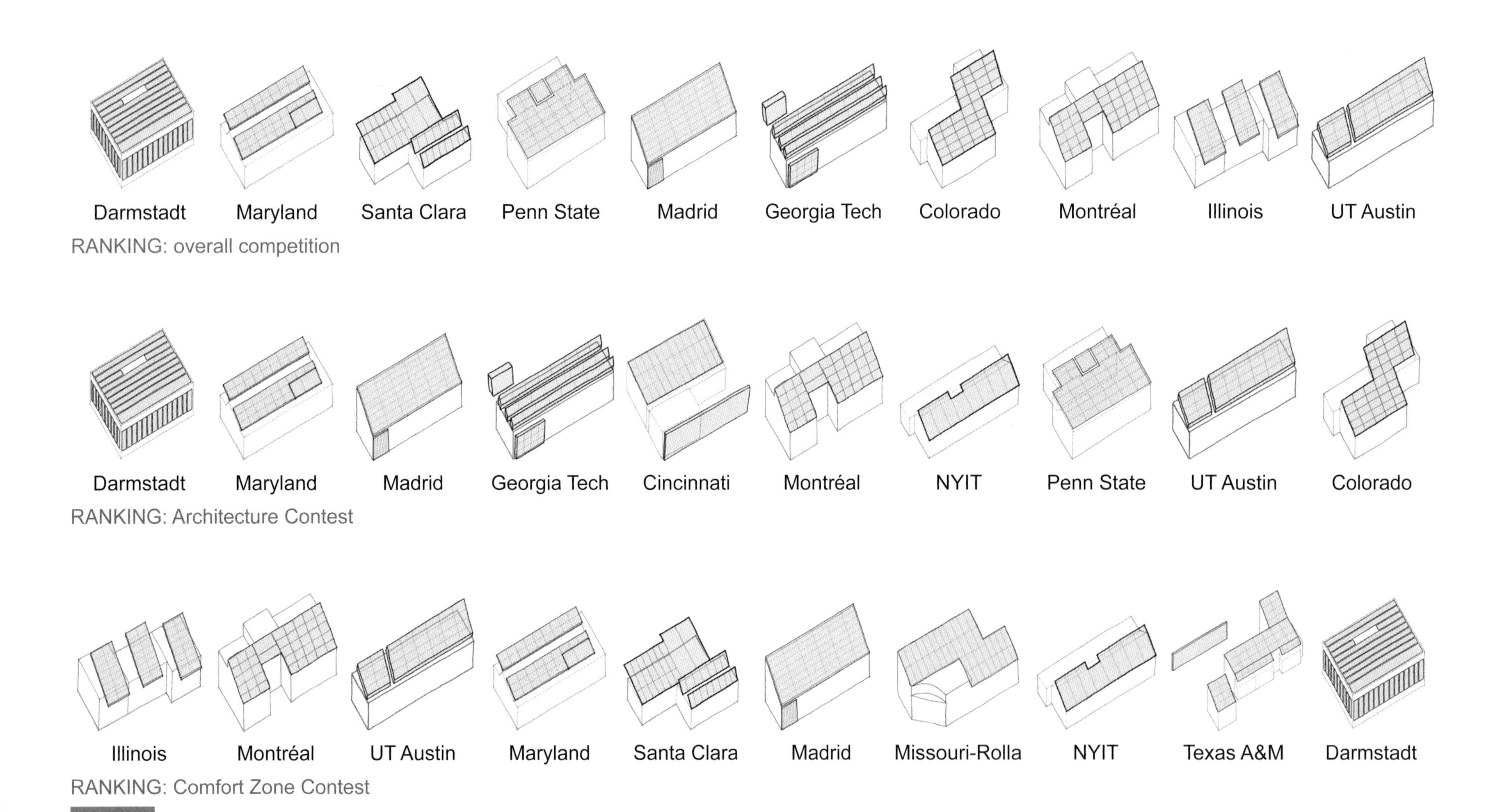

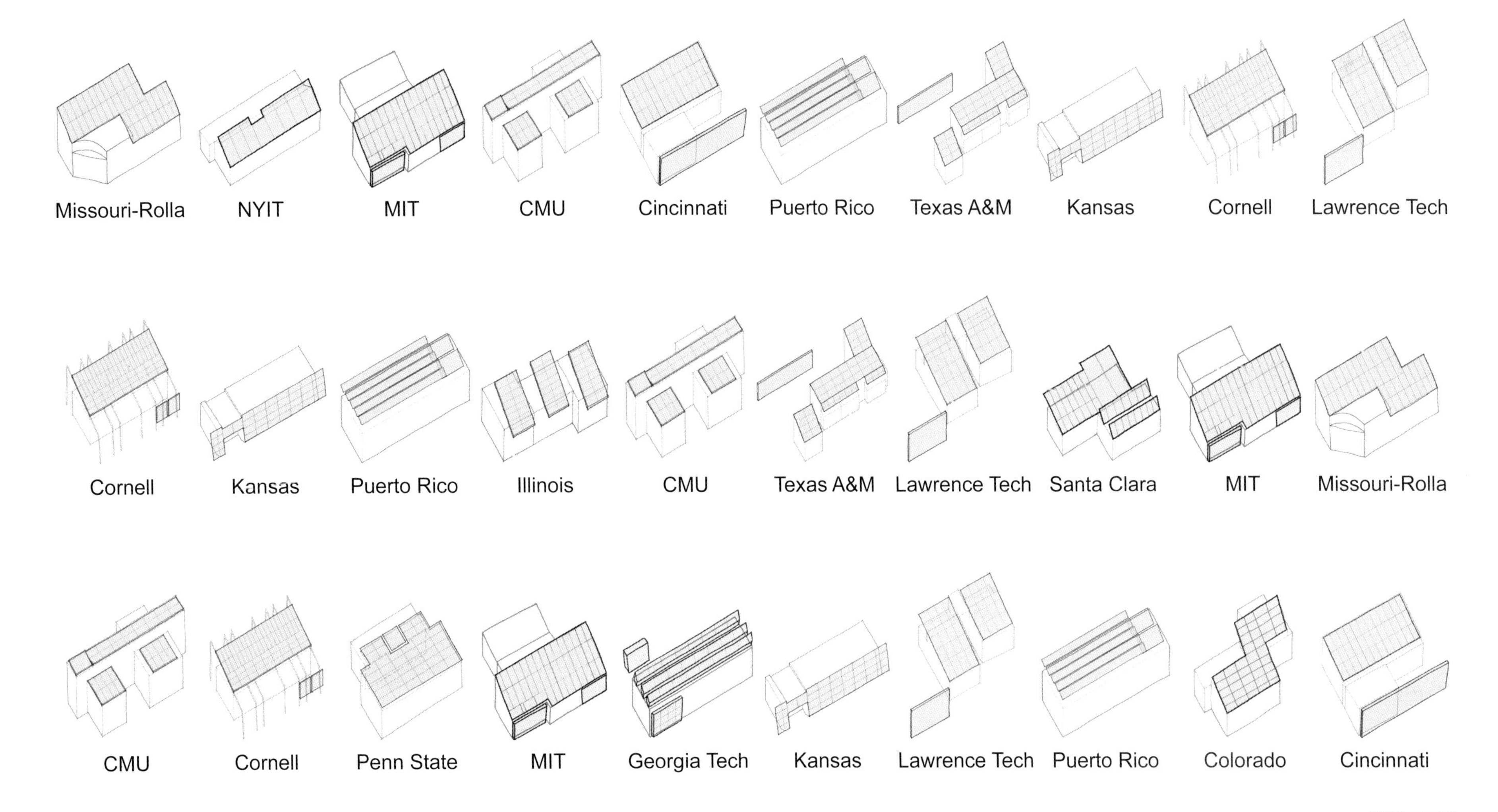
Missouri-Rolla
NYIT
MIT
CMU
Cincinnati
Puerto Rico
Texas A&M
Kansas
Cornell
Lawrence Tech
Cornell
Kansas
Puerto Rico
Illinois
CMU
Texas A&M
Lawrence Tech
Santa Clara
MIT
Missouri-Rolla
CMU
Cornell
Penn State
MIT
Georgia Tech
Kansas
Lawrence Tech
Puerto Rico
Colorado
Cincinnati

Mechanical Systems Layout

These diagrams show the layout of the systems for energy collection, storage, distribution and core. Systems for energy collection include photovoltaic arrays and solar thermal arrays. Systems for storage include batteries for electrical storage and tanks for hot water. The distribution systems include ducts and pipes for distribution of hot and cold water and air. Systems within the core include the air handlers and systems for heating and cooling.

- collection
- storage
- distribution
- core

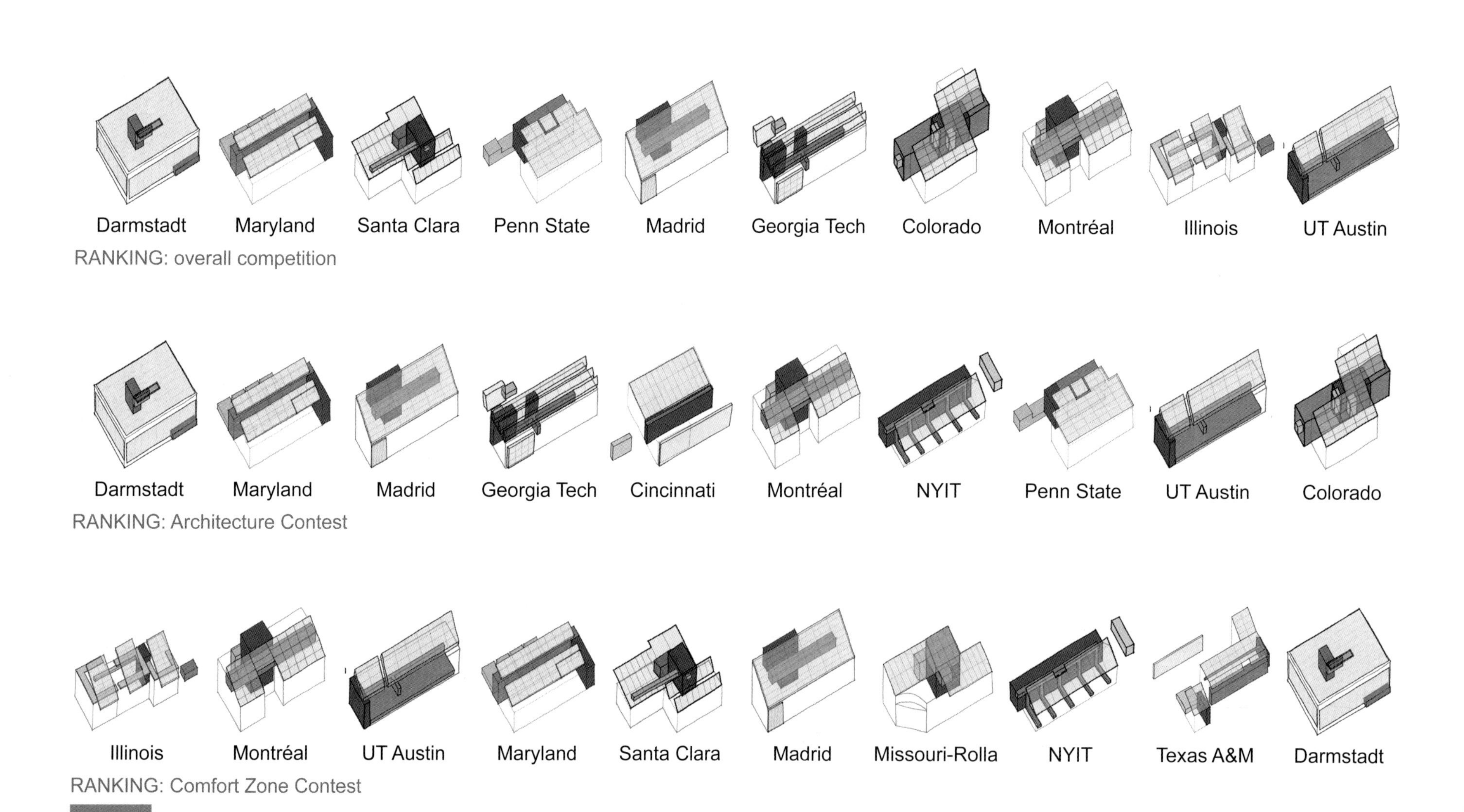

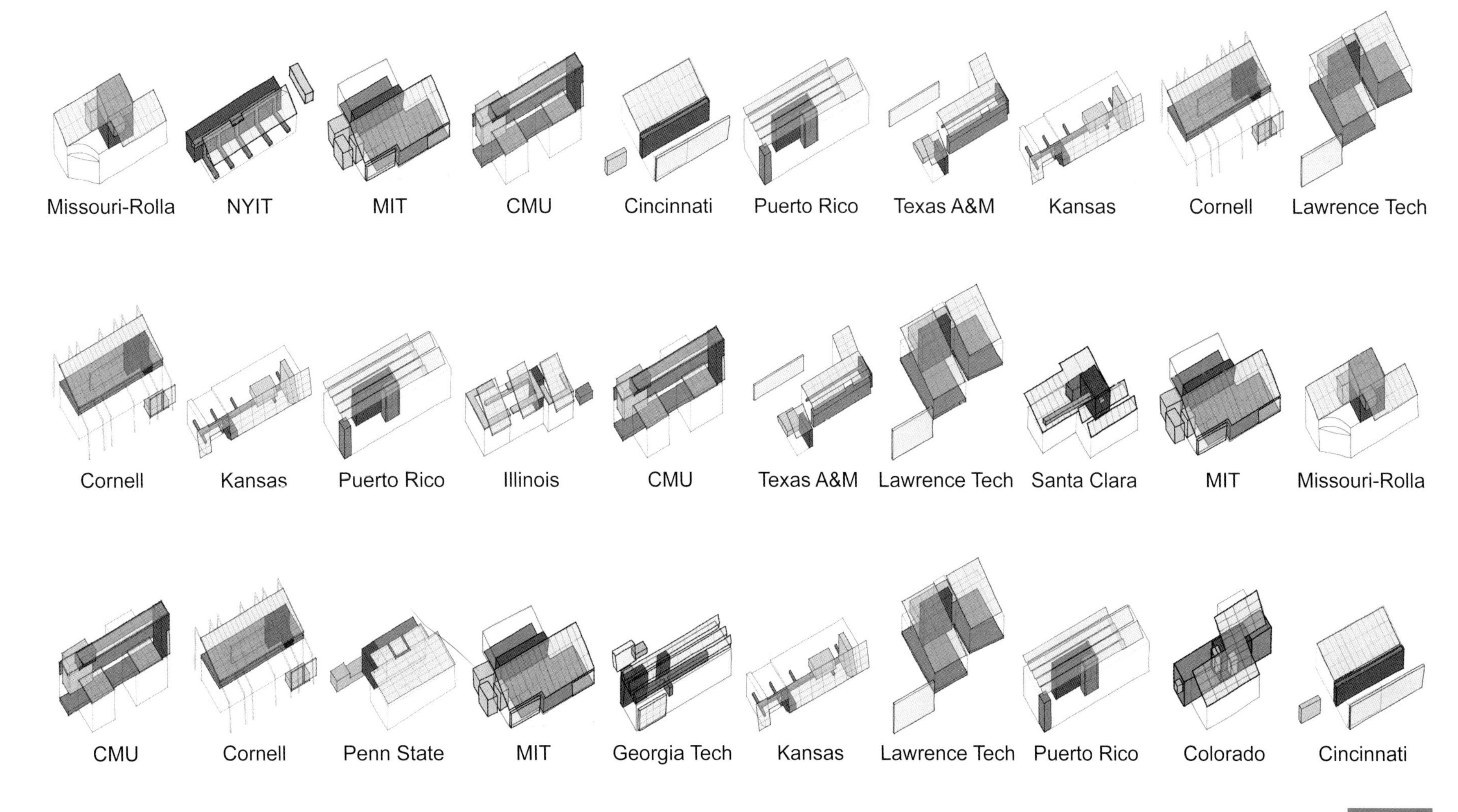
Missouri-Rolla
NYIT
MIT
CMU
Cincinnati
Puerto Rico
Texas A&M
Kansas
Cornell
Lawrence Tech
Cornell
Kansas
Puerto Rico
Illinois
CMU
Texas A&M
Lawrence Tech
Santa Clara
MIT
Missouri-Rolla
CMU
Cornell
Penn State
MIT
Georgia Tech
Kansas
Lawrence Tech
Puerto Rico
Colorado
Cincinnati

Precedents in Zero-Energy Design:

Part 3: Analysis

Introduction to Analyses

The following analyses represent an investigation into the impact of decisions made by the competing Solar Decathlon 2007 teams upon their performance in the Architecture and Comfort Zone Contests as well as in the overall competition rankings.

The categories explored in the analyses were developed from a broad range of issues that would be relevant to anyone interested in architecture, passive design, systems design and the impacts that decisions in each of these areas has upon the overall design of a house.

The categories include: location, architectural design parti, systems design parti, thermal transfer, daylighting, shading, ventilation and the incorporation of photovoltaics.

All of the data represented here came from publicly accessible files, and from data available from the US Department of Energy and the teams that competed in the 2007 competition. The data was interpreted as objectively as possible.

Interpretation of Charts

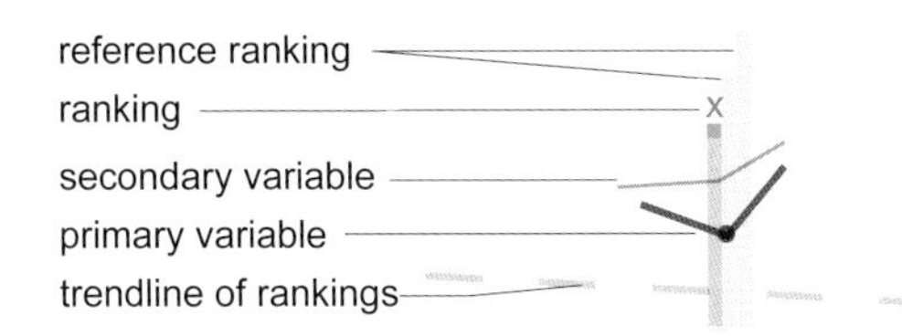

A chart is a visual representation of data. By representing the relationships between the houses in a graphic format, we can recognize patterns and relationships that provide us with additional levels of understanding about the information.

The charts in this section are designed to provide multiple levels of information in a single graphic depiction. In each chart, the horizontal axis represents an ascending order of the houses. In some cases this is the order that they ranked in the overall competition, while in others it is a more subjective ordering (such as ascending levels of formal complexity).

ranking
Architecture
overall
Comfort Zone

In the vertical axis, most charts include a list of the ranking of each house in the Architecture Contest, Comfort Zone Contest or overall competition ranking, as shown by vertical bars. These are often connected with a line so that the relative rankings can be compared to the specific data within the graph, also in the horizontal axis.

For each set of data, the comparison of rankings is given an average trendline. A trendline is a calculated average momentum of the overall data set. This provides a basis for comparison of the overall slope and trajectory of the data.

The slope of the trendline is compared to the primary variable in the chart, to assess whether the specific rankings are increasing or decreasing relative to the variable.

increasing rankings
decreasing rankings

It is important to recognize that a higher number in the rankings leads to a higher bar. However, a higher bar indicates a decrease in the rankings. Any evaluation of rankings with a trendline sloping downwards indicates an increase in rankings, while a trendline sloping upwards indicates a decrease in rankings.

Location: Institution

All 20 competing institutions designed and built their house at or near their institution. They transported the house from their institution to Washington, DC in October 2007 for the three weeks of the competition, before disassembling and shipping the houses back to the institution or to another location.

The hypothesis behind this analysis is that houses from locations closer to Washington, DC would rank higher in the Comfort Zone Contest due to their knowledge of the climate.

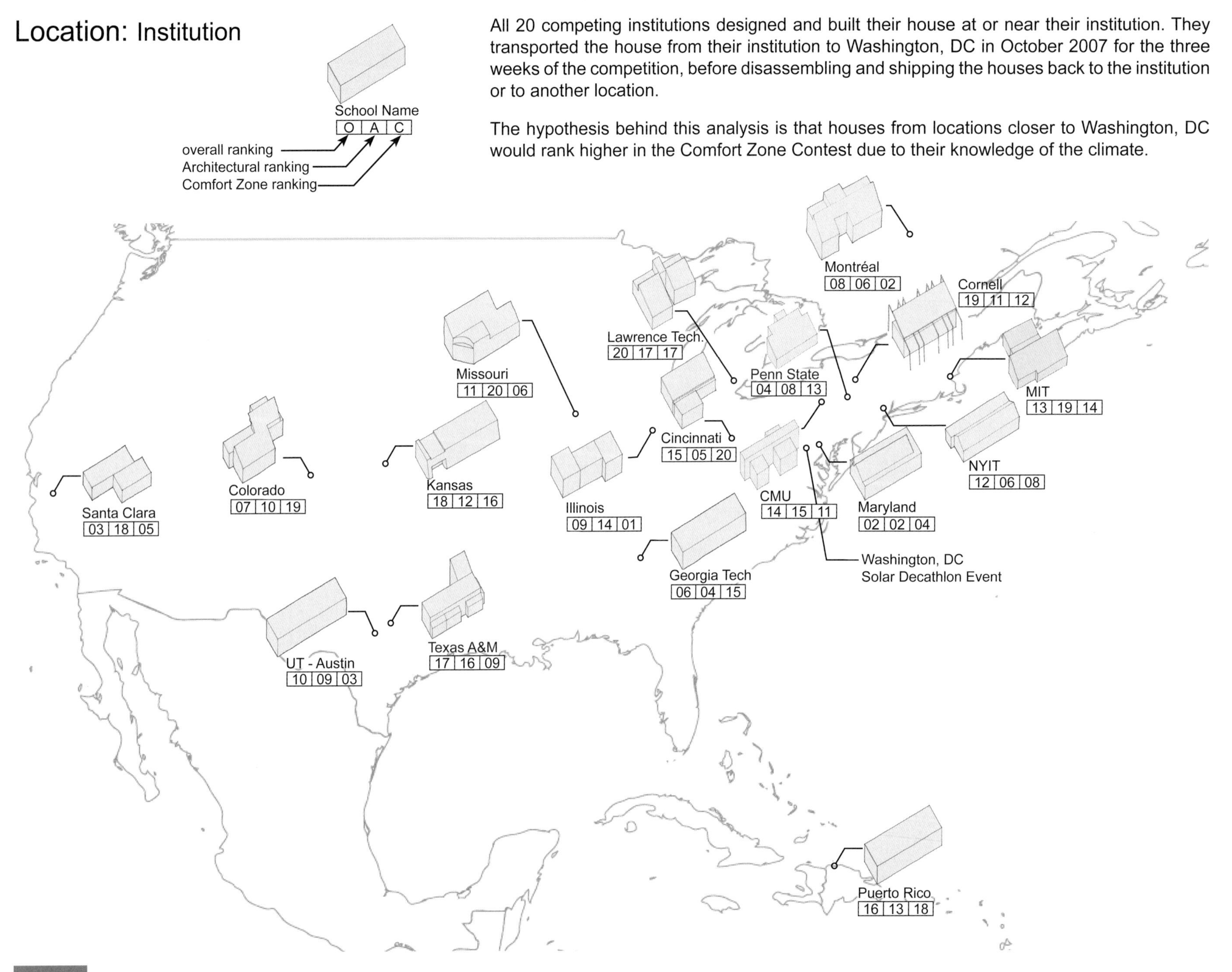

Team Darmstadt won the competition, though they also traveled the farthest. However, they scored tenth in the Comfort Zone Contest. In contrast, Team Maryland, the team closest to Washington, ranked second overall and fourth in the Comfort Zone Contest.

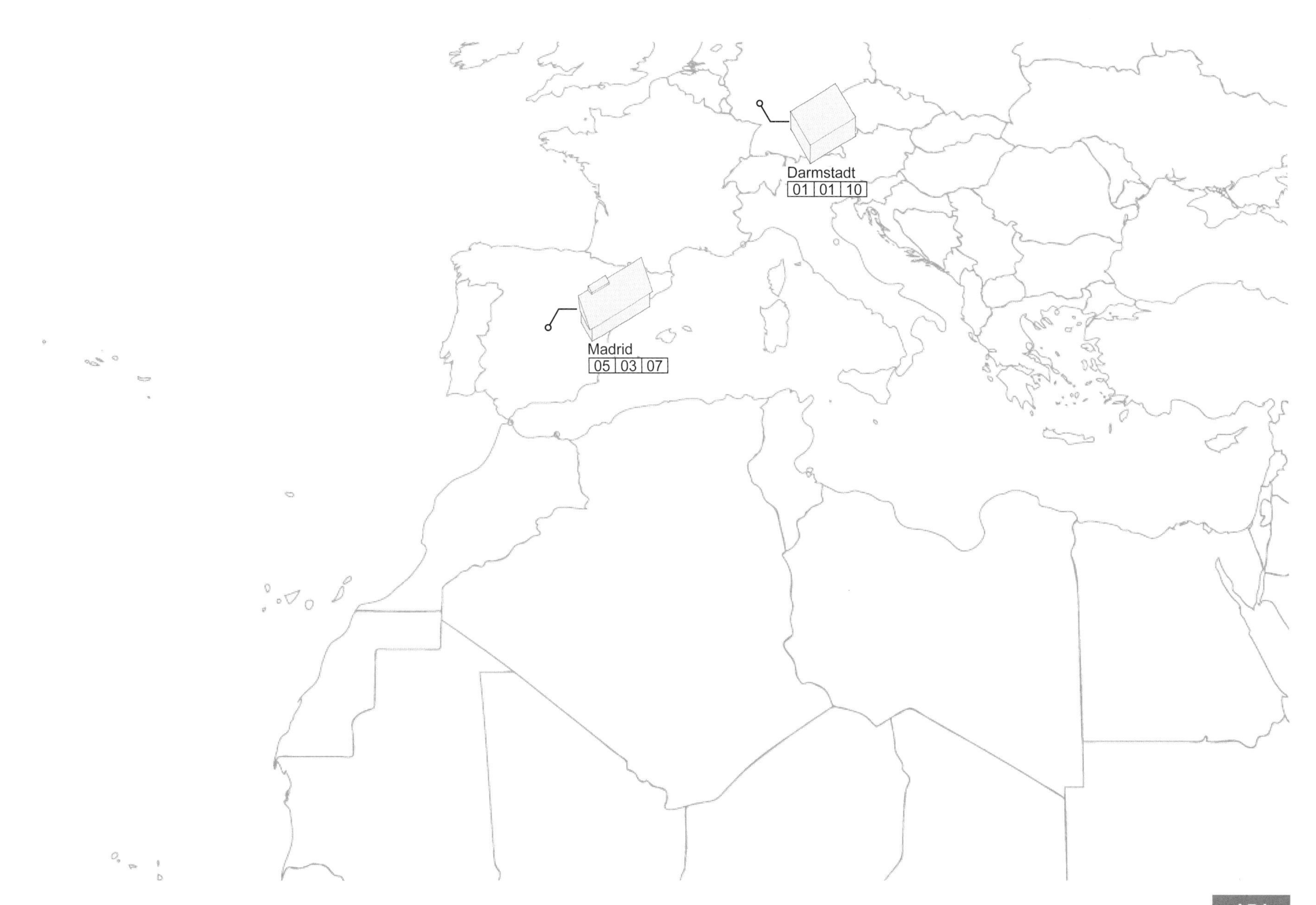

Location: Project Cost and Distance

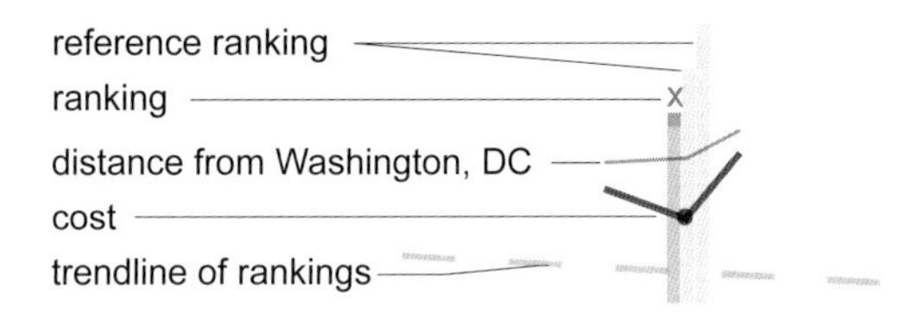

There was no budget imposed by, and no detailed budget information required for, the competition. Each team received a DOE Grant but was responsible for securing all additional funds on their own. As a result, the amount spent varied from $270,000 to $1,378,297 (a factor of 5 between them). The chart also shows the distance from the competition site, which has a clear correlation to increased project cost.

The hypothesis tested here is that there would be a recognizable correlation between the overall cost of the houses and their success in the competition and contest rankings.

The results show a clear increase in overall rankings, as well as in the Architecture and Comfort Zone Contests, as the project cost increased. Looking closely at the trendlines, it is evident that the strongest correlation is between the Architecture Contest rankings and the project cost.

If we assume that passive means of achieving high performance in the Comfort Zone Contest would cost less than active (technological or mechanized) means, then a contest in which every team fully utilized passive options for achieving thermal comfort would not show this trend.

overall competition rankings

1	Darmstadt	$1,378,297
2	Maryland	$448,470
3	Santa Clara	$800,000
4	Penn State	$505,000
5	Madrid	$1,000,000
6	Georgia Tech	$700,000
7	Colorado	$300,000
8	Montréal	$510,000
9	Illinois	$622,000
10	UT Austin	$450,000
11	Missouri-Rolla	$425,000
12	NYIT	$400,000
13	MIT	$270,000
14	CMU	$410,000
15	Cincinnati	$434,900
16	Puerto Rico	$364,902
17	Texas A&M	$550,000
18	Kansas	$412,808
19	Cornell	$675,000
20	Lawrence Tech	$672,000
Average Project Cost		$560,669

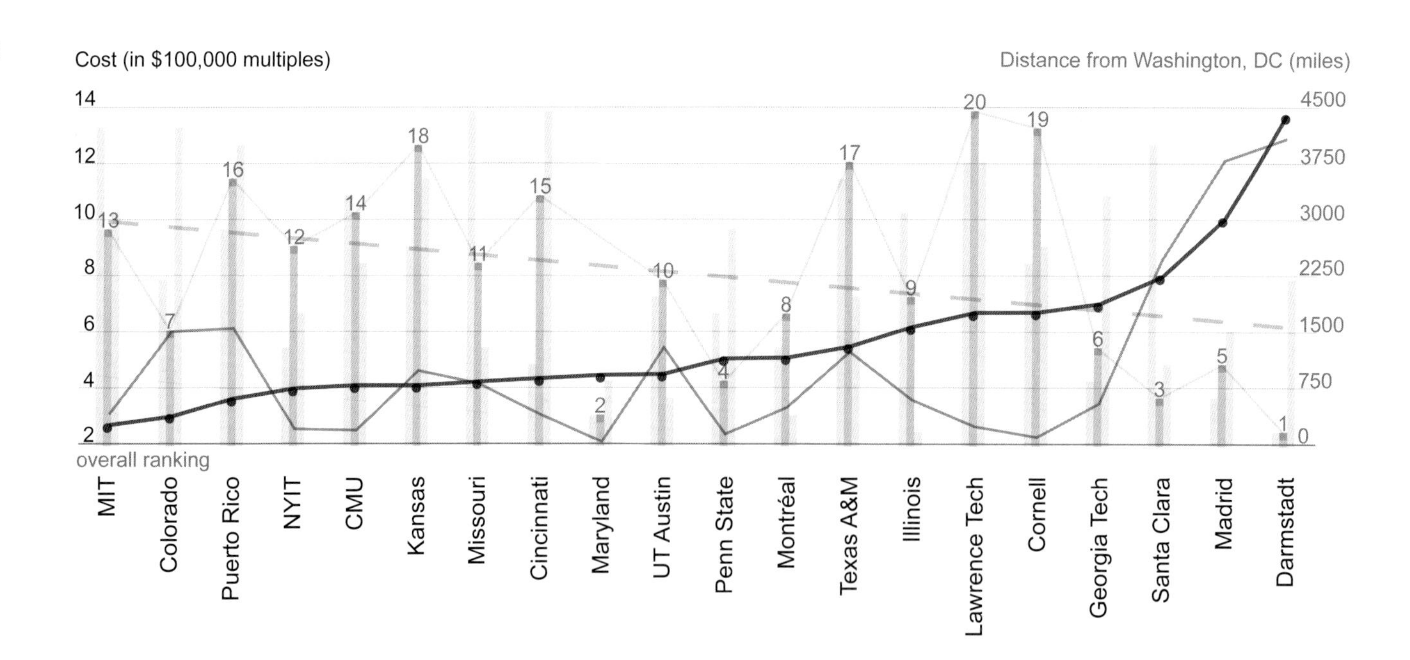

Architecture Contest rankings

1	Darmstadt	$1,378,297
2	Maryland	$448,470
3	Madrid	$1,000,000
4	Georgia Tech	$700,000
5	Cincinnati	$434,900
6	Montréal	$510,000
7	NYIT	$400,000
8	Penn State	$505,000
9	UT Austin	$450,000
10	Colorado	$300,000
11	Cornell	$675,000
12	Kansas	$412,808
13	Puerto Rico	$364,902
14	Illinois	$622,000
15	CMU	$410,000
16	Texas A&M	$550,000
17	Lawrence Tech	$672,000
18	Santa Clara	$800,000
19	MIT	$270,000
20	Missouri-Rolla	$425,000

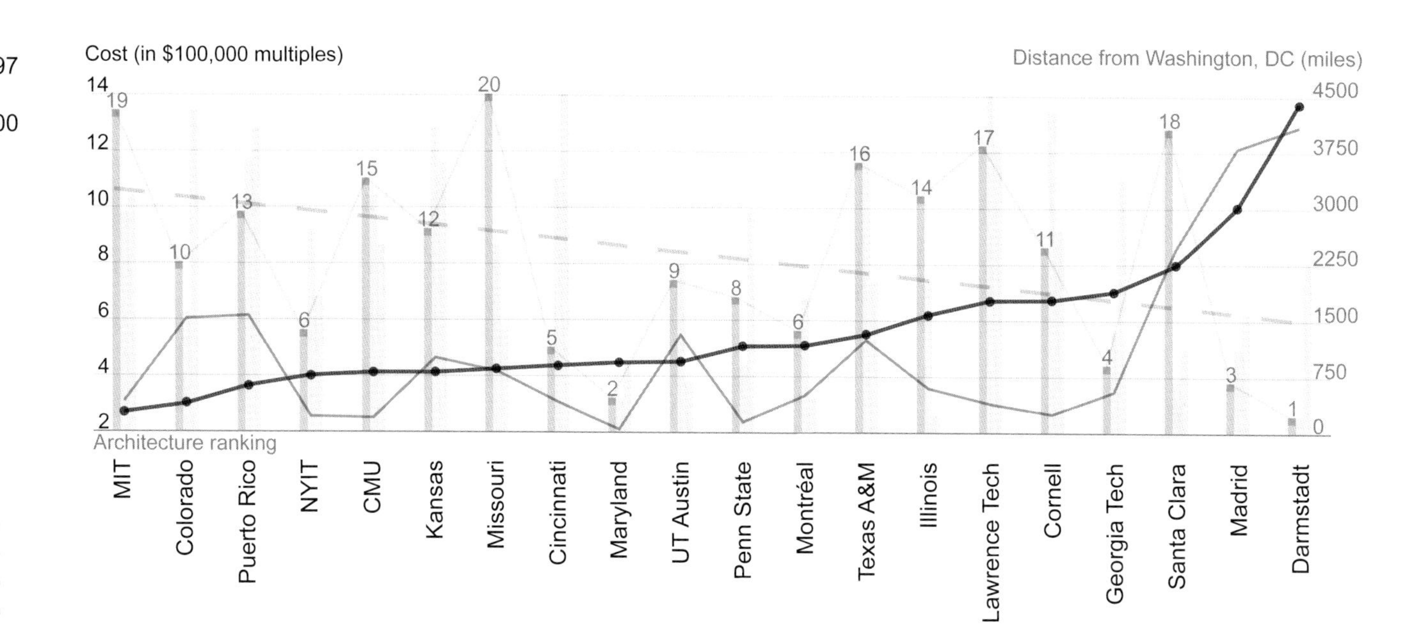

Comfort Zone Contest rankings

1	Illinois	$622,000
2	Montréal	$510,000
3	UT Austin	$450,000
4	Maryland	$448,470
5	Santa Clara	$800,000
6	Madrid	$1,000,000
7	Missouri-Rolla	$425,000
8	NYIT	$400,000
9	Texas A&M	$550,000
10	Darmstadt	$1,378,297
11	CMU	$410,000
12	Cornell	$675,000
13	Penn State	$505,000
14	MIT	$270,000
15	Georgia Tech	$700,000
16	Kansas	$412,808
17	Lawrence Tech	$672,000
18	Puerto Rico	$364,902
19	Colorado	$300,000
20	Cincinnati	$434,900

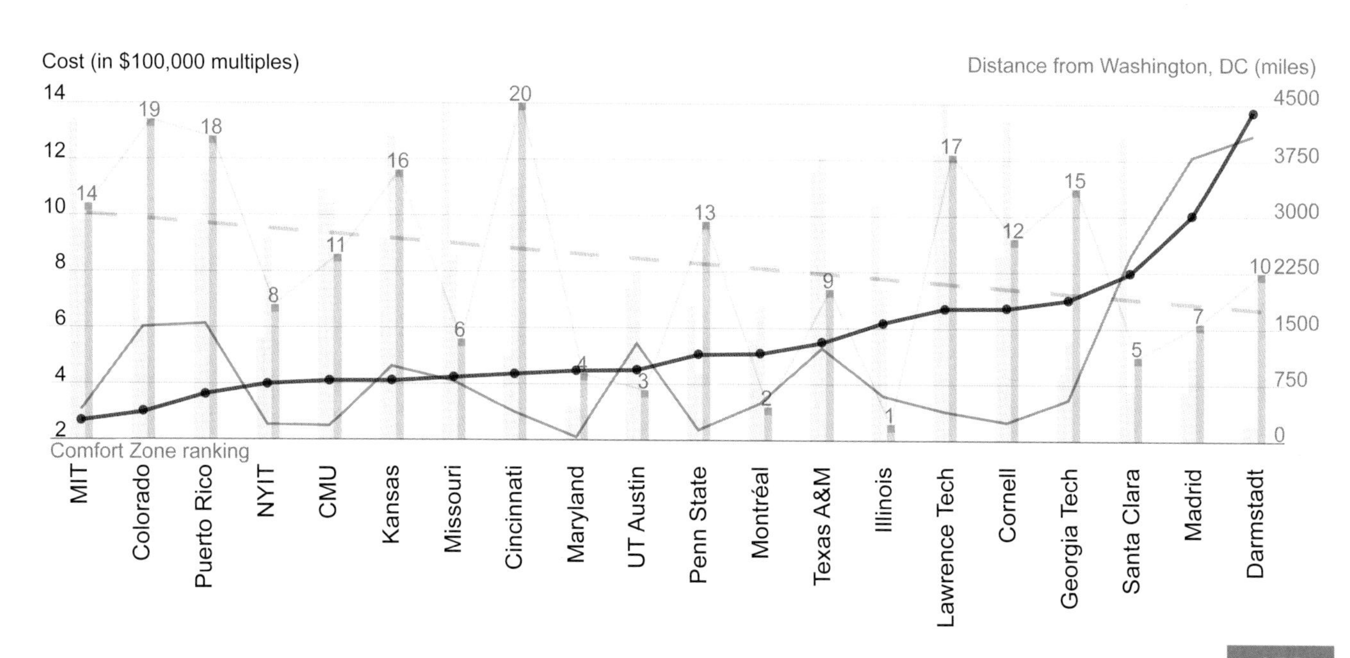

Location: Latitude of Institution

Each of the 20 houses was designed and built in locations near the team's institution, yet the competition took place in Washington, DC. This disparity introduces a critical question facing every team – should the house be designed for the climate in their home location or the climate in Washington for the two weeks of the competition?

In terms of a climatic comparison, latitude is likely more indicative of climatic similarity than distance. These charts compare the latitudinal distances between the institution locations and Washington, DC at 38.5° North latitude.

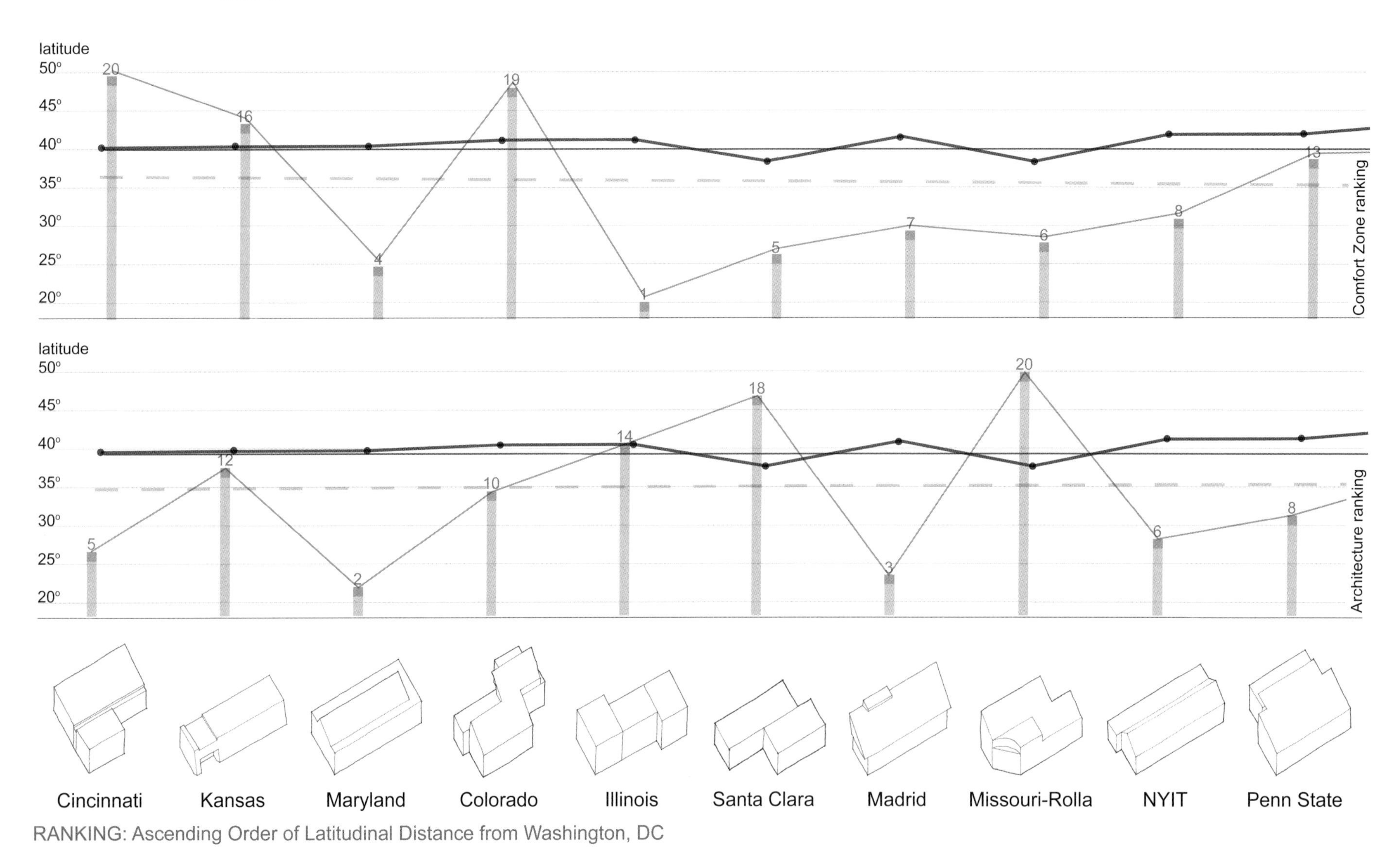

RANKING: Ascending Order of Latitudinal Distance from Washington, DC

The trendlines for the Architecture and Comfort Zone Contests show no significant indication of a tendency to rank higher or lower as a result of increased latitudinal distance from Washington, DC. Comfort Zone rankings show a very slight increase as latitudinal distance increases, while Architecture rankings show a slight decrease.

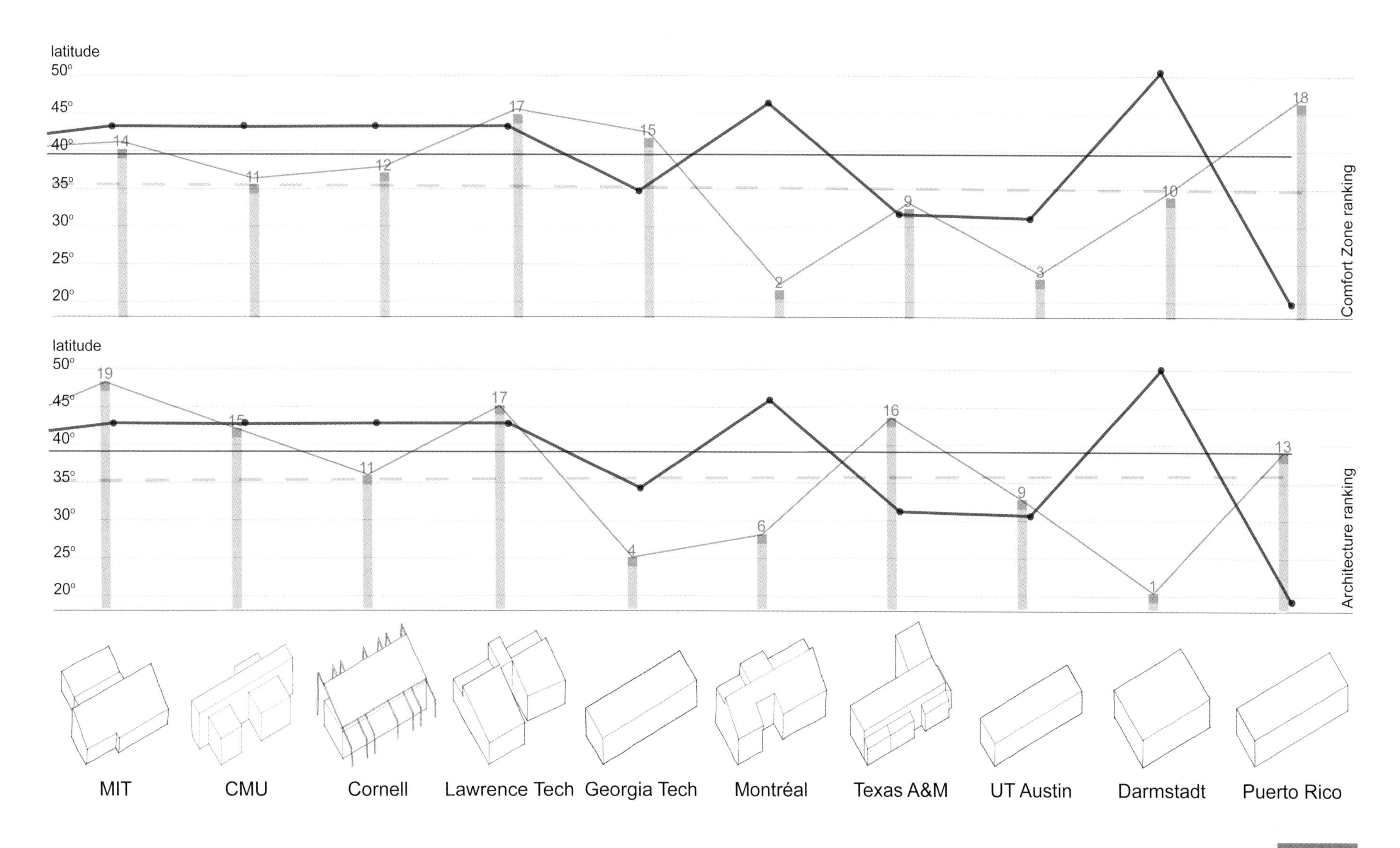

Location: on Solar Village

O overall ranking
A Architecture ranking
C Comfort Zone ranking

Each team competing in the Solar Decathlon competition was assigned a specific location along "Main Street" in the "Solar Village" on the National Mall. Ten houses were on the north side of Main Street and ten houses were on the south side. The hypothesis behind these investigations is that the arbitrary assignment of location on the north or south side of the street would have an impact on competition rankings. Houses that are reliant on solar energy for their power have an inherent bias towards the equator (the south side of the building in Washington, DC).

The results show that there is a correlation between which side of the "Main Street" of the Solar Village a house is located and its success in the overall competition ranking.

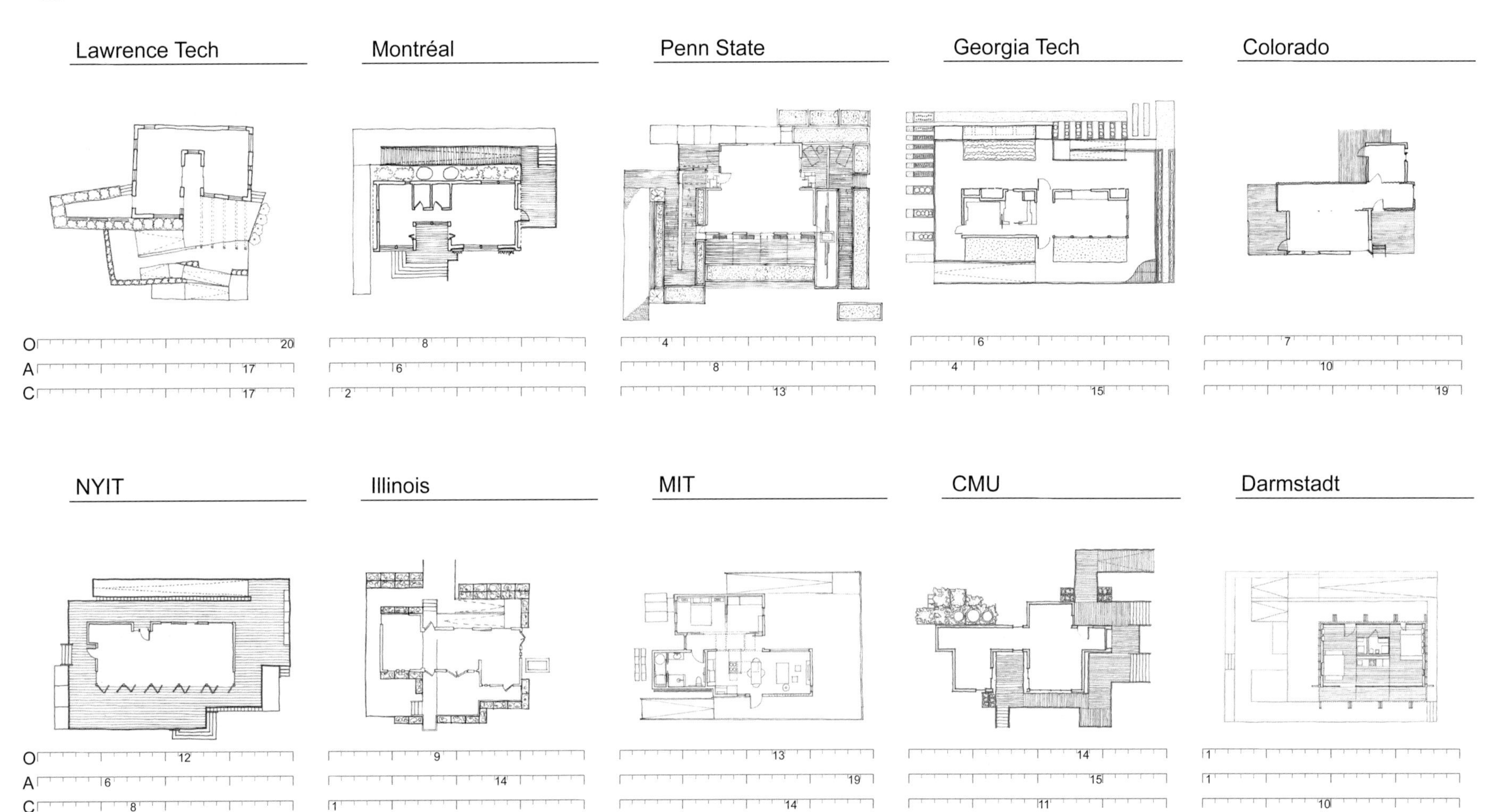

LAYOUT: Location in Solar Village on National Mall

Teams located on the north side of the axis ranked significantly higher in the overall competition, though there was no significant difference in the Architecture or Comfort Zone Contest rankings between north and south.

	overall average:	Architecture average:	Comfort Zone average:
north	**8.7**	10.7	**10.2**
south	12.3	**10.2**	10.8

Team	O	A	C
Missouri-Rolla	11	20	06
Maryland	2	2	4
UT Austin	10	9	3
Puerto Rico	16	13	18
Santa Clara	3	18	5
Kansas	18	12	16
Cornell	19	11	12
Cincinnati	15	5	20
Texas A&M	17	16	9
Madrid	5	3	7

Location: Site Circulation

O overall ranking
A Architecture ranking
C Comfort Zone ranking

Every team had to balance their design response to the "Main Street" entrance with the need to address solar access on the south face of their house. In addition, each team made critical decisions about the entry sequence from Main Street into their house. There were multiple, competing requirements for the facades of these small buildings. This series of diagrams shows the percentage of glazing on the south facade, as well as the entry sequence.

The intent of this analysis was to investigate how the combined factors of public access and southern glazing were impacted by location on either side of Main Street.

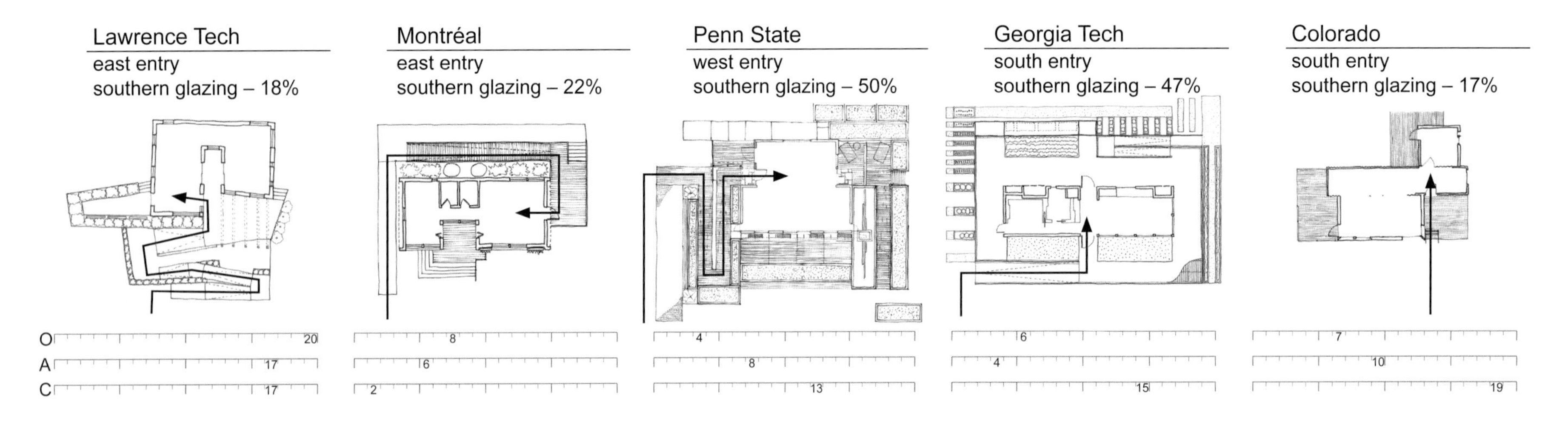

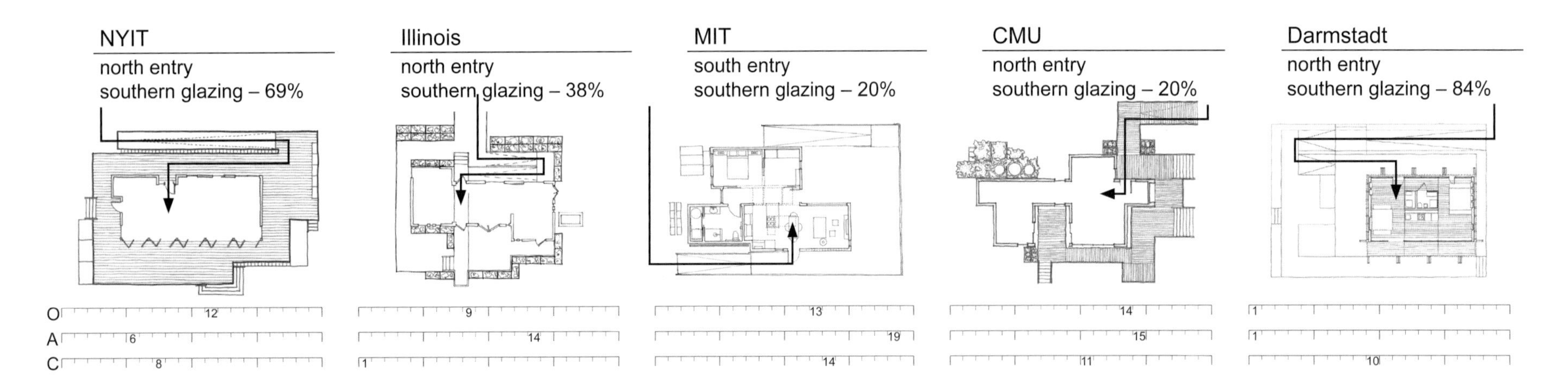

LAYOUT: Location in Solar Village on National Mall

There was a significant difference in the amount of southern glazing when a house was located on the north side of Main Street as compared to being on the south side. Those houses whose south facade didn't face Main Street had an average of 10% more glazing on the south facade.

north of axis - avg south glazing: **30.8%**
south of axis - avg south glazing: **39.8%**

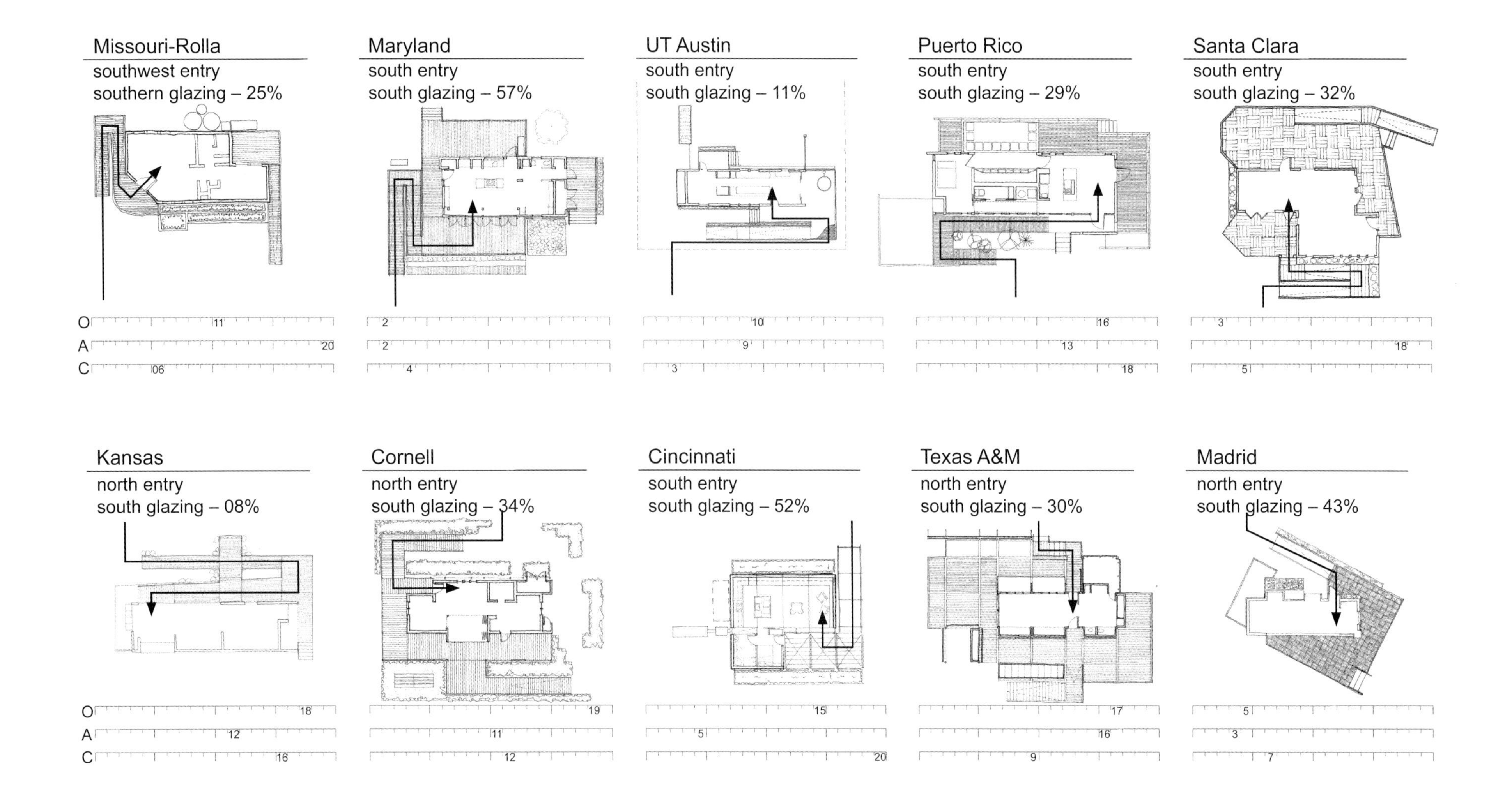

Concept Parti: Evolution

Conceptual clarity is a critical aspect of contemporary architecture and design. This series of diagrams provides a basis for comparison of the relationship between the exterior parti, the interior parti and the overall concept. This sequence provides an opportunitity to assess the recognition of an evolution that links the interior and exterior parti at a conceptual level.

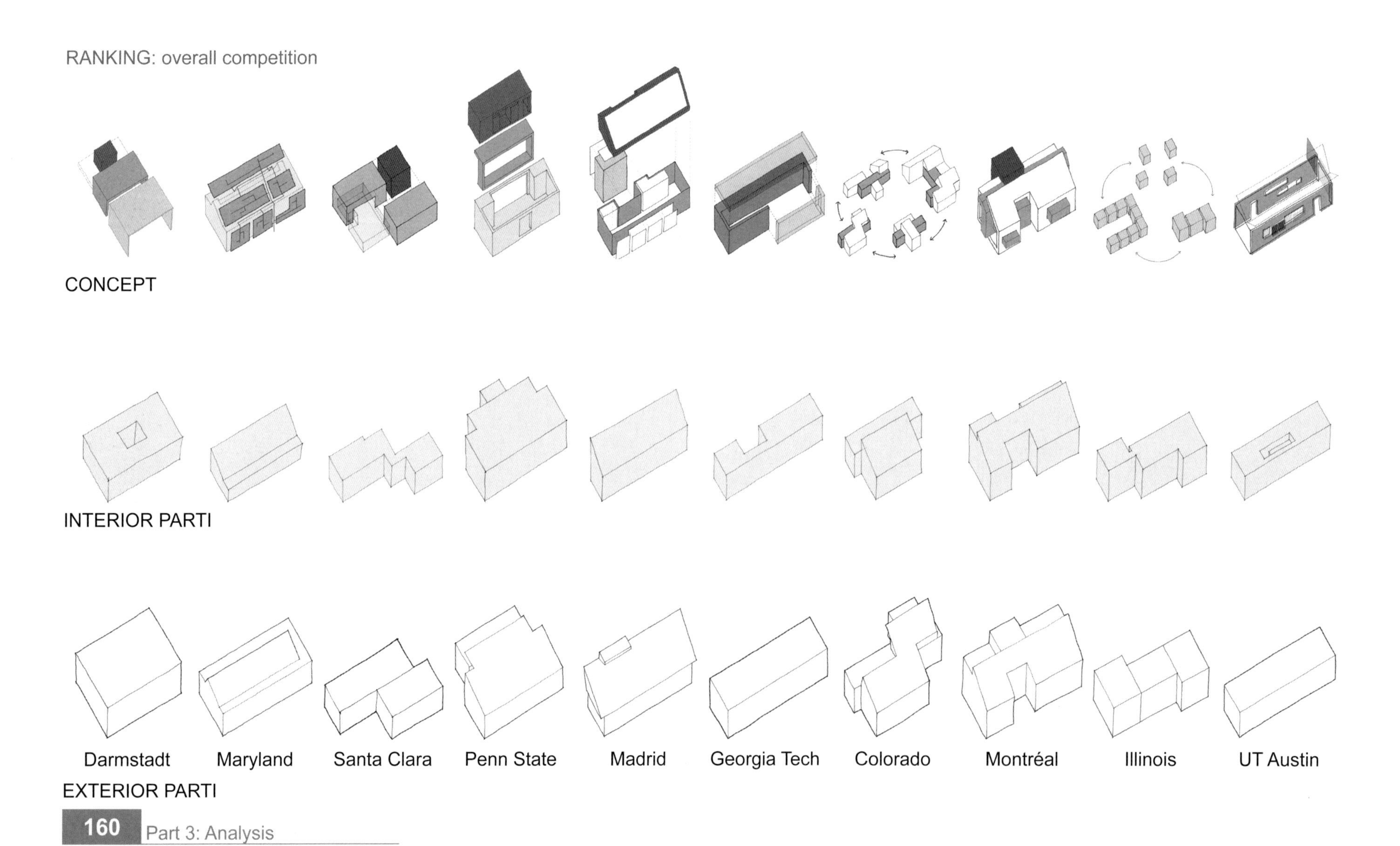

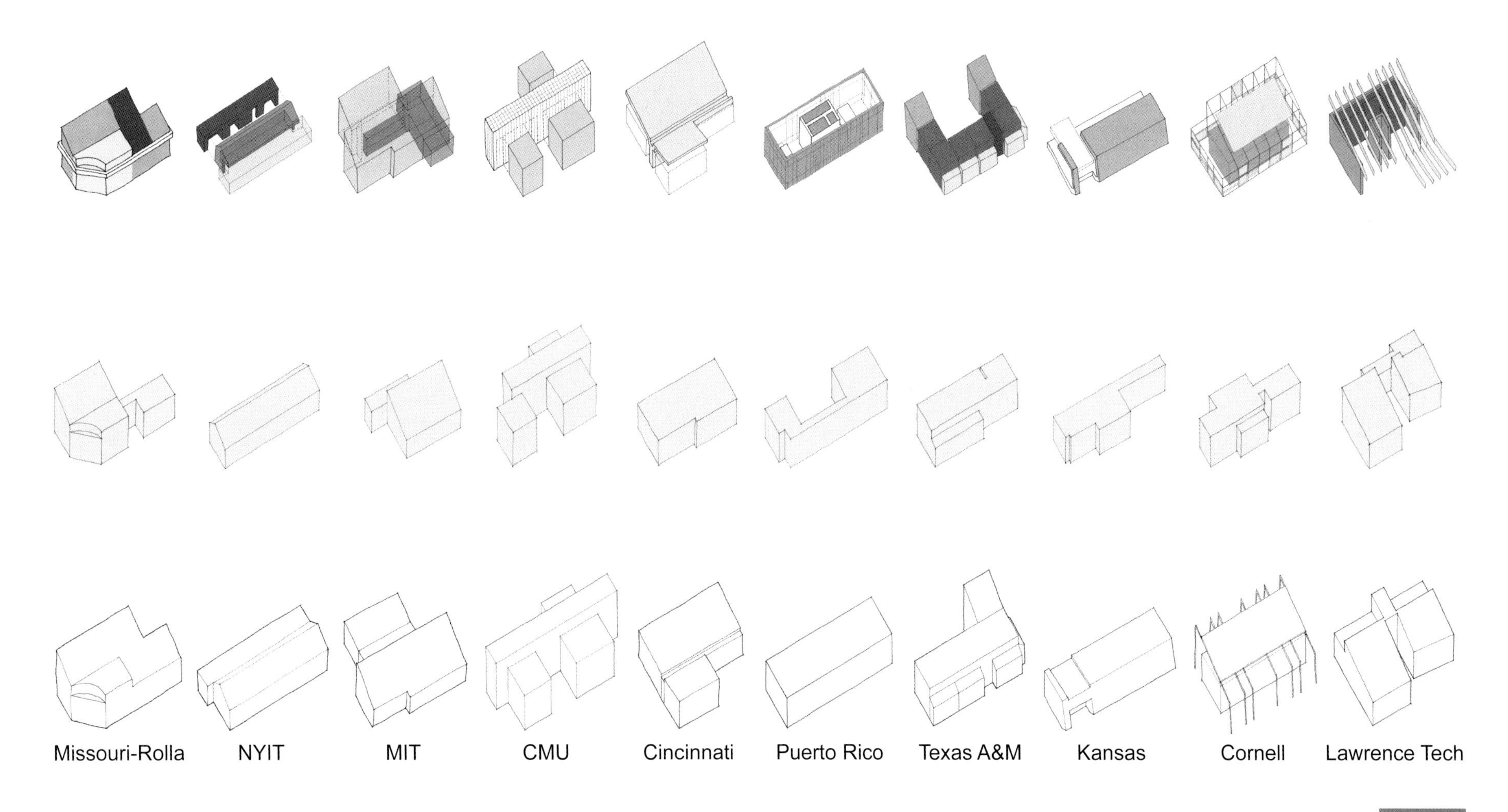
Missouri-Rolla
NYIT
MIT
CMU
Cincinnati
Puerto Rico
Texas A&M
Kansas
Cornell
Lawrence Tech

Exterior Parti: Ascending Complexity

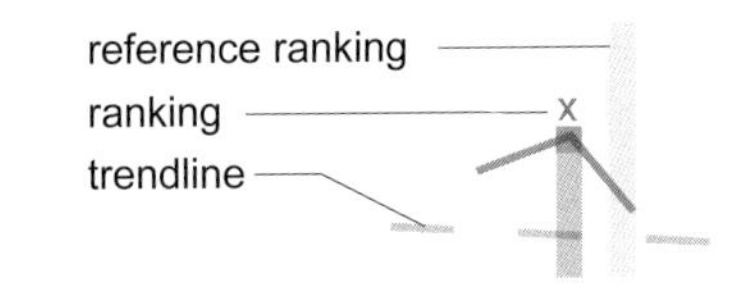

This sequence represents an ascending order of simplicity to complexity in the exterior form parti for all 20 houses. There are some houses that have clearly defined, unified forms; some that have multiple forms; and others that are less legible in a formal analysis. The variants include: single mass or bar-shaped form, offset bar form, central spine, offset masses and two separate masses.

This chart compares the ascending complexity of the exterior form to the Architecture Contest and overall rankings.

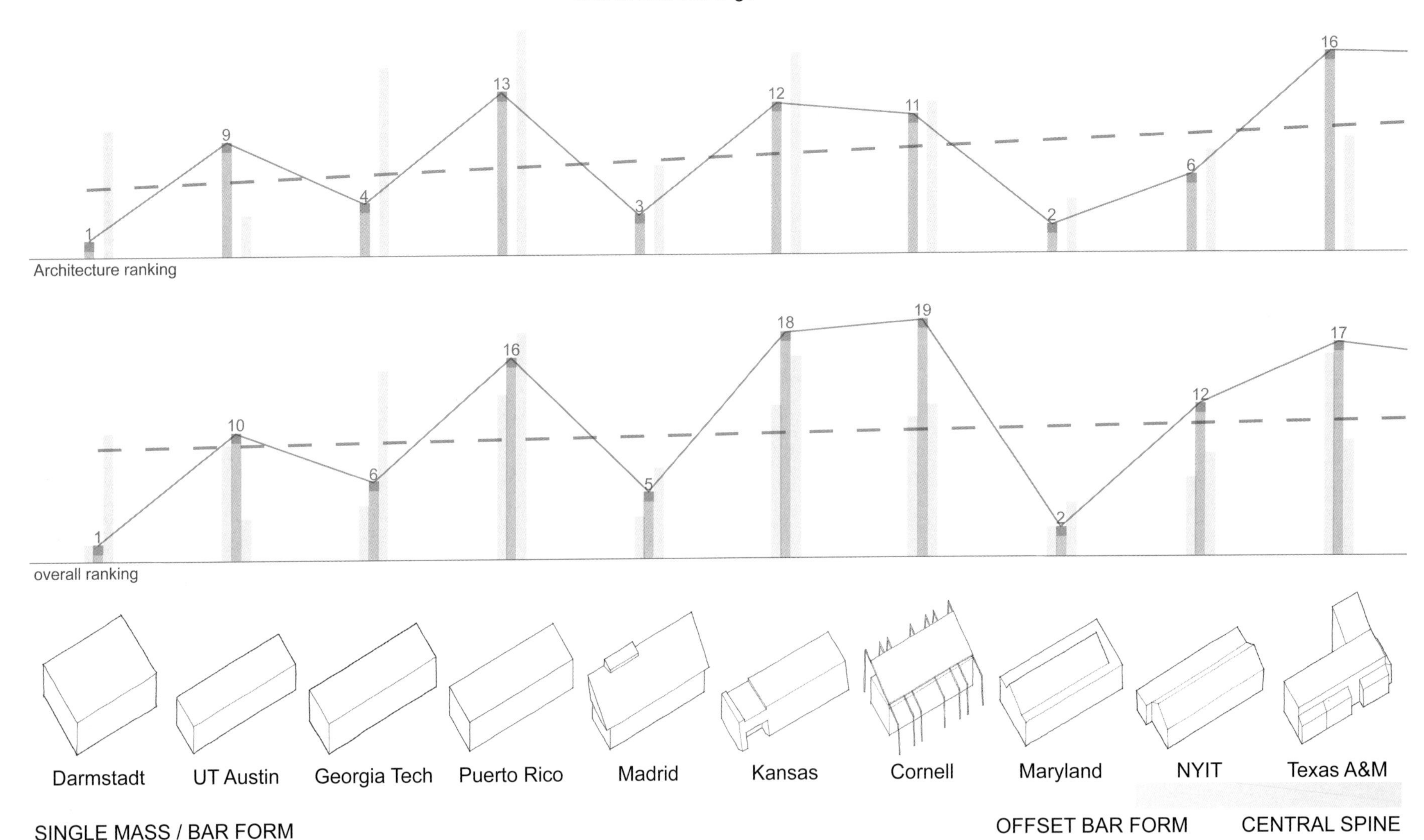

RANKING: Ascending Complexity of Exterior Form Parti

There is a correlation between increasing complexity and decreased rankings in the Architecture Contest and overall competition. The correlation is stronger with the Architecture rankings, which suggests that a simpler exterior form parti correlates to higher rankings in the Architecture Contest.

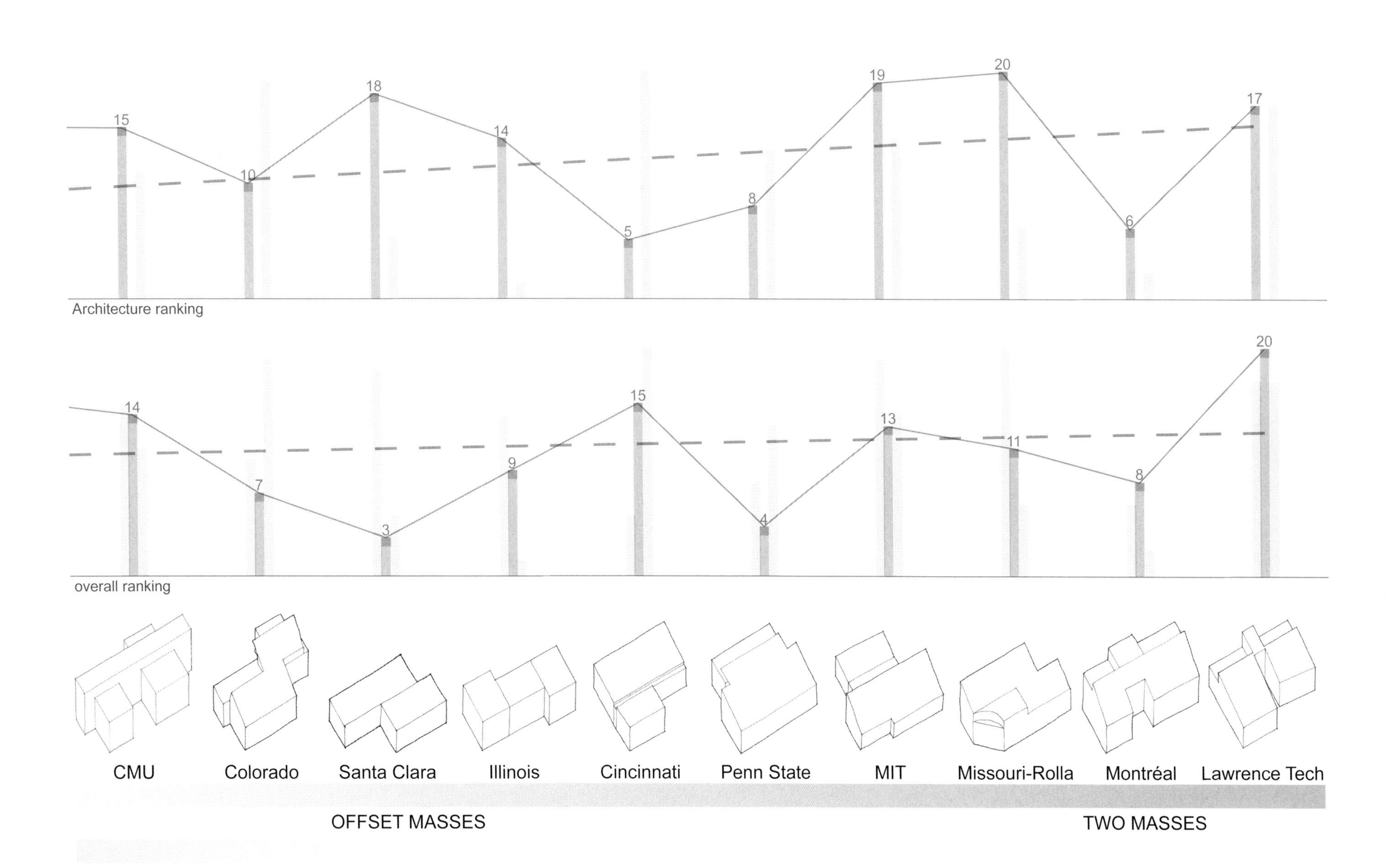

Interior Parti: Ascending Complexity

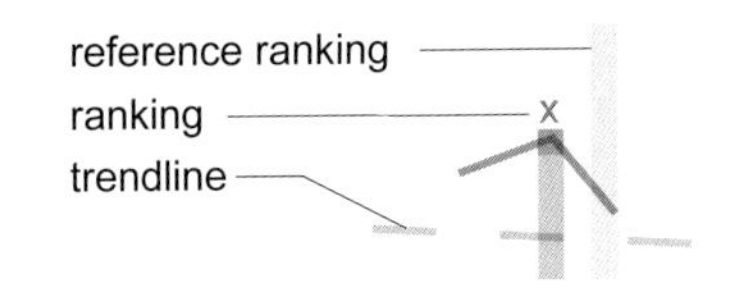

This sequence represents an ascending order of simplicity to complexity in the interior spatial parti for the 20 houses. There are some houses that have clearly defined spatial forms and others that have multiple readings. The variants include: single mass or bar-shaped space, ring-shaped space, subtracted spaces, additive spaces, multiple spaces and houses with two distinct spaces.

This chart compares the ascending complexity of the interior spatial parti to the Architecture and overall rankings.

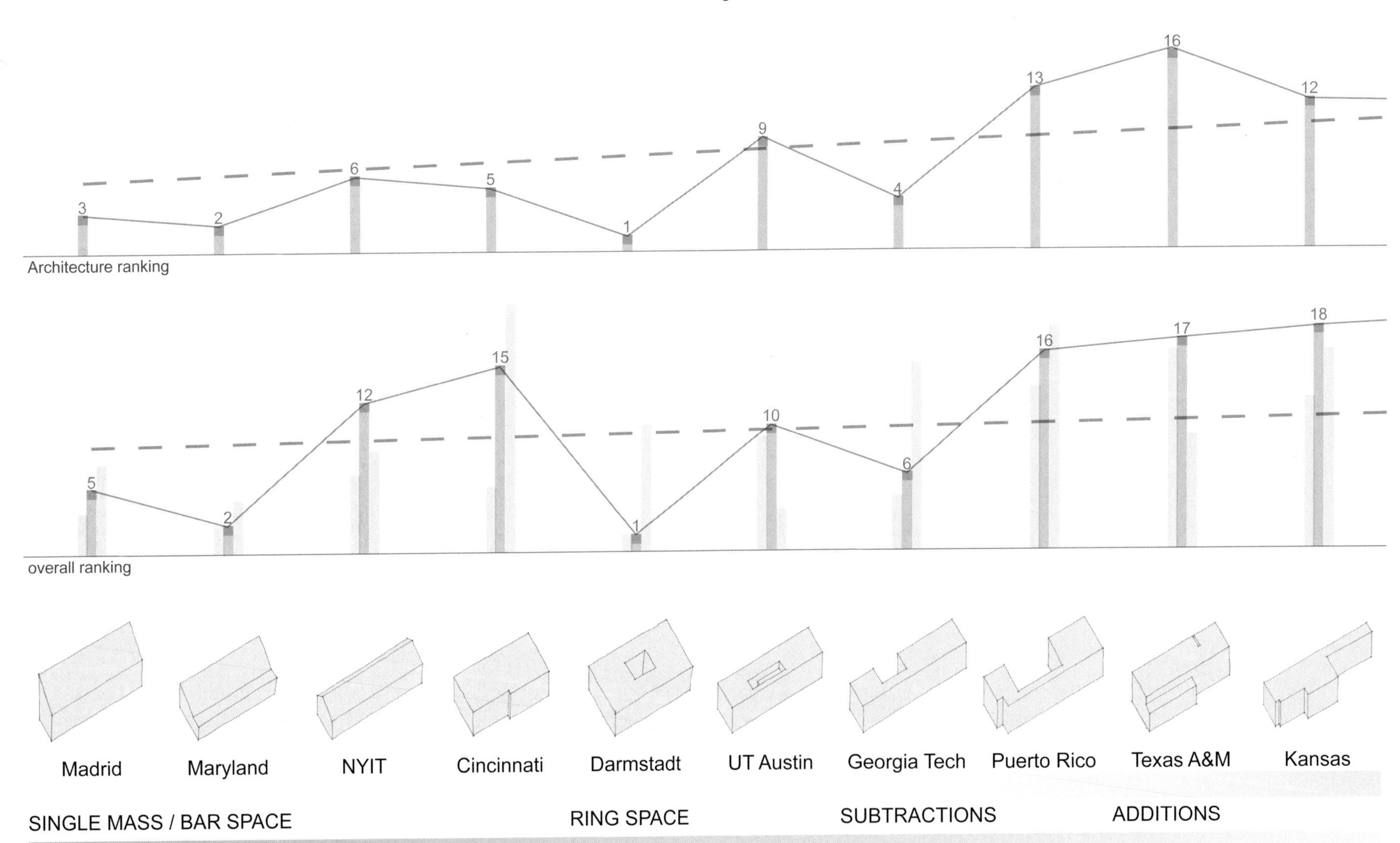

RANKING: Ascending Complexity of Interior Spatial Parti

As with the exterior form parti, the trendlines show that there is a correlation between the increase in interior spatial complexity and lower overall rankings, while there is a strong correlation between increasing complexity and decreased Architecture Contest rankings. Therefore, those teams that designed a simpler interior spatial parti tended to rank higher in the Architecture Contest.

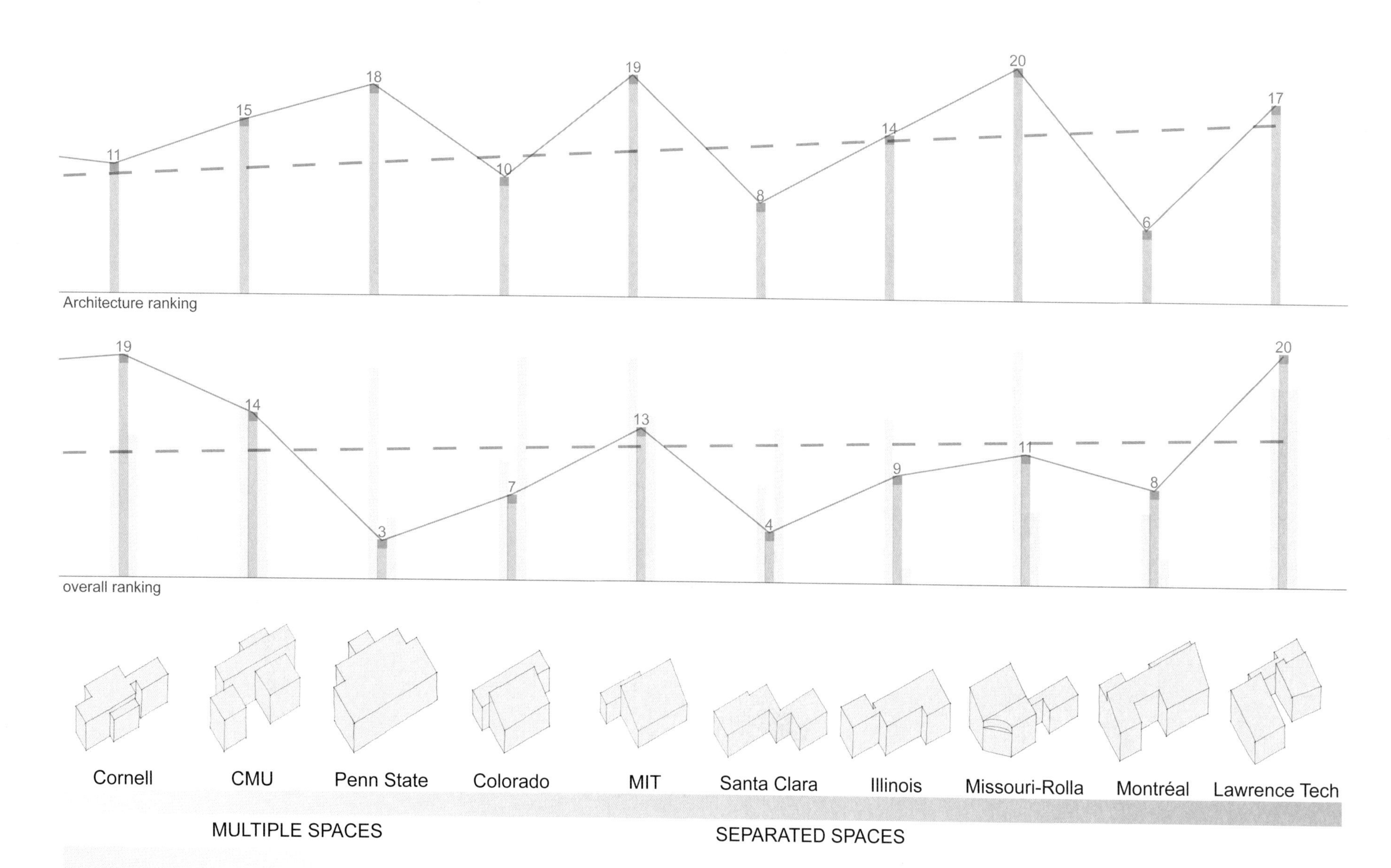

Systems Parti: Core Evolution

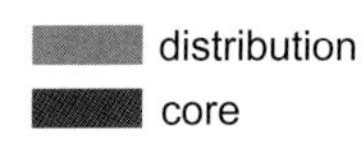

This series of diagrams explores the relationship between the exterior form parti, the interior spatial parti and the location of the core and distribution elements of the mechanical systems of each house. The diagrams have been ordered based on the complexity of the mechanical systems core in relation to overall building layout. The intent is to discover to what degree the design of the mechanical system is integrated with the design of the interior and exterior parti of the house.

RANKING: Ascending Complexity of Systems Core Organization

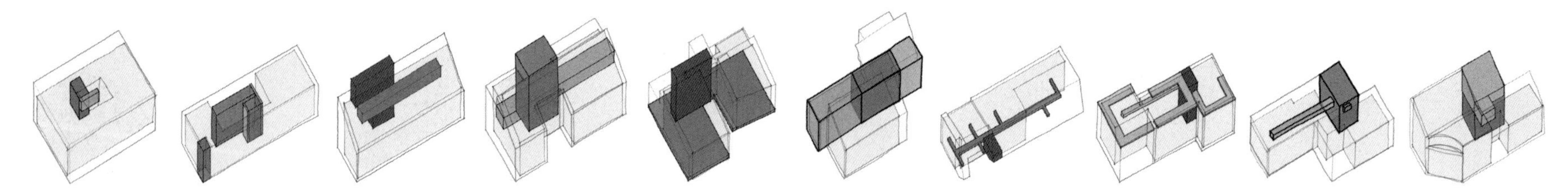

INTERIOR PARTI WITH MECHANICAL CORE AND DISTRIBUTION

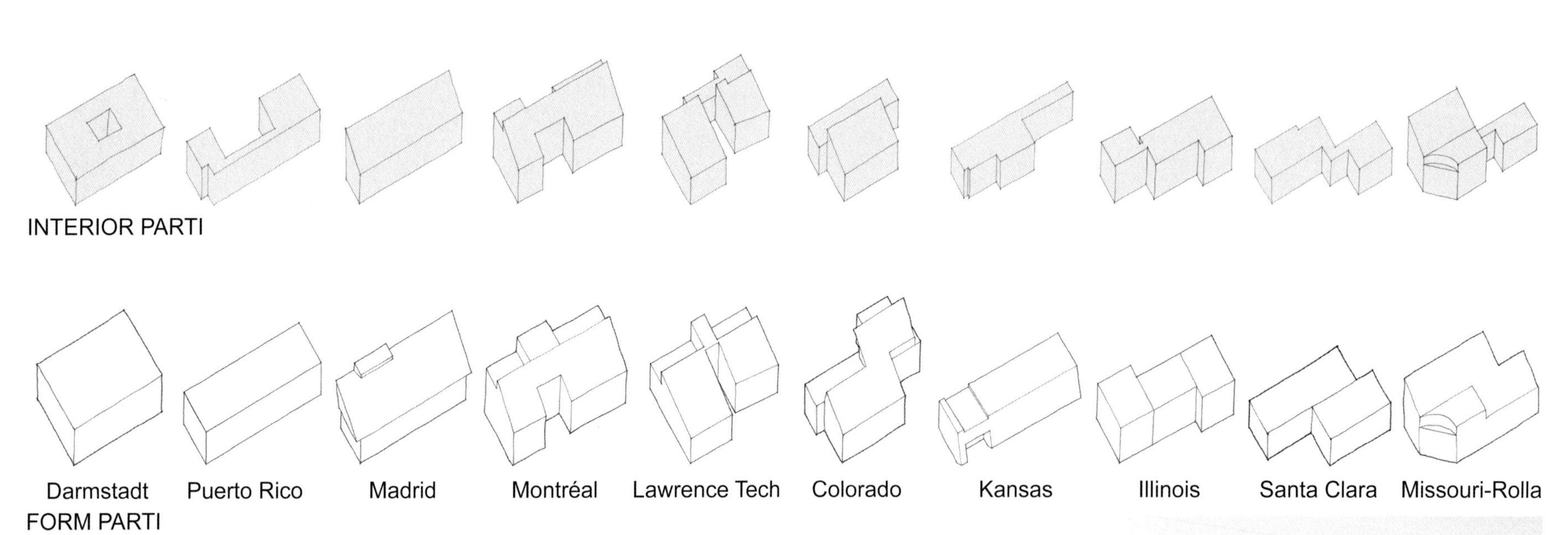

CENTRAL CORE

EDGE CORE

Hypotheses about the relationship between the layout of the mechanical systems and the interior and exterior parti led to a series of diagrammatic studies that parse the mechanical systems into four parts:

collection – the photovoltaic and solar thermal arrays for the collection of solar energy
storage – batteries and water tanks for storage of electric and thermal energy
distribution – the ducts, pipes or panels used for distribution of heating or cooling
core – the location of the mechanical equipment for heating and cooling

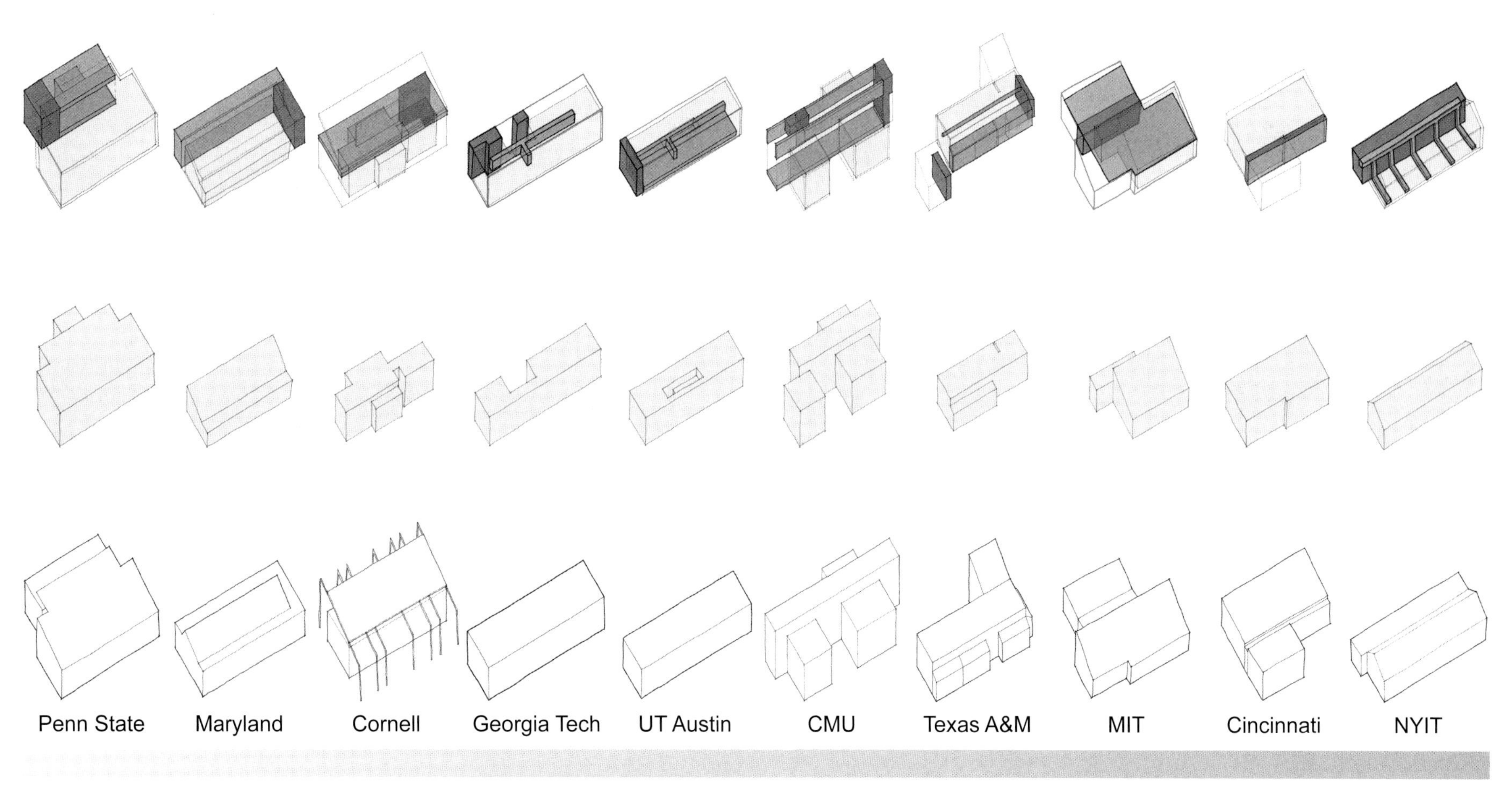

Systems Parti: Systems Evolution

This series of diagrams addresses the relationship between the location of systems devoted to the collection and storage of energy and to the location of the mechanical core and distribution systems for heating and cooling the houses.

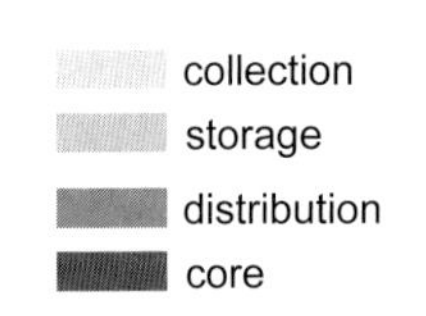

RANKING: overall competition

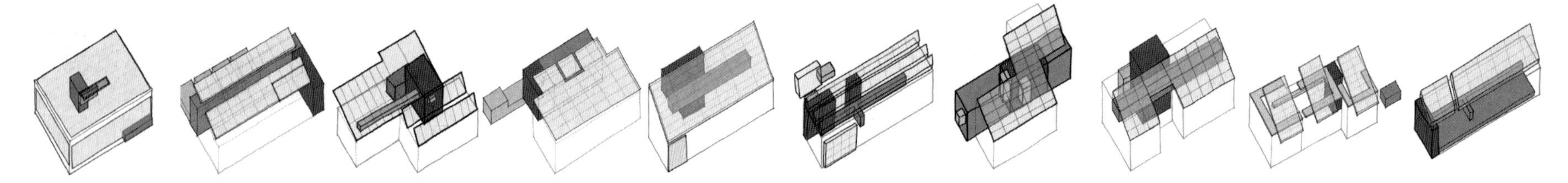

COMBINED ENERGY COLLECTION, STORAGE AND DISPERSION

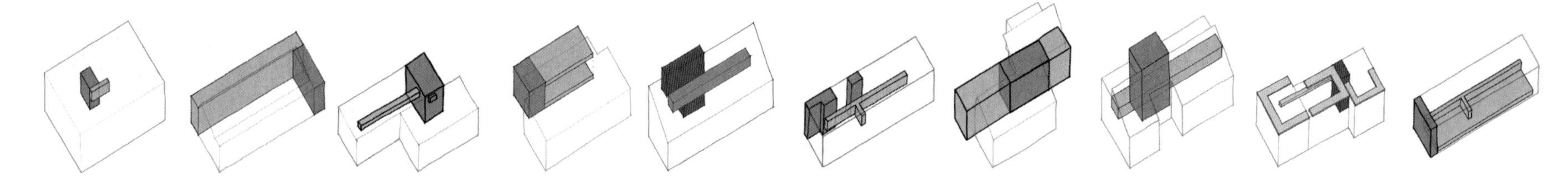

MECHANICAL CORE AND DISTRIBUTION

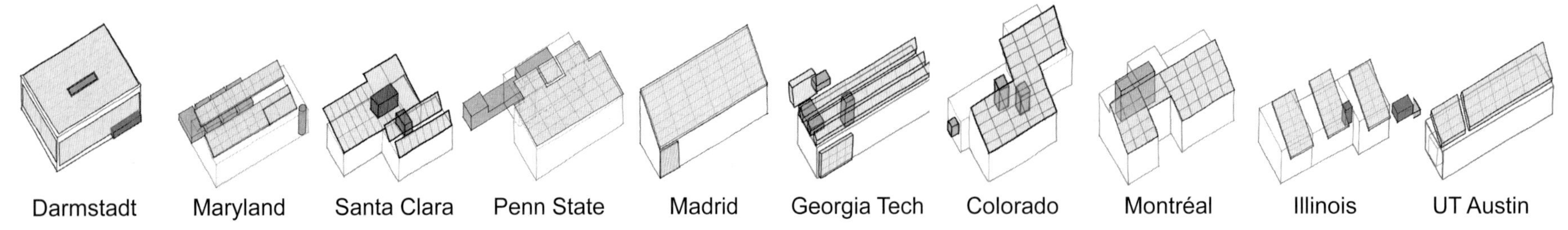

ENERGY COLLECTION AND STORAGE

Hypotheses about the relationship between the layout of the mechanical systems and the interior and exterior parti led to a series of diagrammatic studies that parse the mechanical systems into four parts:

collection – the photovoltaic and solar thermal arrays for the collection of solar energy
storage – batteries and water tanks for storage of electric and thermal energy
distribution – the ducts, pipes or panels used for distribution of heating or cooling
core – the location of the mechanical equipment for heating and cooling

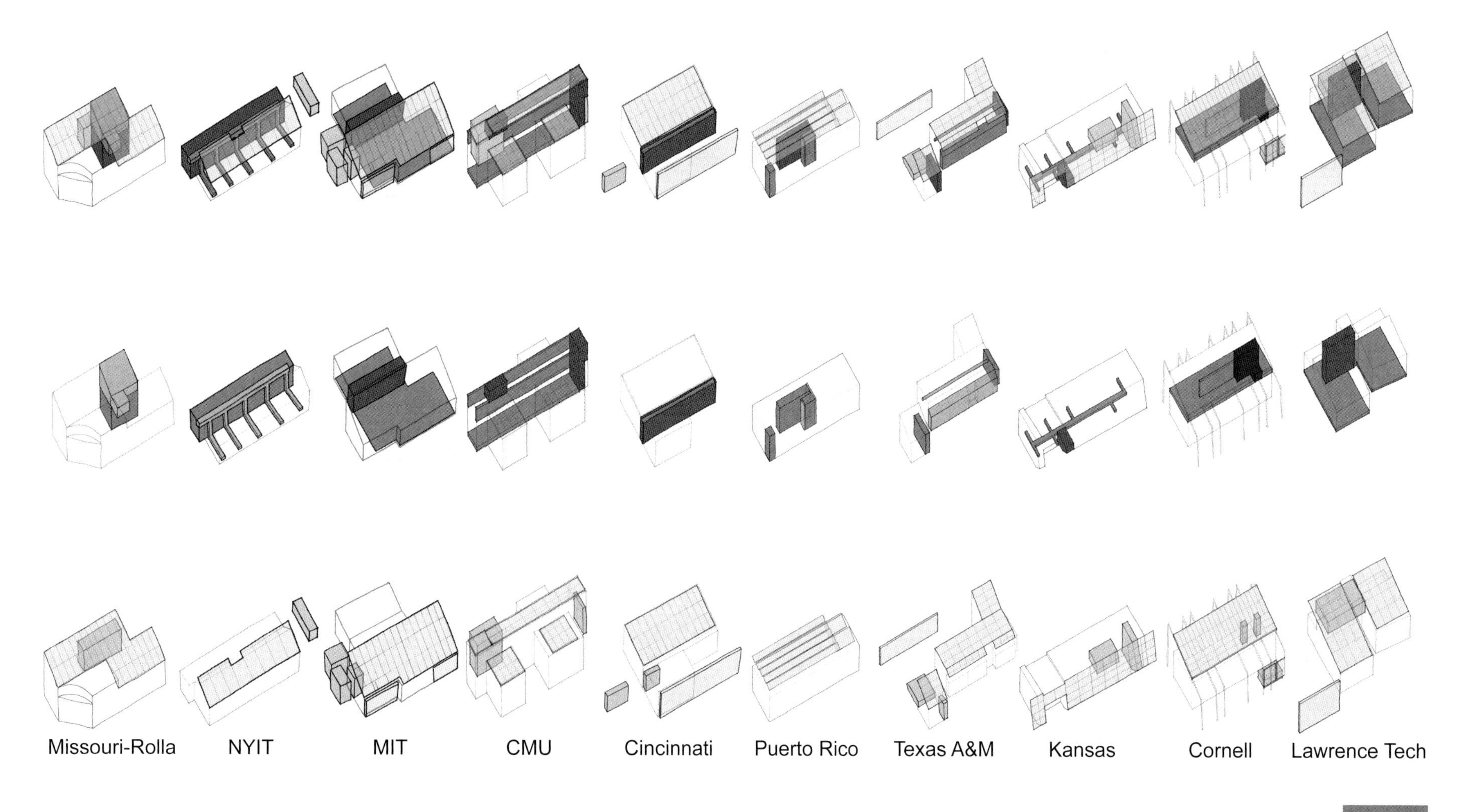

Thermal Transfer

"Insulate before you insolate" is a statement well known to those practicing passive solar design. This refers to the need to "insulate" (design wall assemblies that will resist heat transfer) before "insolating" (introducing direct solar heat gain through glazing). Each team took a different strategy towards insulating their house. Most glazing assemblies are much less effective at resisting heat loss than an opaque insulated wall, although some teams utilized contemporary technolgies and materials that allow light to transfer through the wall while retaining high levels of resistance (limiting heat transfer).

Lawrence Tech

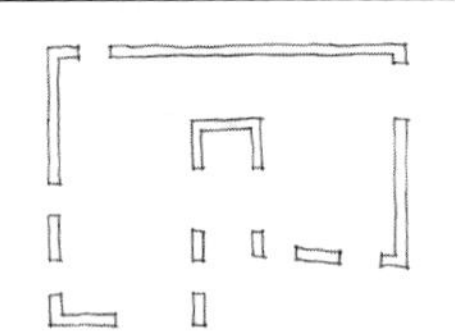

floor area:	**612 sf**
surface:	**2722 sf**
volume:	**5896 cf**
surface : volume	**.462**
thermal conductance:	**148**

Montréal

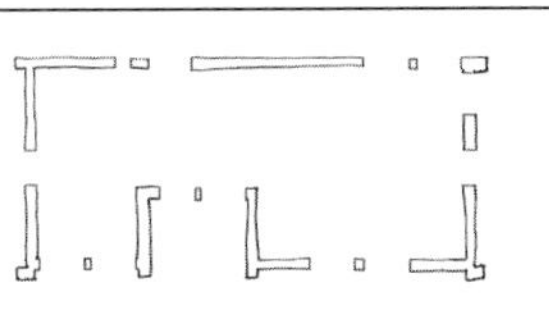

floor area:	**580 sf**
surface:	**2228 sf**
volume:	**3930 cf**
surface : volume	**.567**
thermal conductance:	**104**

Penn State

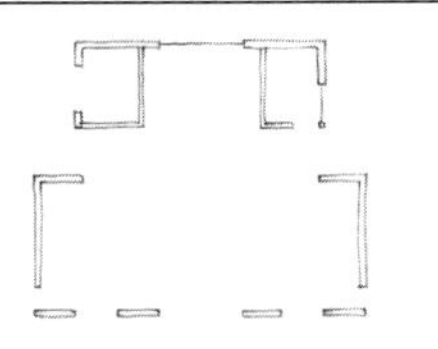

floor area:	**608 sf**
surface:	**1960 sf**
volume:	**6436 cf**
surface : volume	**.305**
thermal conductance:	**214**

Georgia Tech

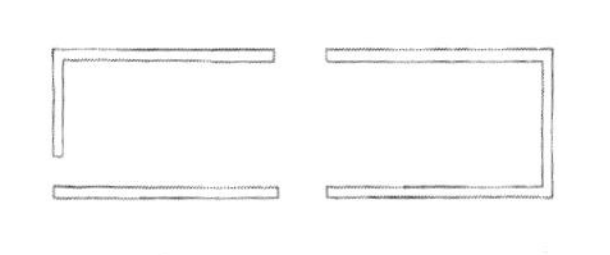

floor area:	**564 sf**
surface:	**1974 sf**
volume:	**7056 cf**
surface : volume	**.280**
thermal conductance:	**90**

Colorado

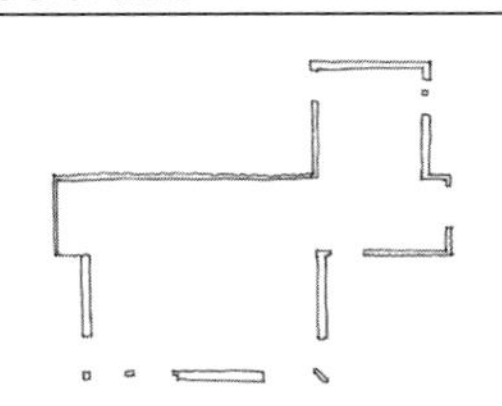

floor area:	**659 sf**
surface:	**2597 sf**
volume:	**8567 cf**
surface : volume	**.303**
thermal conductance:	**194**

NYIT

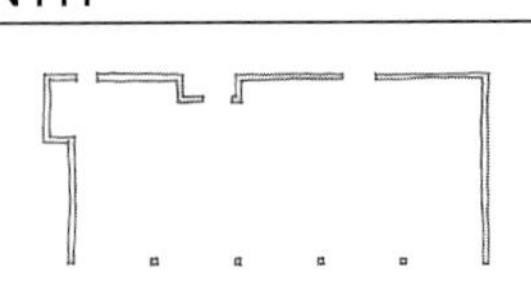

floor area:	**595 sf**
surface:	**1917 sf**
volume:	**5998 cf**
surface : volume	**.320**
thermal conductance:	**129**

Illinois

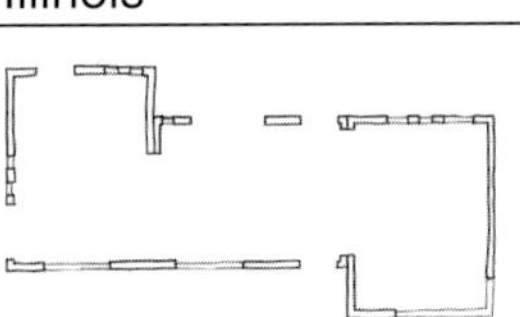

floor area:	**504 sf**
surface:	**1524 sf**
volume:	**4284 cf**
surface : volume	**.356**
thermal conductance:	**106**

MIT

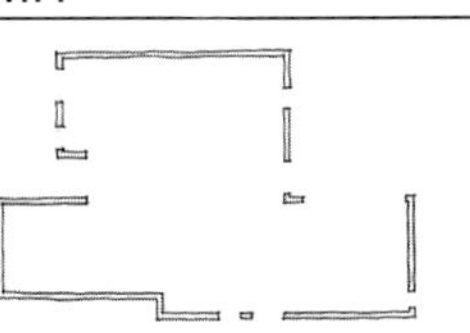

floor area:	**655 sf**
surface:	**2155 sf**
volume:	**7364 cf**
surface : volume	**.293**
thermal conductance:	**145**

CMU

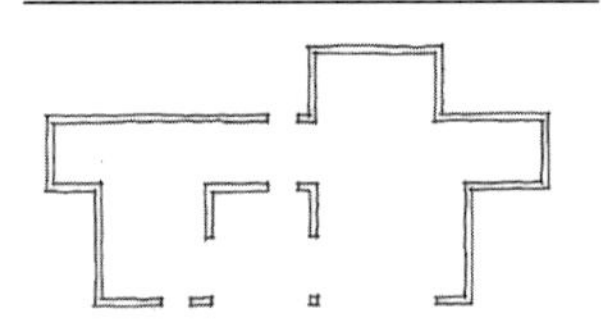

floor area:	**651 sf**
surface:	**3311 sf**
volume:	**7563 cf**
surface : volume	**.438**
thermal conductance:	**295**

Darmstadt

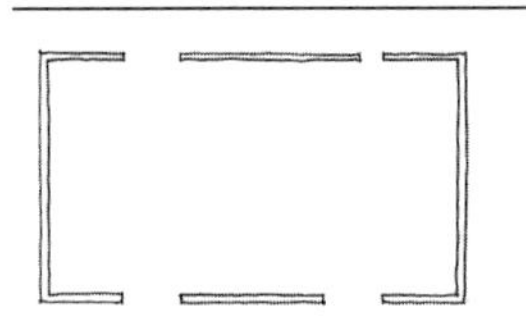

floor area:	**638 sf**
surface:	**1657 sf**
volume:	**5234 cf**
surface : volume	**.317**
thermal conductance:	**66**

floor area:	refers to the conditioned floor area of each house in square feet
surface:	refers to overall exterior surface area of each house in square feet
volume:	refers to interior volume of conditioned space in cubic feet
surface to volume:	refers to the ratio of exterior surface to volume [lower number = less surface area per volume = less heat loss]
thermal conductance:	unitless number that is the product of the areas of exterior surface assemblies multiplied by their specific conductances (U-values) [lower number = less thermal transfer through the envelope] [lower number = better insulating capacity]

Missouri-Rolla

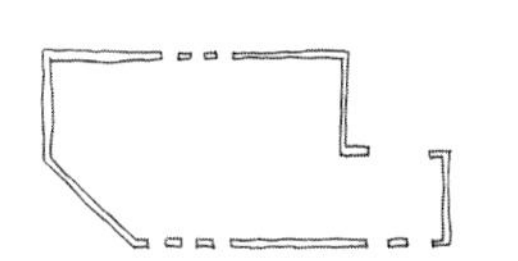

floor area:	**593 sf**
surface:	**2331 sf**
volume:	**5270 cf**
surface : volume	**.442**
thermal conductance:	**184**

Maryland

floor area:	**652 sf**
surface:	**2135 sf**
volume:	**5356 cf**
surface : volume	**.399**
thermal conductance:	**154**

UT Austin

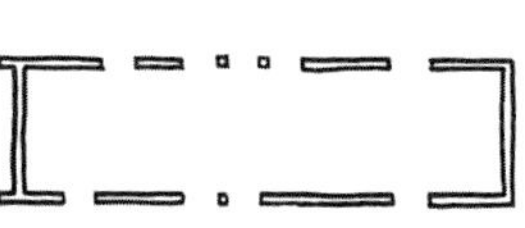

floor area:	**632 sf**
surface:	**1917 sf**
volume:	**6481 cf**
surface : volume	**.296**
thermal conductance:	**131**

Puerto Rico

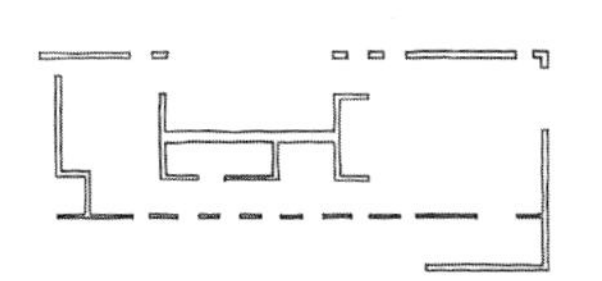

floor area:	**459 sf**
surface:	**2005 sf**
volume:	**3900 cf**
surface : volume	**.514**
thermal conductance:	**142**

Santa Clara

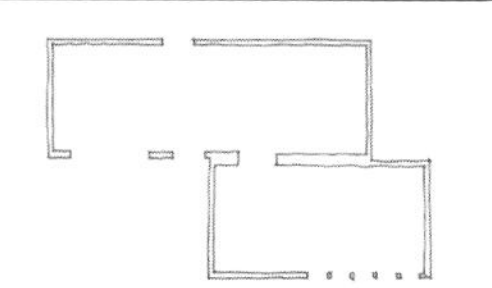

floor area:	**544 sf**
surface:	**1774 sf**
volume:	**4352 cf**
surface : volume	**.408**
thermal conductance:	**90**

Kansas

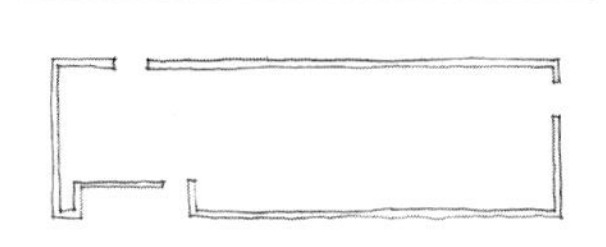

floor area:	**595 sf**
surface:	**2283 sf**
volume:	**5664 cf**
surface : volume	**.403**
thermal conductance:	**152**

Cornell

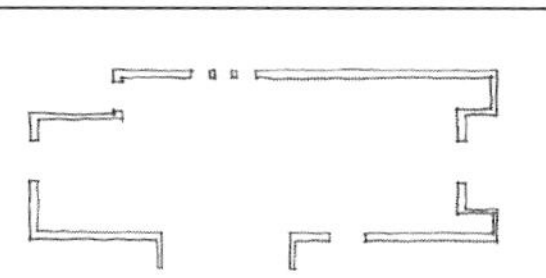

floor area:	**540 sf**
surface:	**1524 sf**
volume:	**6422 cf**
surface : volume	**.237**
thermal conductance:	**114**

Cincinnati

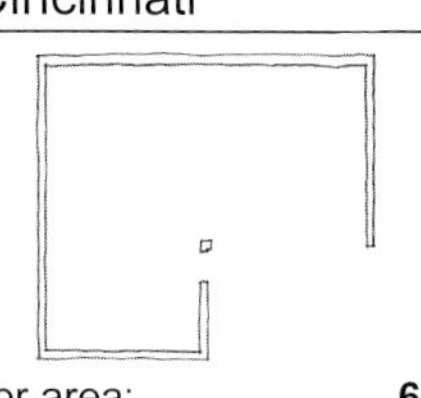

floor area:	**676 sf**
surface:	**1893 sf**
volume:	**6032 cf**
surface : volume	**.314**
thermal conductance:	**158**

Texas A&M

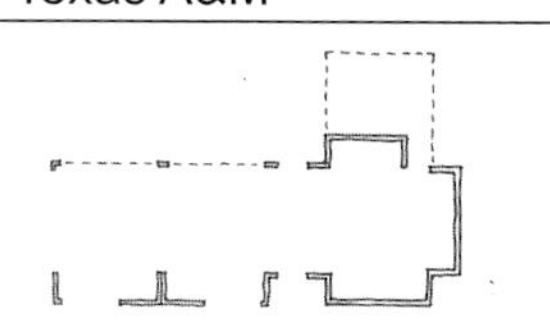

floor area:	**465 sf**
surface:	**1900 sf**
volume:	**5752 cf**
surface : volume	**.330**
thermal conductance:	**220**

Madrid

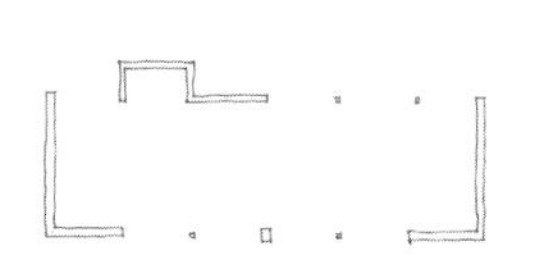

floor area:	**512 sf**
surface:	**1999 sf**
volume:	**6020 cf**
surface : volume	**.332**
thermal conductance:	**132**

Thermal Transfer: Conditioned Floor Area

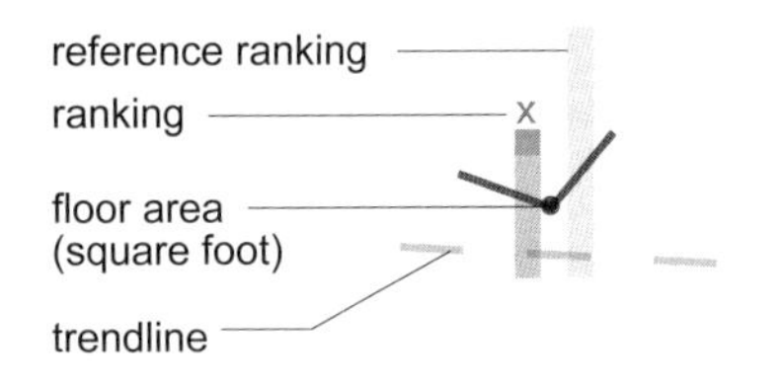

According to the National Association of Home Builders, the average home size in the United States was 2,330 square feet in 2004. For the one-bedroom, one-bathroom houses of the Solar Decathlon there was a minimum of 450 conditioned square feet and a maximum of 800 conditioned square feet of interior space allowed for each house. Many of the teams chose to create less conditioned space, because smaller spaces require less energy to heat and cool while utilizing fewer resources.

The hypothesis behind this analysis was that there would be a correlation between an increase in conditioned square footage and a decrease in the Architecture or Comfort Zone ranking.

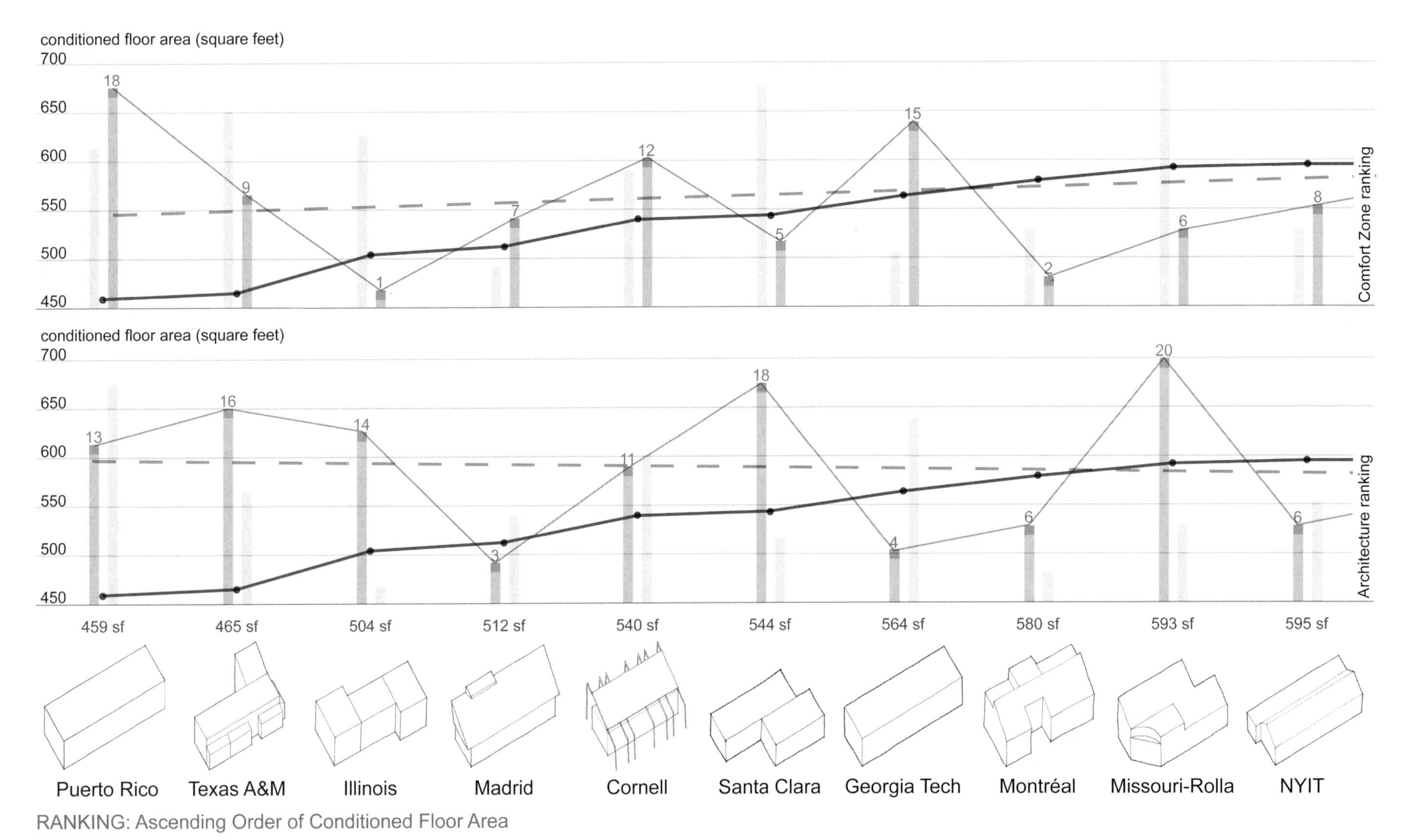

RANKING: Ascending Order of Conditioned Floor Area

There is a correlation, with larger houses ranking better in the Architecture Contest while in the Comfort Zone Contest increased conditioned floor area correlated slightly to decreased rankings. Those judging Architecture responded positively to larger spaces, but the result was a decrease in Comfort Zone rankings.

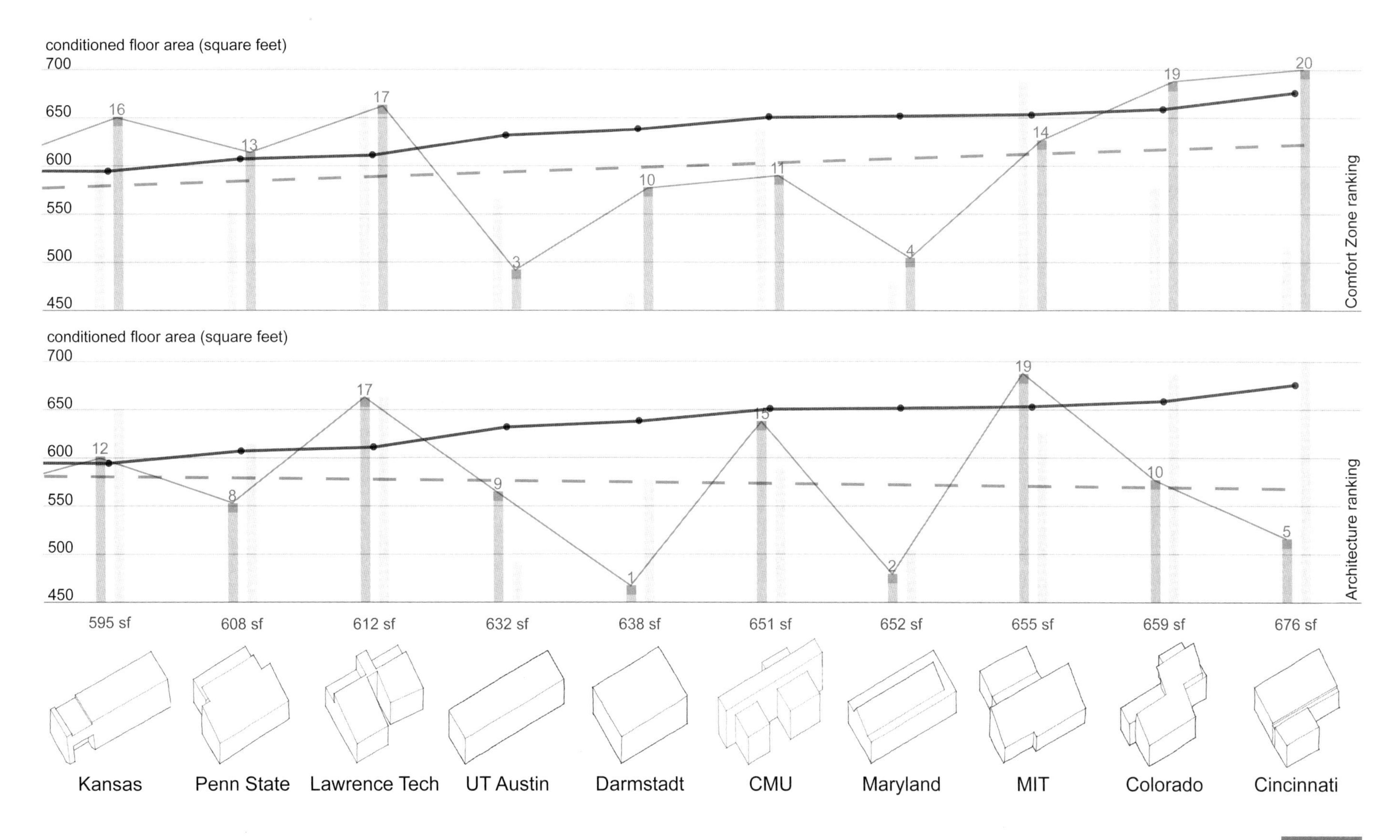

Thermal Transfer: Surface Area

Surface area refers to the amount of exterior wall (building skin) that separates the interior from the exterior. The shape and proportions of each house led to greater or lesser exposed surface areas. Simpler forms tend to have less exposed surface area. This chart compares the exposed surface area of each house to its rankings in the Architecture and Comfort Zone Contests.

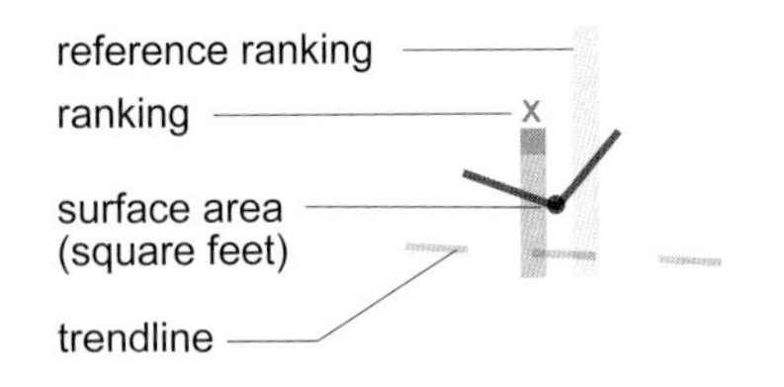

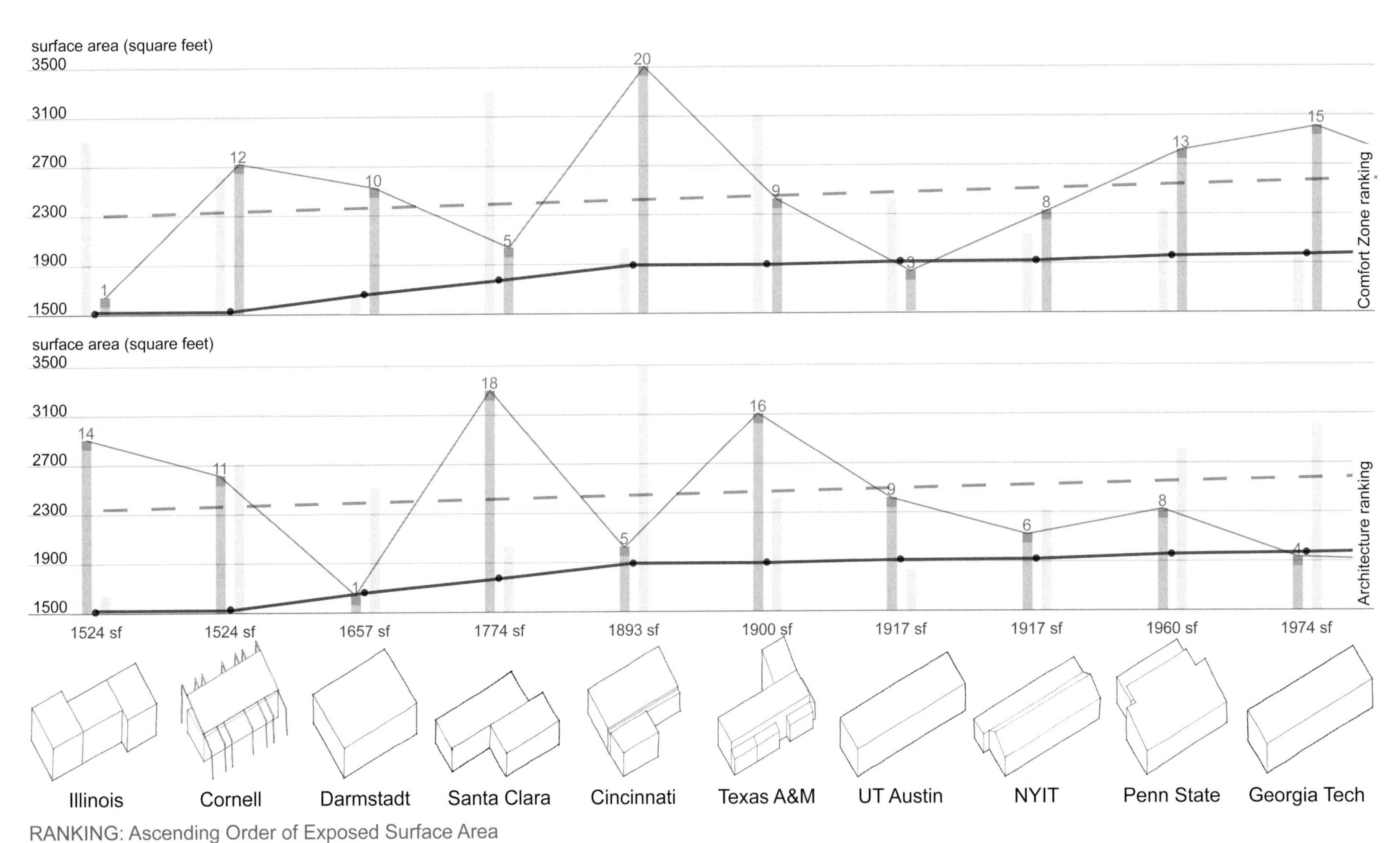

RANKING: Ascending Order of Exposed Surface Area

Though the trendline slope is not significant, there is a direct correlation between an increase in exposed surface area and a decrease in the Architecture and Comfort Zone rankings.

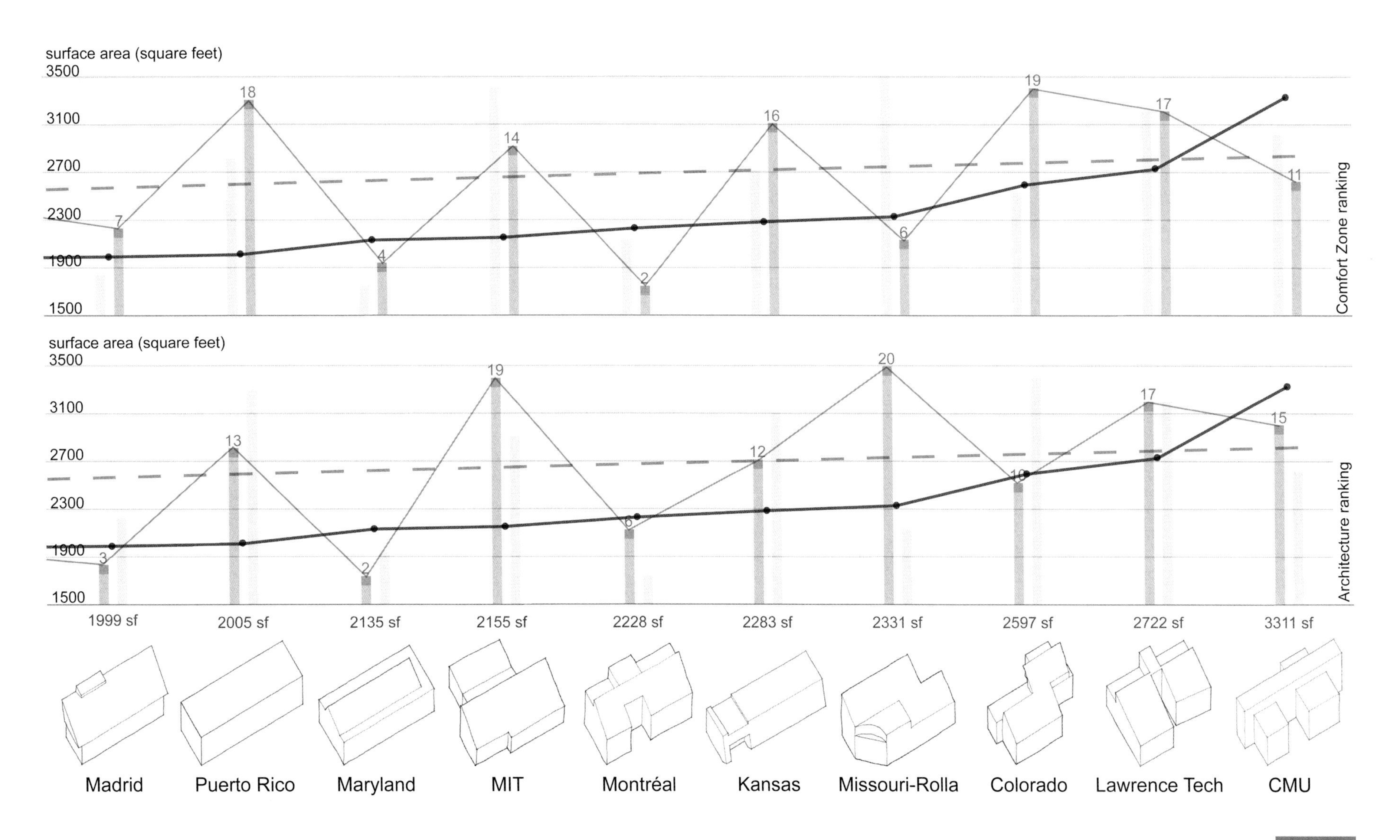

Thermal Transfer: Interior Volume

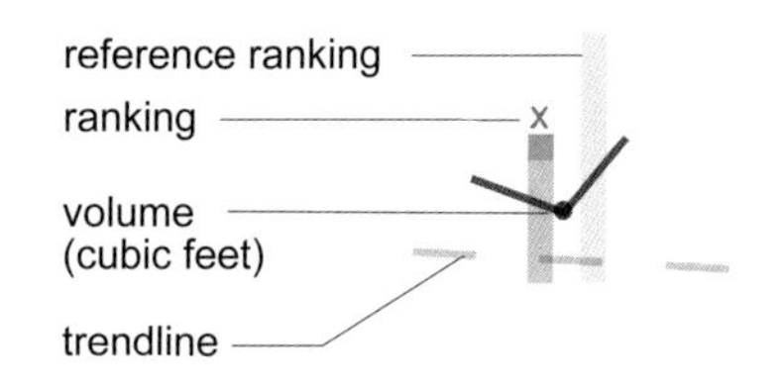

Though there were limitations on the square footage of each house, as well as height limitations, there was no specific limitation of volume. Design choices about shape and proportion impact the overall volume of a building. Greater volume may have architectural benefits, but increased volume leads to an increased amount of air that needs to be conditioned.

This chart compares the impact of conditioned interior volume on the rankings in the Architecture and Comfort Zone Contests.

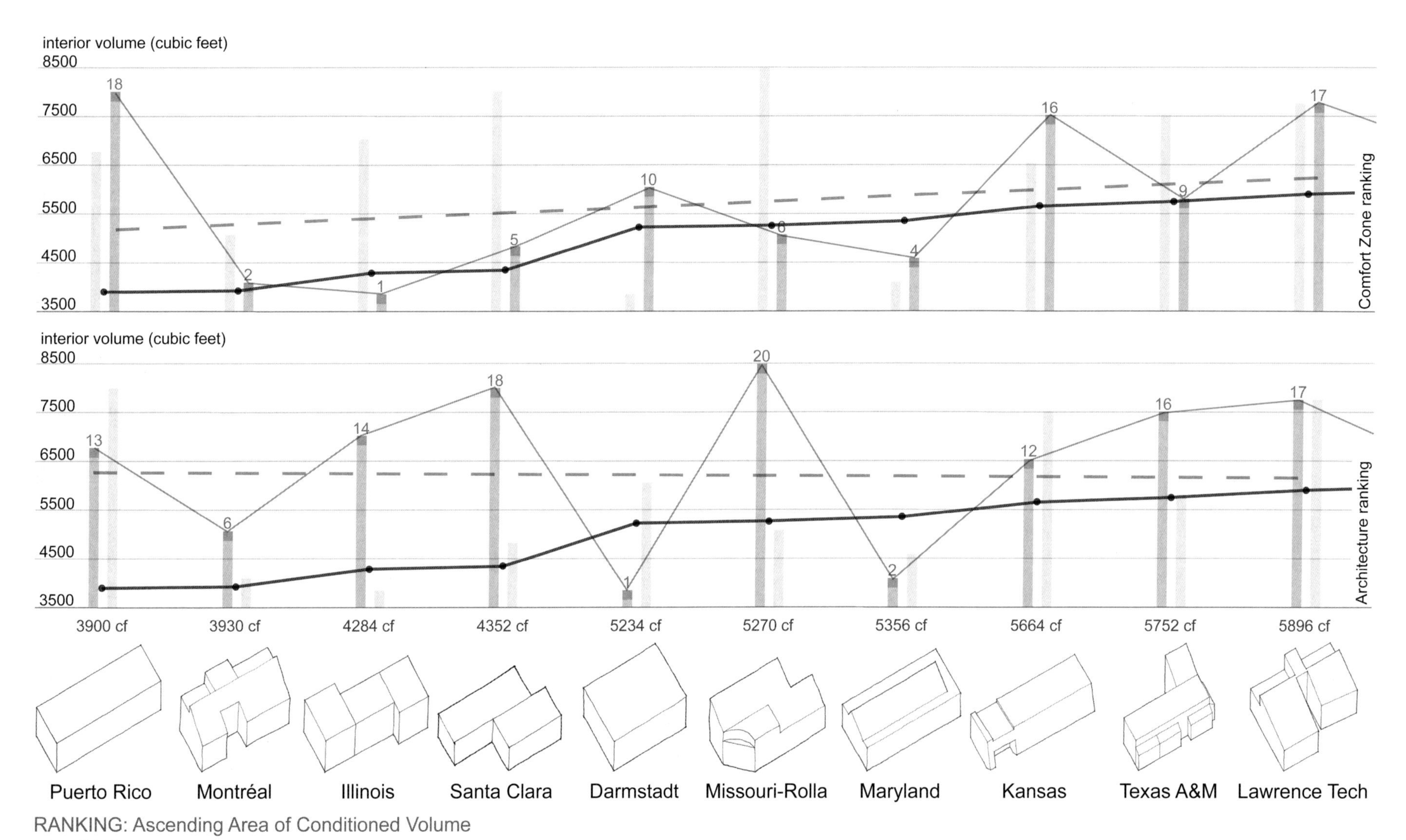

RANKING: Ascending Area of Conditioned Volume

This comparison shows a direct correlation between volume and rankings in the Architecture and Comfort Zone Contests. As volume increased, the teams had lower rankings in the Comfort Zone Contest but slightly higher rankings in the Architecture Contest.

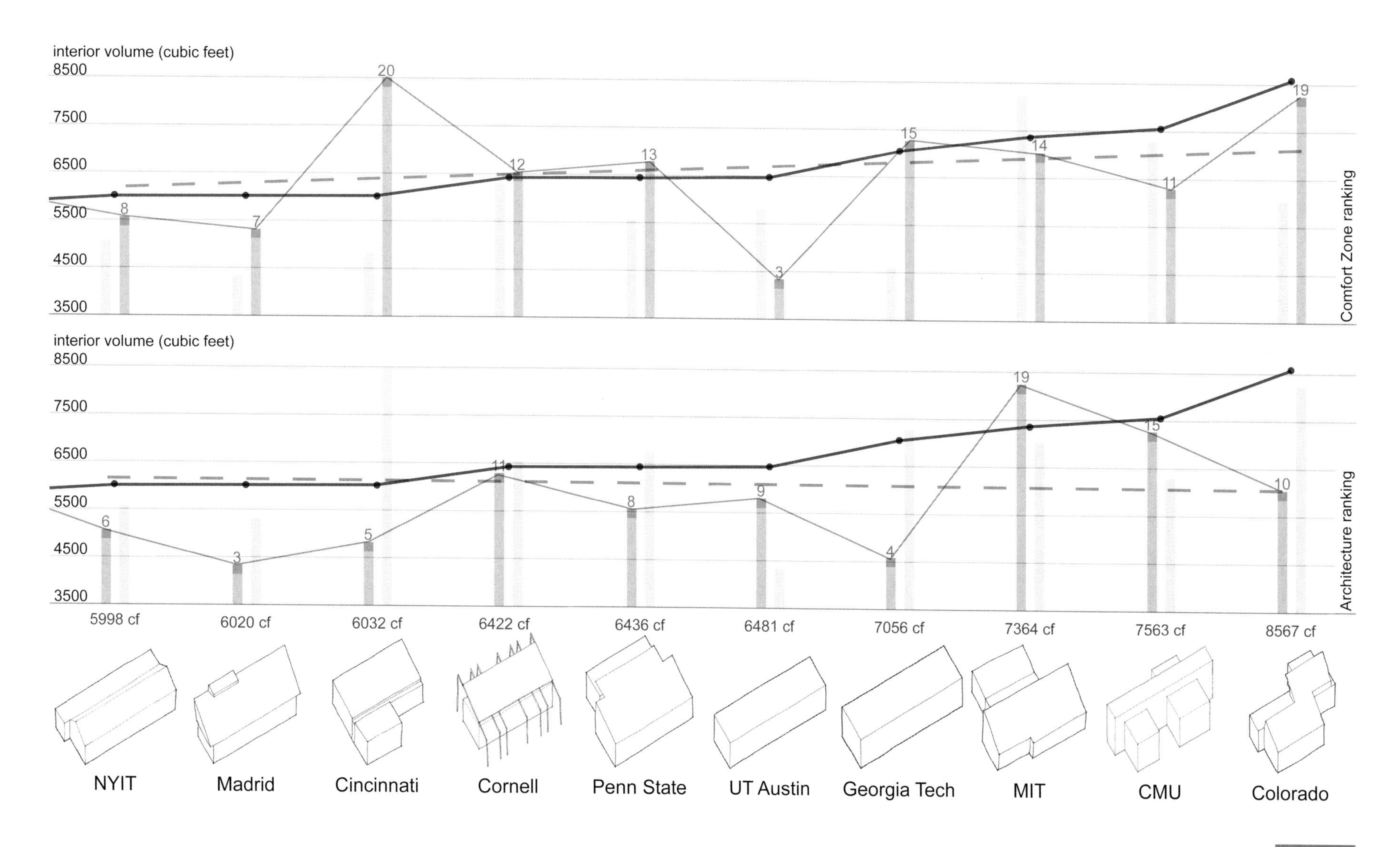

Thermal Transfer: Surface to Volume

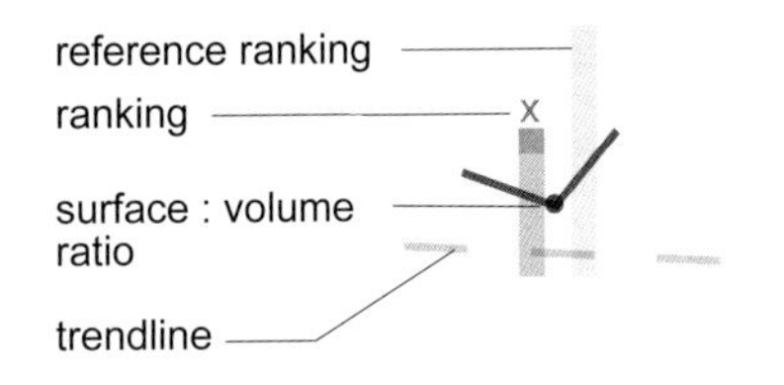

For a given volume, as the surface area increases the heating or cooling load also increases. The relationship of exterior exposed surface to interior conditioned volume is therefore a potential indicator of interior thermal comfort. The greater the exposed surface area, the greater opportunity for heat transfer through the building skin. The hypothesis was that an increased ratio of surface to volume would lead to a decrease in Comfort Zone rankings.

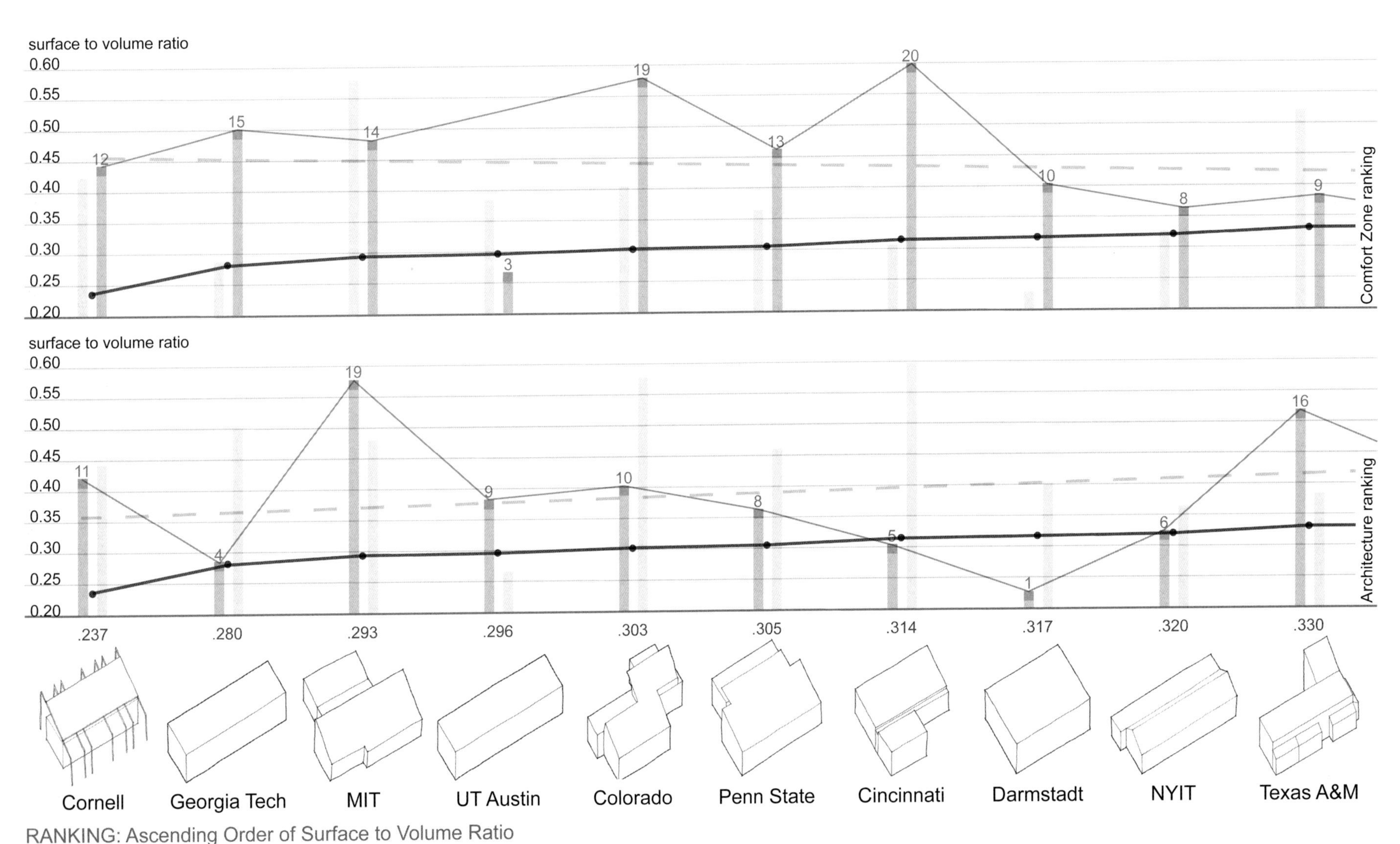

RANKING: Ascending Order of Surface to Volume Ratio

There is a direct correlation between the increase of surface area as compared to volume and the rankings in the Architecture and Comfort Zone Contests. The Architecture rankings decreased as the ratio of surface to volume increased while the Comfort Zone rankings increased.

The increase in surface to volume ratio typically correlates to an increasingly complex form (such as CMU and Colorado). Other houses with simple forms and minimal extrusions have higher than expected surface to volume ratios as a result of their comparatively low ceiling heights (decreased volume).

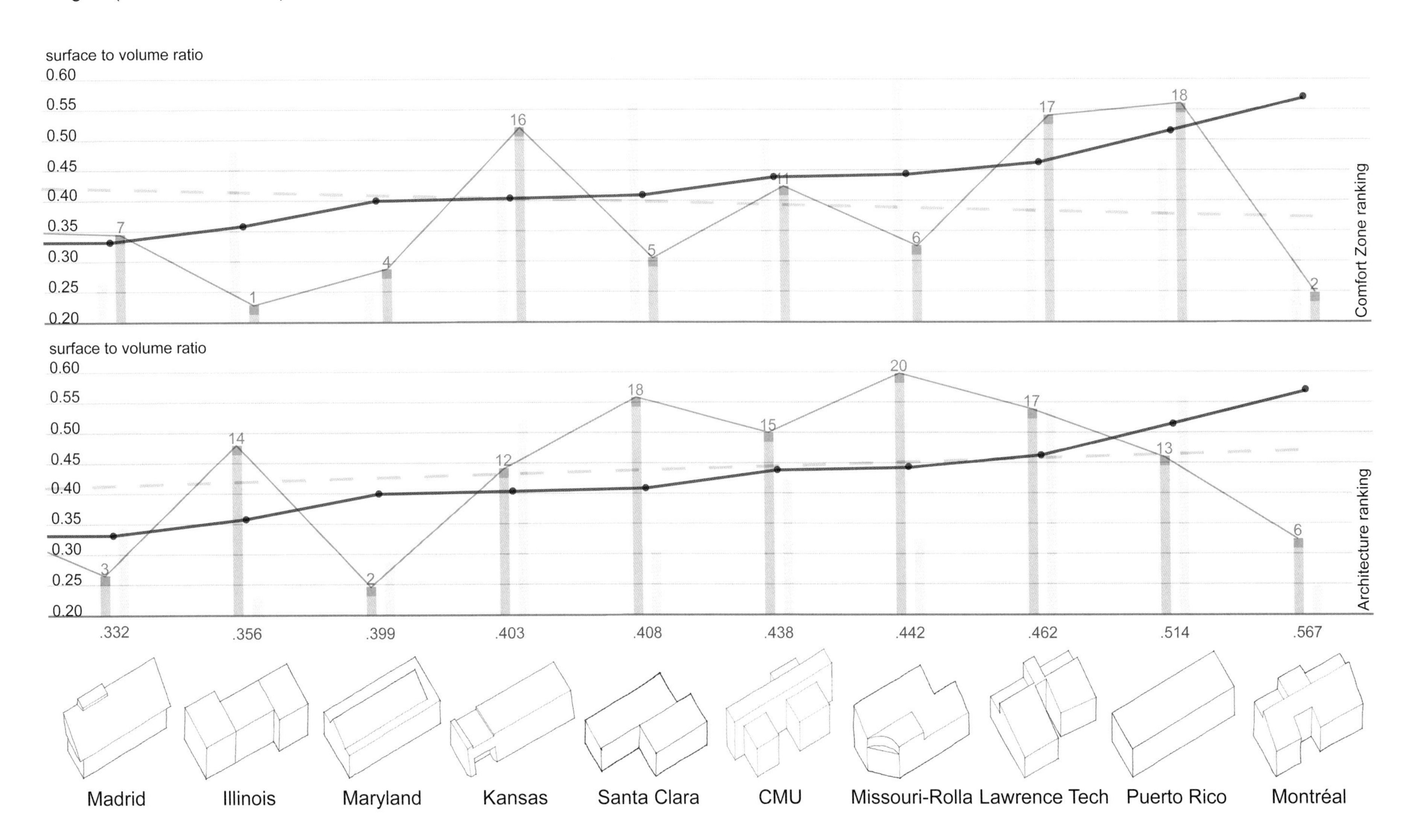

Thermal Transfer: Conductance

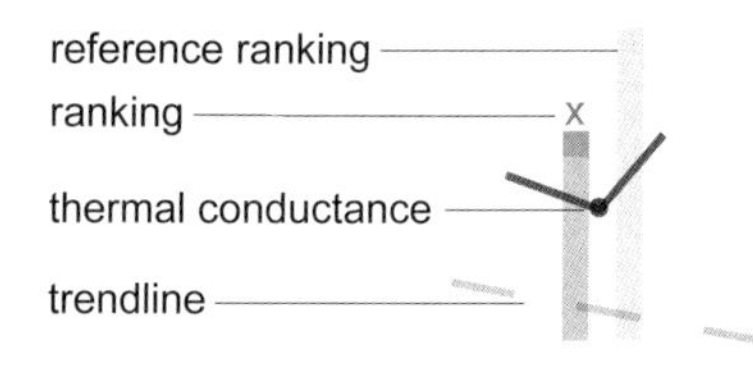

The materials that comprise the assemblies of the exterior walls and roof of each house each have specific capacities to permit or resist the conductance of heat. The *thermal conductance* rating is a numerical compilation of the conductance values of all exterior surfaces of a house multiplied by its total exterior square footage (U-value x Area). A higher U-value correlates to higher conductance or lower resistance. A lower U-value indicates an assembly that is better insulated than one with a higher U-value (or thermal conductance number).

This chart shows the thermal conductance rating for each house as compared to its rankings in the Architecture and Comfort Zone Contests.

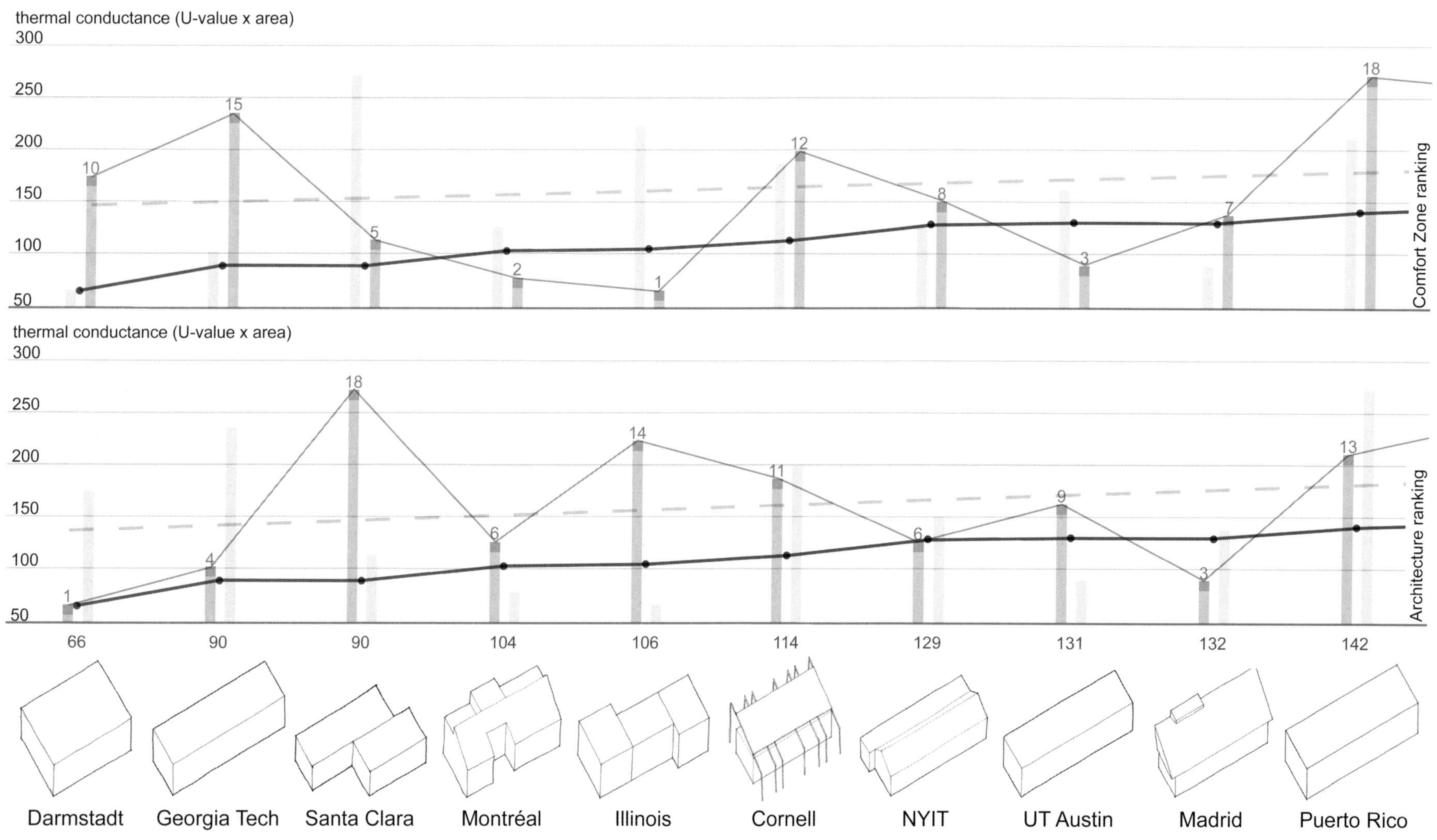

RANKING: Ascending Order of Thermal Conductance

Greater conductance equates to a decrease in resistance (greater potential heat loss). There is a direct correlation between increased levels of thermal conductance (decreased insulation capacity) and a decrease in rankings in the Architecture and Comfort Zone Contests.

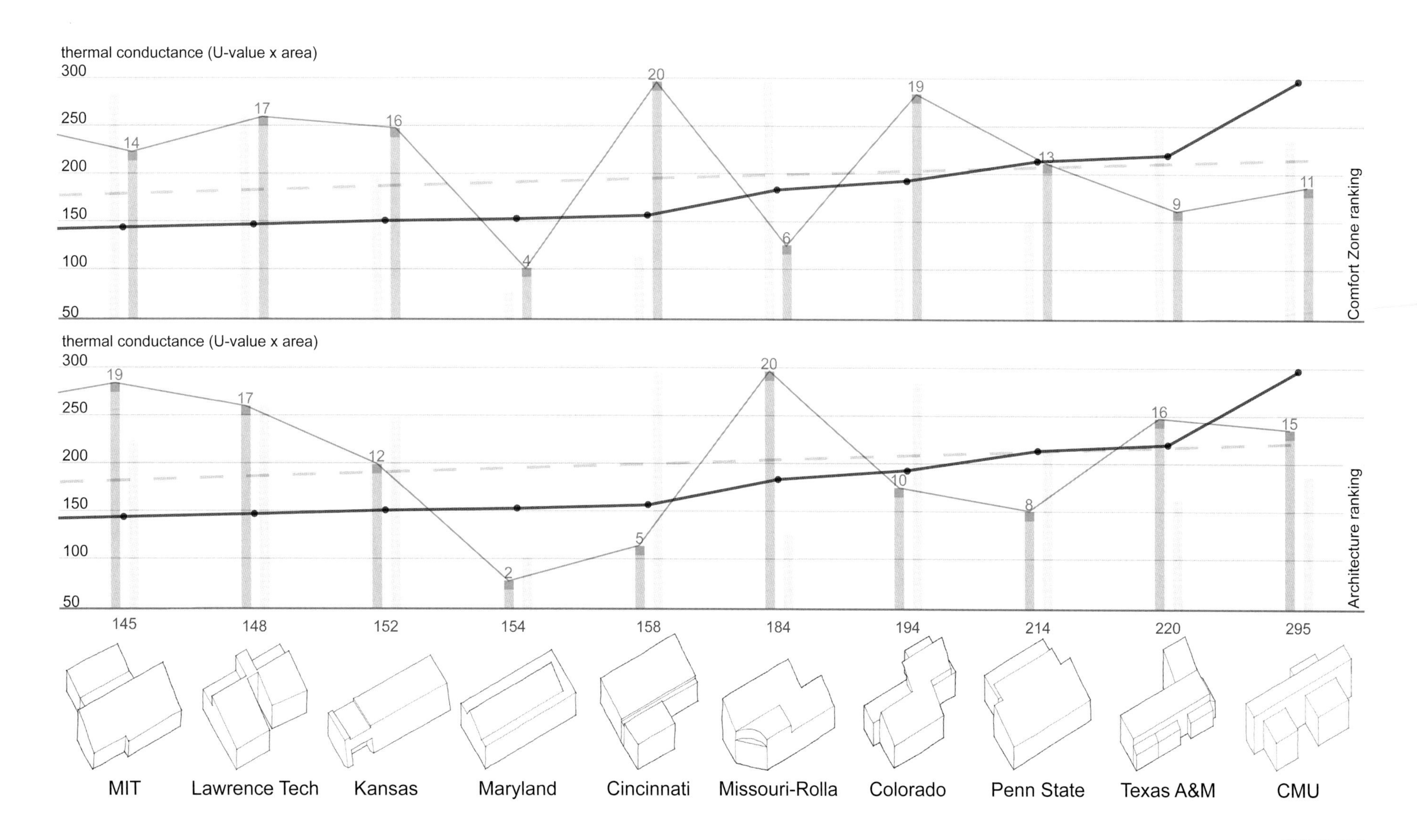

Thermal Storage

Thermal storage refers to the utilzation of thermal mass materials that have an increased capacity to store heat energy as a result of their specific heat and density. Thermal storage is achieved with the use of thermal mass materials (such as stone) used to either store heat as a means to mitigating sensible temperature increases or to slow the flow of heat. Thermal storage materials include phase change materials (PCM) that remove heat through the latent heat of phase change. Though these were extremely effective ways of mitigating temperature swings in the houses, only seven of the 20 teams used thermal storage materials in their designs. This was probably a result of another challenge facing every team – the need to transport the house to the National Mall in Washington, DC. The transportation requirement likely impacted the decision to utilize thermal storage materials.

Internal Thermal Storage

When utilizing thermal storage for passive solar design, there are rule-of-thumb ratios of south-facing glazing to internal thermal storage material surface:

With direct solar gain on the thermal storage material:
ratio of 1 : 3 (glazing to material).

Without direct solar gain on the thermal storage material:
ratio of 1 : 6 (glazing to material).

Penn State

O	04
A	08
C	13

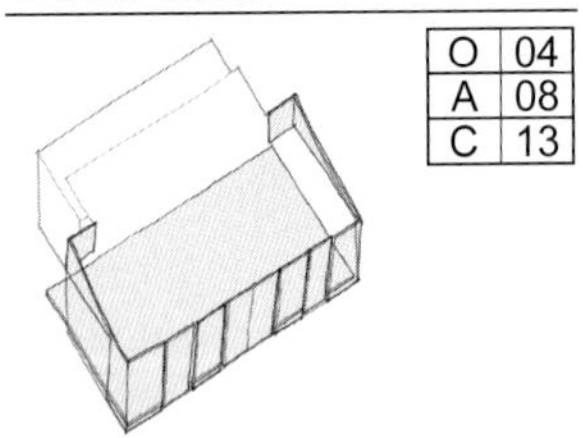

material 1: **milk-bottle wall**
location: **south interior**
area: **128 sf**

material 2: **slate flooring**
location: **interior**
area: **560 sf**

south glazing : mass: **1 : 5.4**

Cornell

O	19
A	11
C	12

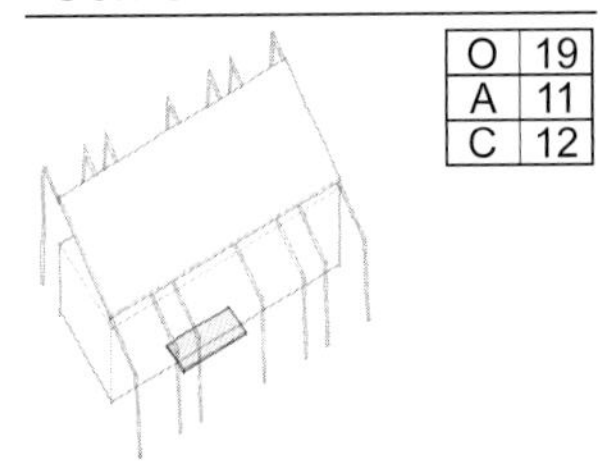

material: **bluestone floor**
location: **south interior**
area: **95 sf**

south glazing : mass: **1 : 1.0**

Darmstadt

O	01
A	01
C	10

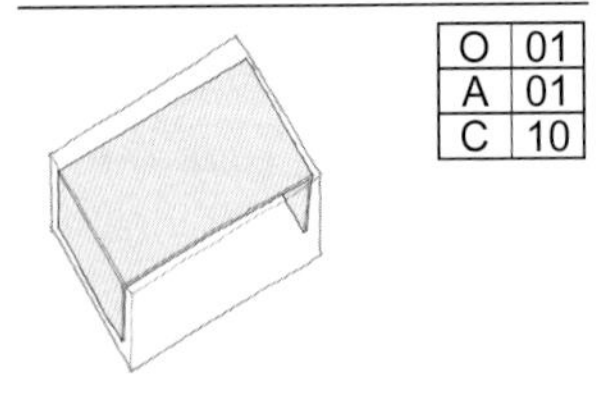

material: **phase change matrls**
location: **int. walls and ceiling**
area: **888 sf**

south glazing : mass: **1 : 3.8**

Madrid

O	05
A	03
C	07

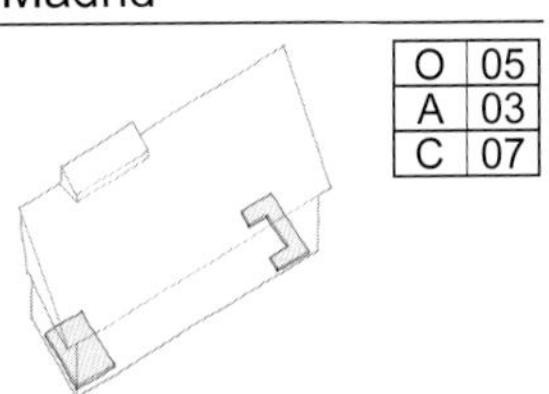

material: **phase change matrls**
location: **int. east and west flr**
area: **104 sf**

south glazing : mass: **1 : 0.5**

External Thermal Storage

Penn State

O	04
A	08
C	13

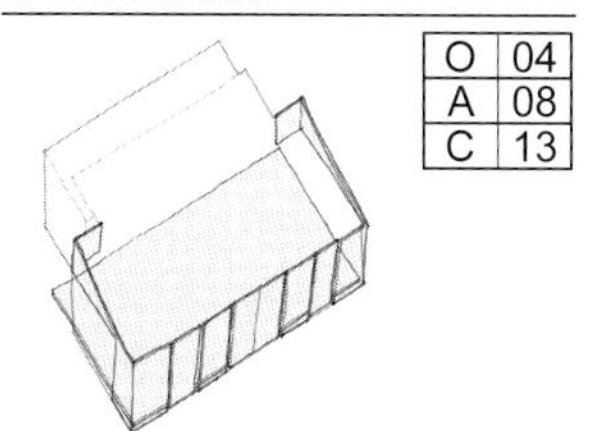

material 1: **slate shingles**
location: **n, s, e, w exterior**
area: **224 sf**
thickness: **0.5 in**

material 2: **fiber-cement board**
location: **e, w exterior**
area: **72 sf**
thickness: **0.375 in**

Colorado

O	07
A	10
C	19

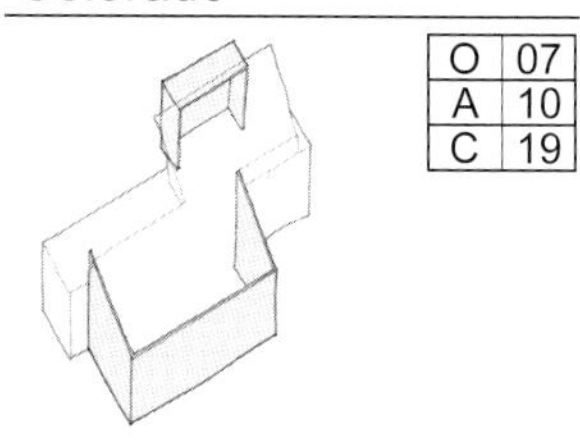

material: **fiber-cement board**
location: **n, s, e, w exterior**
area: **537 sf**
thickness: **0.375 in**

MIT

O	13
A	19
C	14

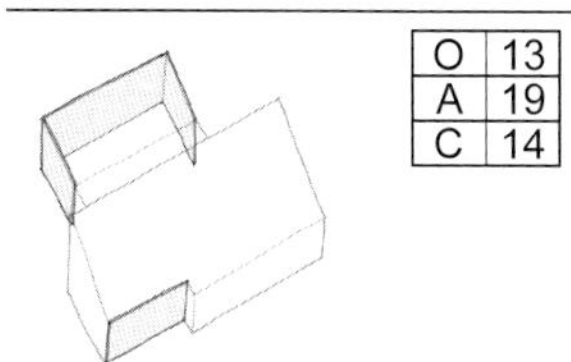

material: **fiber-cement board**
location: **n, s, e, w exterior**
area: **490 sf**
thickness: **0.375 in**

Santa Clara

O	03
A	18
C	05

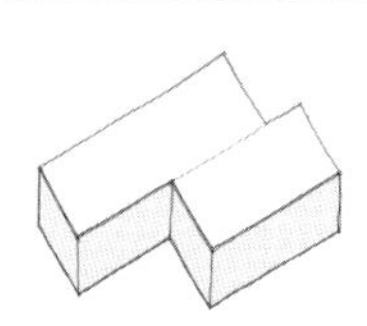

material: **fiber-cement board**
location: **s, w exterior**
area: **454 sf**
thickness: **0.375 in**

Every team chose particular exterior cladding materials for their house for a multitude of reasons. For a climate with significant heat gain, such as Washington, DC, a thermal storage material on the exterior has significant advantages. Only four teams in the 2007 competition chose thermal storage materials for their cladding.

Assessing the impacts of the thermal storage material as a result of Comfort Zone rankings, it appears that this may not have been as effective as anticipated. Most likely, it is a result of having insufficient thickness of the material to make a significant reduction in the speed at which heat was transferred through the wall.

Thermal Storage: Mass Materials

The utilization of thermal storage materials would be expected to have positive impacts on the ranking in the Comfort Zone Contest, but of the seven houses that incorported thermal mass internally or externally, the highest rank in the Comfort Zone Contest was 5 and four of the seven houses were in bottom half of the rankings (12, 13, 14 and 19).

Assessing the impacts of an increased use of thermal storage, one could expect to find increased use of thermal storage correlating to an increase in Comfort Zone rankings. The charts on the right reflect the opposite trend – an increase in thermal storage led to worse Comfort Zone rankings. Ironically, there was a sharp increase in rankings for the Architecture Contest and overall score as the amount of thermal storage increased.

The Penn State team utilized a unique thermal storage feature: a series of sliding milk-bottle walls. Though a fascinating and innovate idea for thermal storage, the bottles were left empty during the competition. For the purposes of these charts, the milk-bottles were considered thermal storage (i.e. filled with water).

The results show that there is a direct relationship between the amount of glazing on a given facade and the location of thermal mass within a home or on an exterior surface. By comparing those homes that utilized thermal mass with those that did not, it appears that the homes without thermal mass have a greater percentage of total glazing. Perhaps the homes that incorporate thermal mass have less glazing in part to help ensure storage of the heat trapped by thermal mass. The results show that the homes with thermal mass utilize a greater percentage of glazing in order for sunlight to reach the thermal mass, but overall incorporate less glazing, resulting in less overall heat loss.

This analysis illuminates the fact that either thermal storage was not being utilized effectively in these projects, or that there were other factors that occurred in the houses with thermal storage which impacted their Comfort Zone rankings.

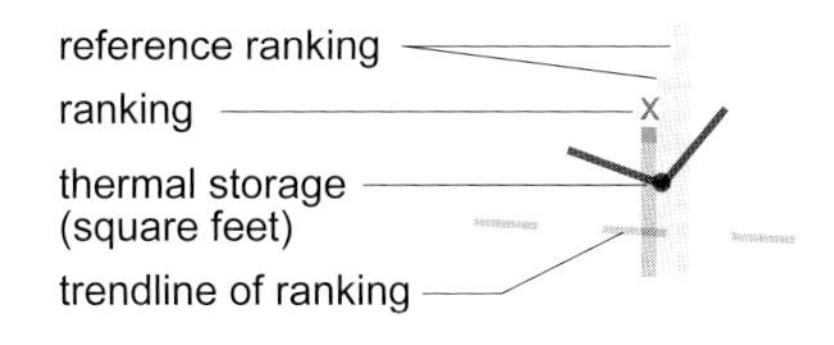
reference ranking
ranking
x
thermal storage
(square feet)
trendline of ranking

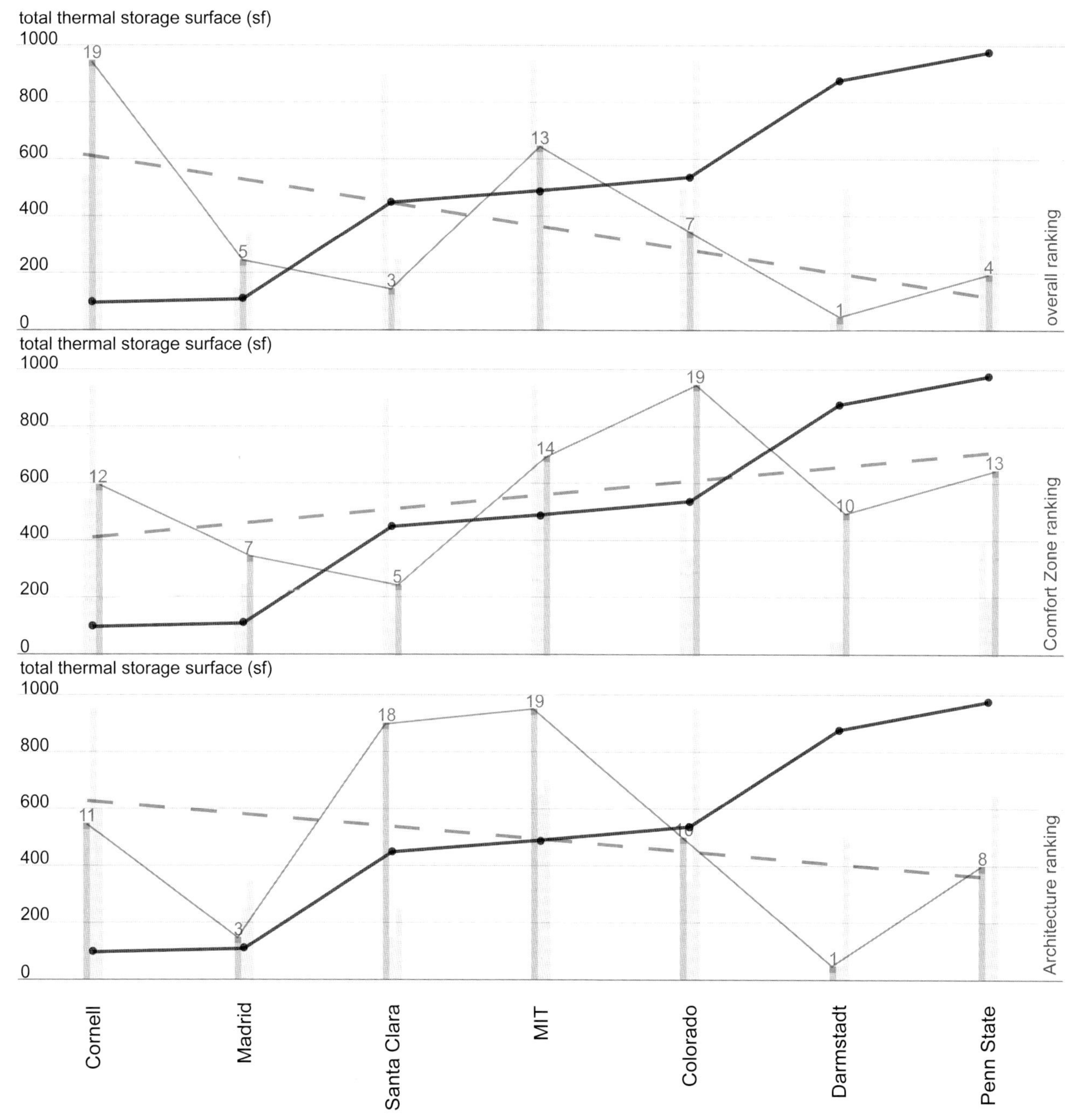
total thermal storage surface (sf)
1000
800
600
400
200
0
19
5
3
13
7
1
4
overall ranking
total thermal storage surface (sf)
1000
800
600
400
200
0
12
7
5
14
19
10
13
Comfort Zone ranking
total thermal storage surface (sf)
1000
800
600
400
200
0
11
3
18
19
10
1
8
Architecture ranking
Cornell
Madrid
Santa Clara
MIT
Colorado
Darmstadt
Penn State

Daylighting

glzg (sf)
N: North
E: East
S: South
W: West
R: Roof
T: Total

Each team attempted to balance the daylight, ventilation and heat through glazing of windows and doors in the walls and roofs of their houses. The data on these pages gives the amounts of glazing on all facades and the roof of each house, as well as a total amount of glazing (in square feet). This data is correlated to rankings in the Architecture and Comfort Zone Contests in charts on the following pages.

Lawrence Tech

glzg (sf)
N: 53.4
E: 120.1
S: 87.6
W: 100.0
R: 14.5
T: 376

Montréal

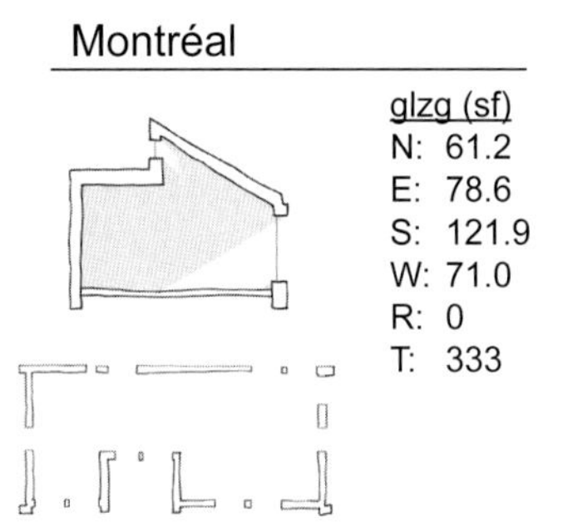

glzg (sf)
N: 61.2
E: 78.6
S: 121.9
W: 71.0
R: 0
T: 333

Penn State

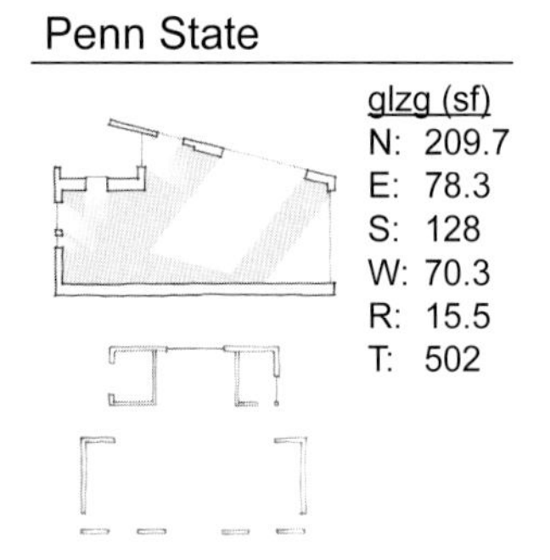

glzg (sf)
N: 209.7
E: 78.3
S: 128
W: 70.3
R: 15.5
T: 502

Georgia Tech

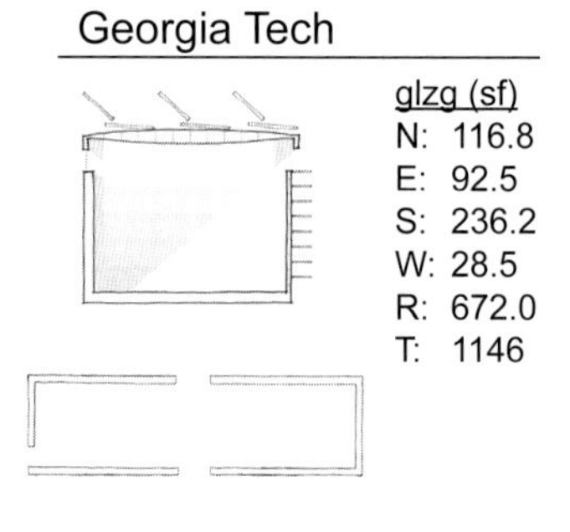

glzg (sf)
N: 116.8
E: 92.5
S: 236.2
W: 28.5
R: 672.0
T: 1146

Colorado

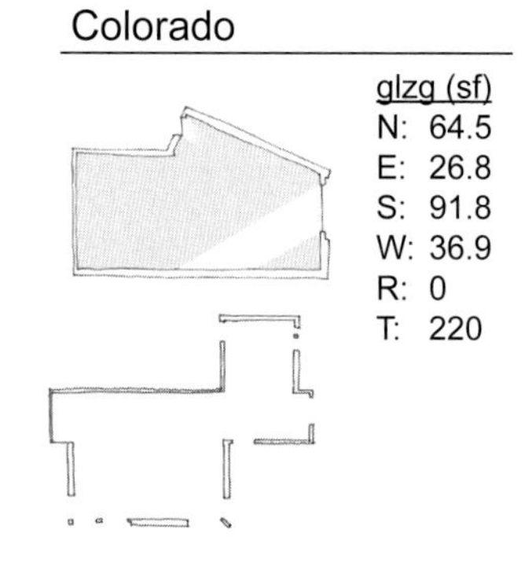

glzg (sf)
N: 64.5
E: 26.8
S: 91.8
W: 36.9
R: 0
T: 220

NYIT

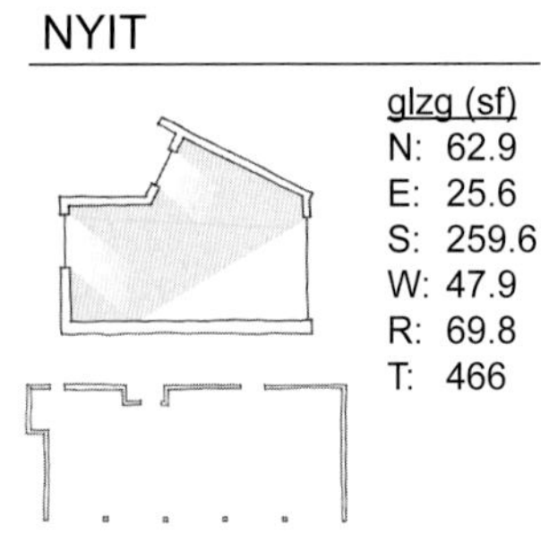

glzg (sf)
N: 62.9
E: 25.6
S: 259.6
W: 47.9
R: 69.8
T: 466

Illinois

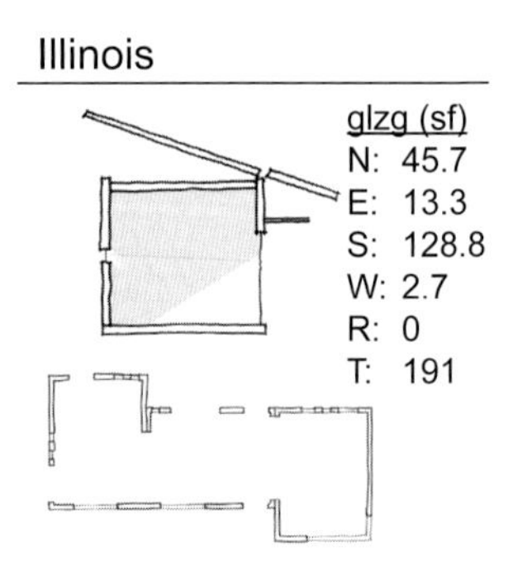

glzg (sf)
N: 45.7
E: 13.3
S: 128.8
W: 2.7
R: 0
T: 191

MIT

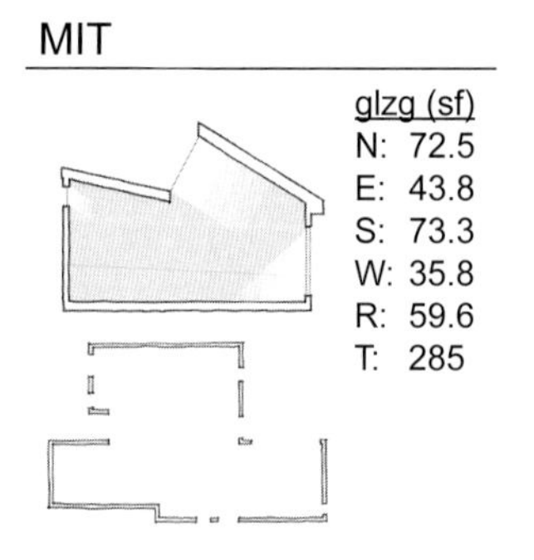

glzg (sf)
N: 72.5
E: 43.8
S: 73.3
W: 35.8
R: 59.6
T: 285

CMU

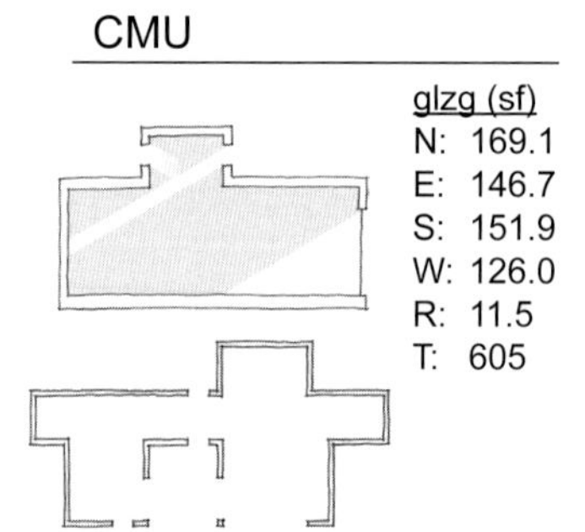

glzg (sf)
N: 169.1
E: 146.7
S: 151.9
W: 126.0
R: 11.5
T: 605

Darmstadt

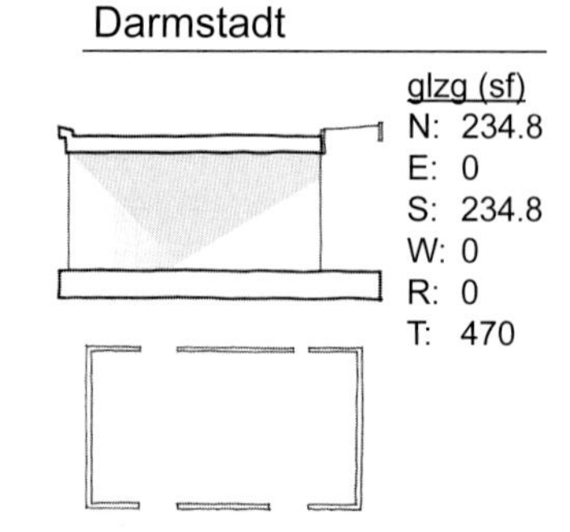

glzg (sf)
N: 234.8
E: 0
S: 234.8
W: 0
R: 0
T: 470

LAYOUT: Location in Solar Village on National Mall

Missouri-Rolla

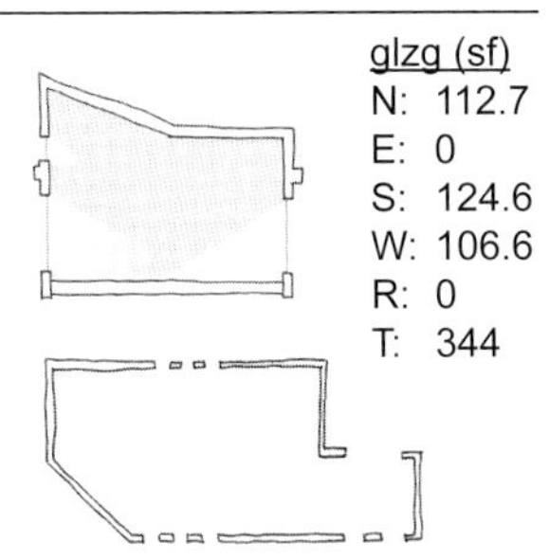

Maryland

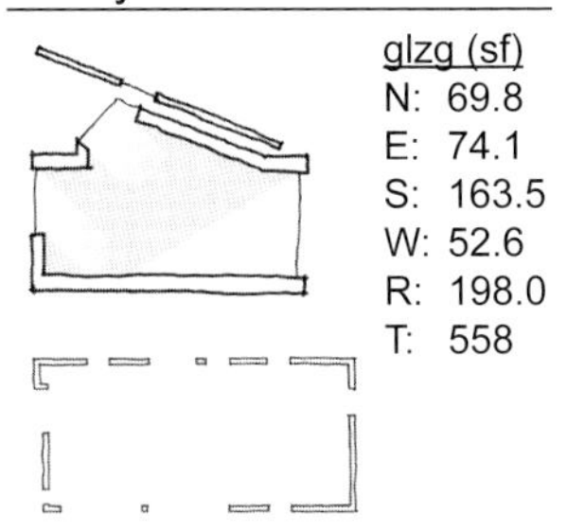

UT Austin

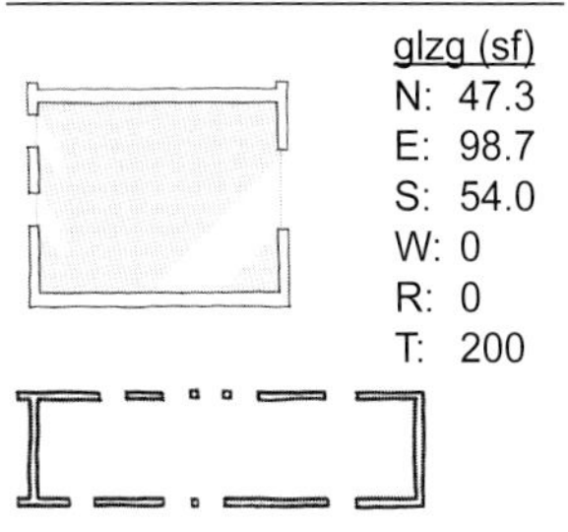

Puerto Rico

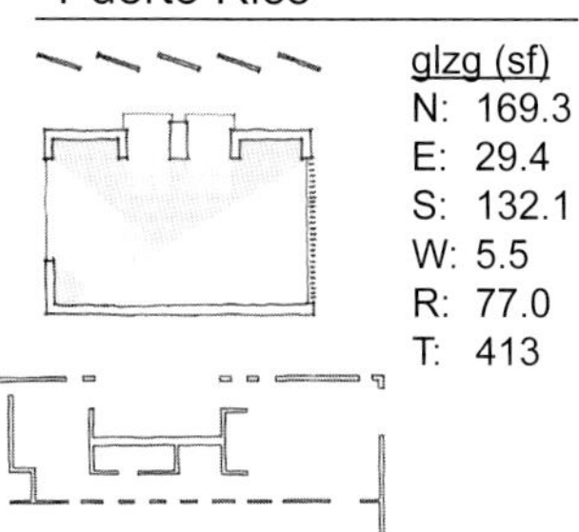

Santa Clara

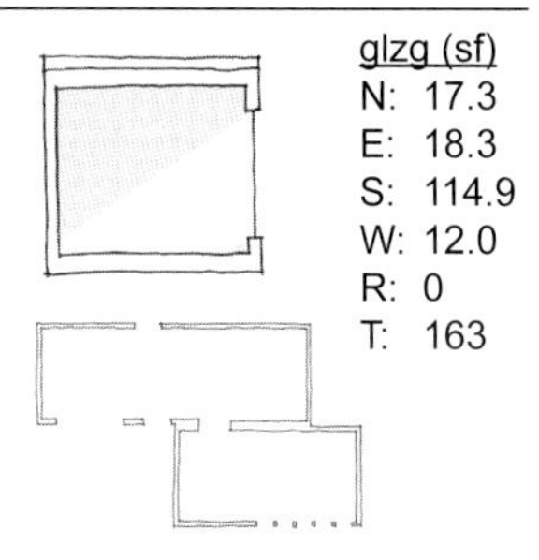

Kansas

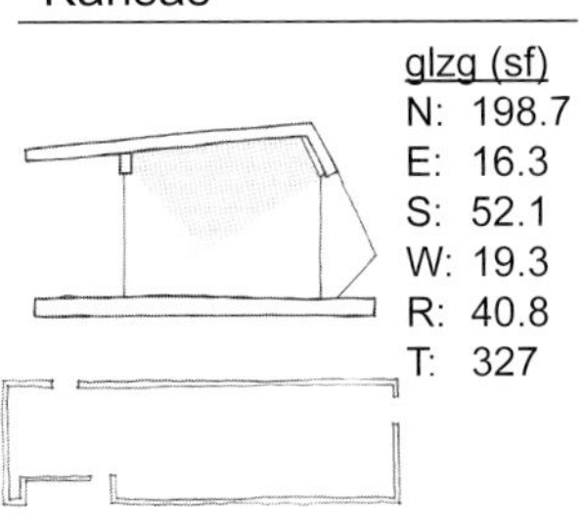

Cornell

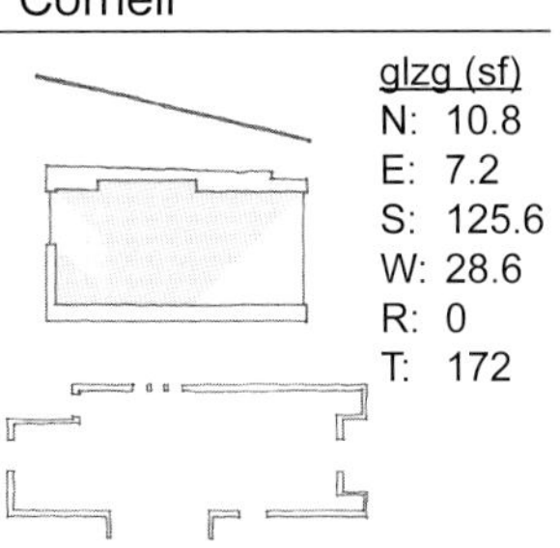

Cincinnati

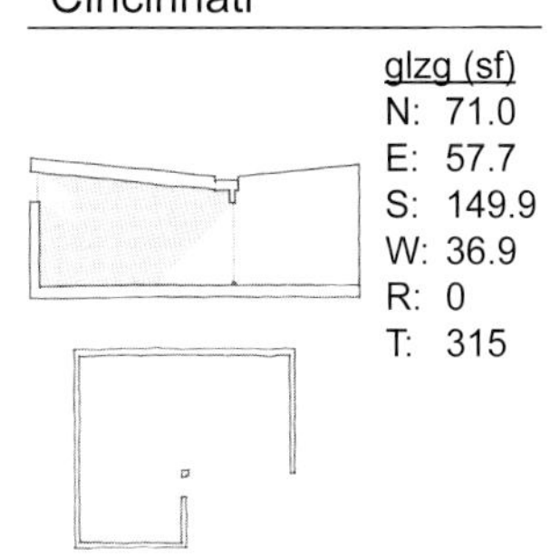

Texas A&M

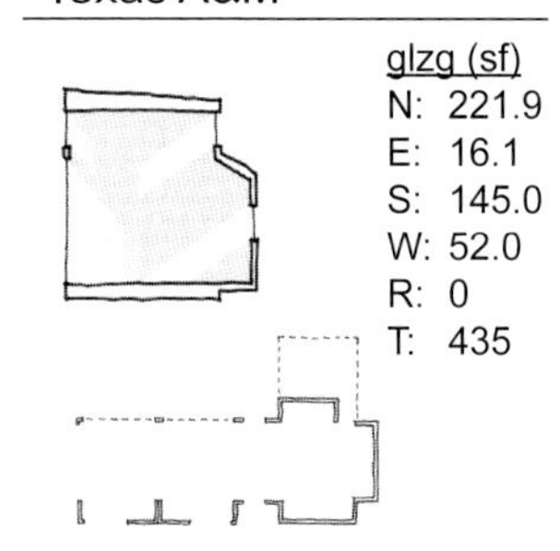

Madrid

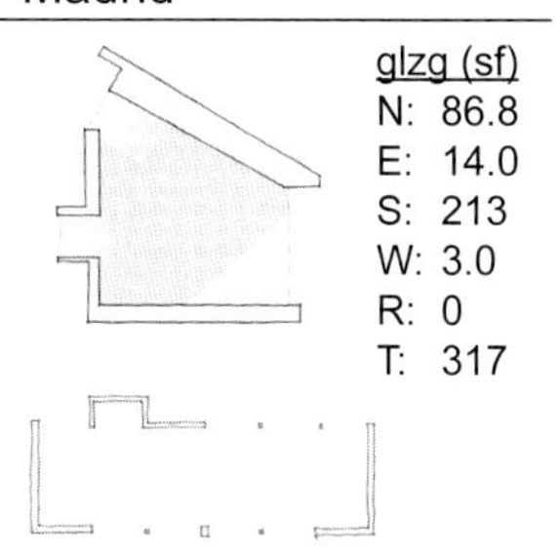

Daylighting: Total Glazing Area

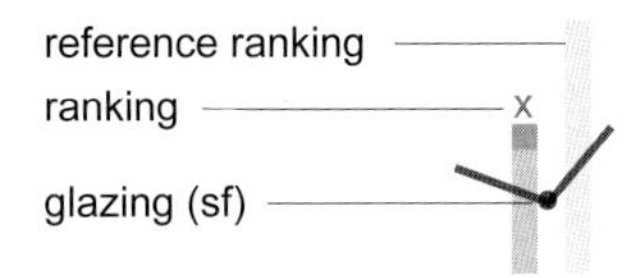

The following information comes from the Lighting Contest data from the 2007 Solar Decathlon Scores and Standings (available from the DOE website). The data represents the footcandle levels averaged over the daylight hours of the competition from an illuminance meter located on the surface of the desk in the office space.

Achieving consistent levels of daylight without overlighting is a challenge. In the lighting competition, the teams received full credit when the lighting levels were between 50–100 footcandles. As is evident by the averages (shown in the Average Illumnance charts), CMU was only house with an average footcandle level within this range. The other houses all have a higher average footcandle level than that which is recommended.

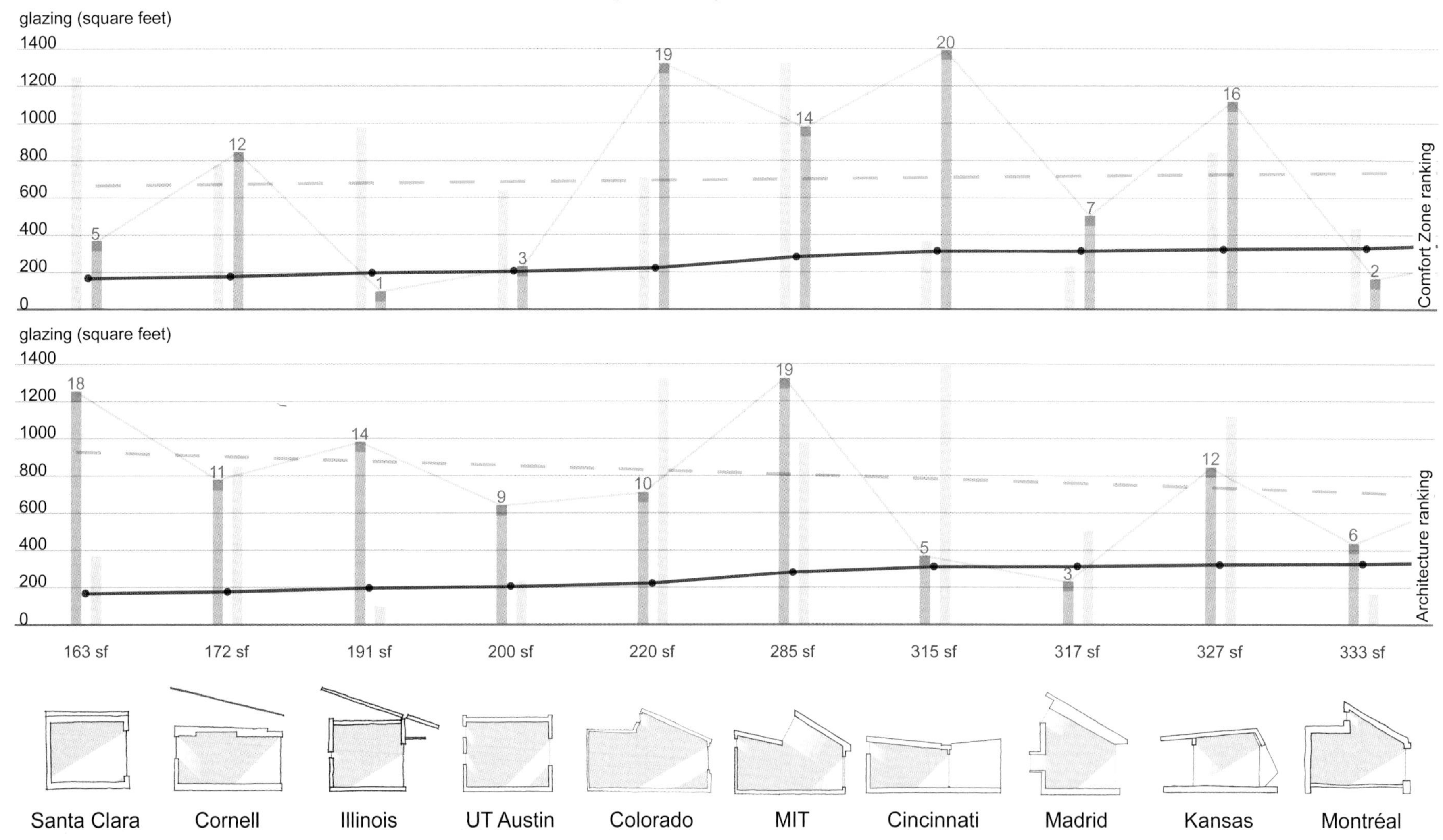

RANKING: Ascending Order of Total Glazing (sf)

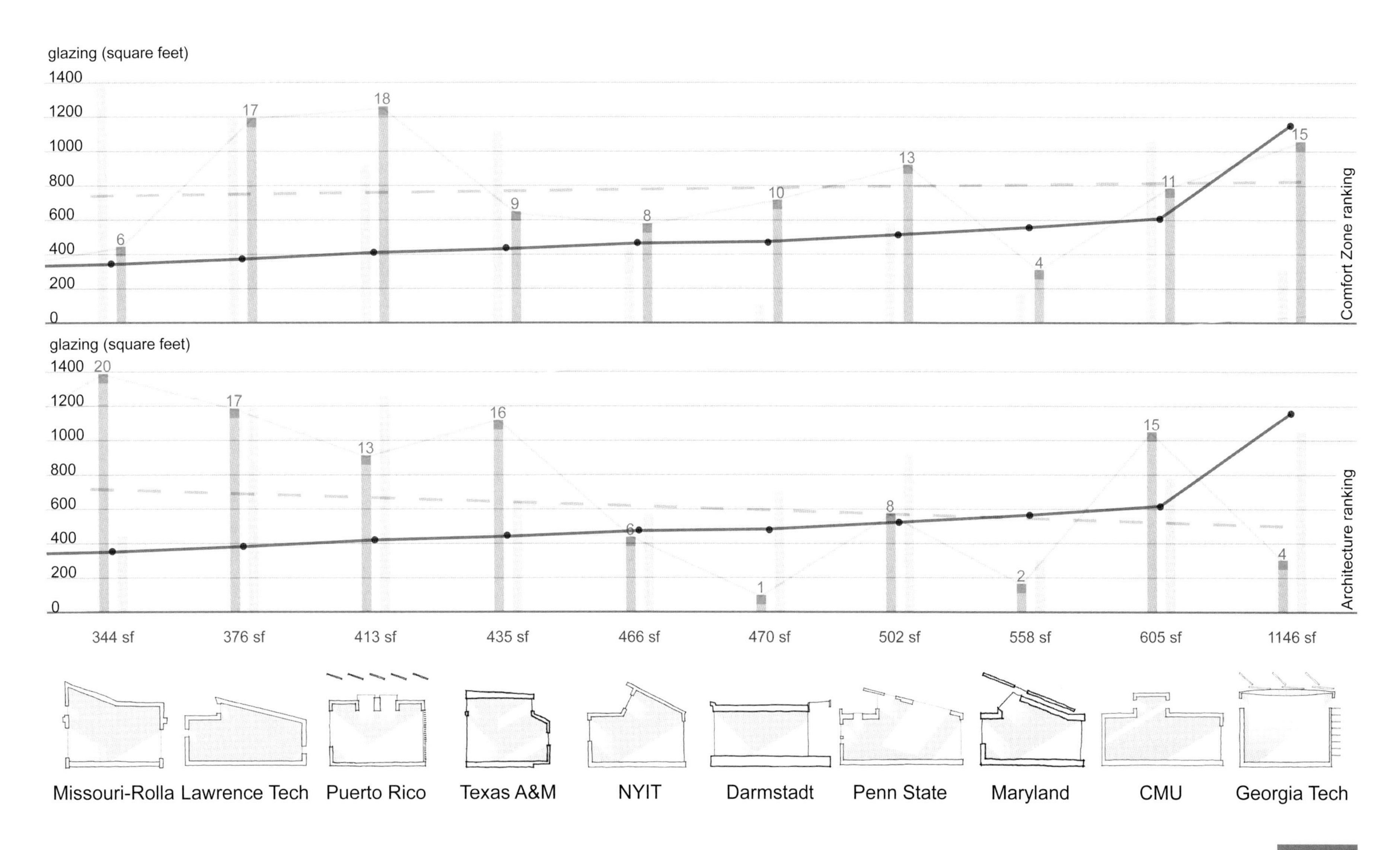
glazing (square feet)
1400
1200
1000
800
600
400
200
0
18
17
15
13
11
10
9
8
6
4
Comfort Zone ranking
glazing (square feet)
1400
1200
1000
800
600
400
200
0
20
17
16
15
13
8
6
4
2
1
Architecture ranking
344 sf
376 sf
413 sf
435 sf
466 sf
470 sf
502 sf
558 sf
605 sf
1146 sf
Missouri-Rolla
Lawrence Tech
Puerto Rico
Texas A&M
NYIT
Darmstadt
Penn State
Maryland
CMU
Georgia Tech

Daylighting: Average Illuminance

There is no specific assessment of the effectiveness of daylighting in the houses of the Solar Decathlon competition. One of the ten contests is "Lighting," and 10 of the 100 points possible in this contest come from "Daytime Workstation Lighting". In this part of the Lighting Contest, a photometer (light level sensor) is placed on the surface of a desk in the office space of each house. Teams are not allowed to have an electric lighting source within 18 inches of the meter. The levels are measured from 9am to 5pm on four days of the competition, and full points are awarded when the lighting levels are at 50 footcandles or above.

Daylighting interior spaces has significant benefits in terms of quality of light and a decrease in energy needs as a result of the reduction in heat from electric lights. However, there are also problems if a space receives too much daylight. Overly daylit spaces can receive too much heat, and are susceptible to glare and overly contrasting lighting conditions.

The 50-footcandle level minimum for this competition is sufficient for typical office activities. Having significantly more than 100 footcandles is not only highly consumptive of energy, but can lead to glare and overlighting issues. According to the information contained in the Lighting Contest data spreadsheet from the 2007 Solar Decathlon Scores and Standings (available from the DOE Solar Decathlon website), only one of the 20 teams remained within the 50 – 100 footcandle range during the daylit hours of the competition. The other 19 teams had more than 100 footcandles of illumination on the photometer, on average with Madrid averaging 1,279 footcandles. The data for Madrid is twice as high as any other team, suggesting that the photometer may have been faulty. The next highest average footcandle reading was Colorado with 648 average footcandles.

The data shows a minor increase in overall rankings as a result of increased footcandles, and a significant increase in Architecture Contest rankings. There is a significant decrease in Comfort Zone rankings as average footcandle levels increase.

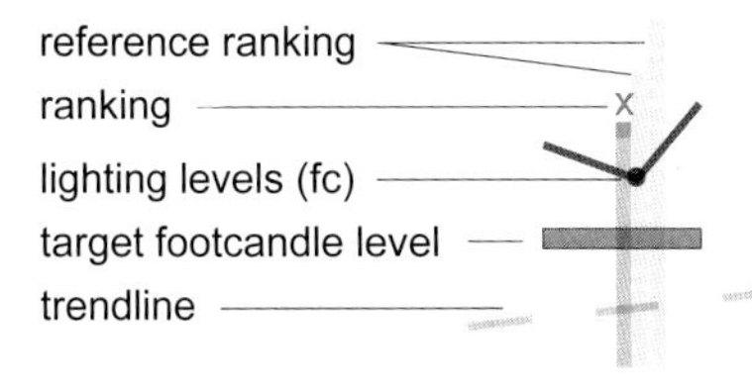
reference ranking
ranking
x
lighting levels (fc)
target footcandle level
trendline

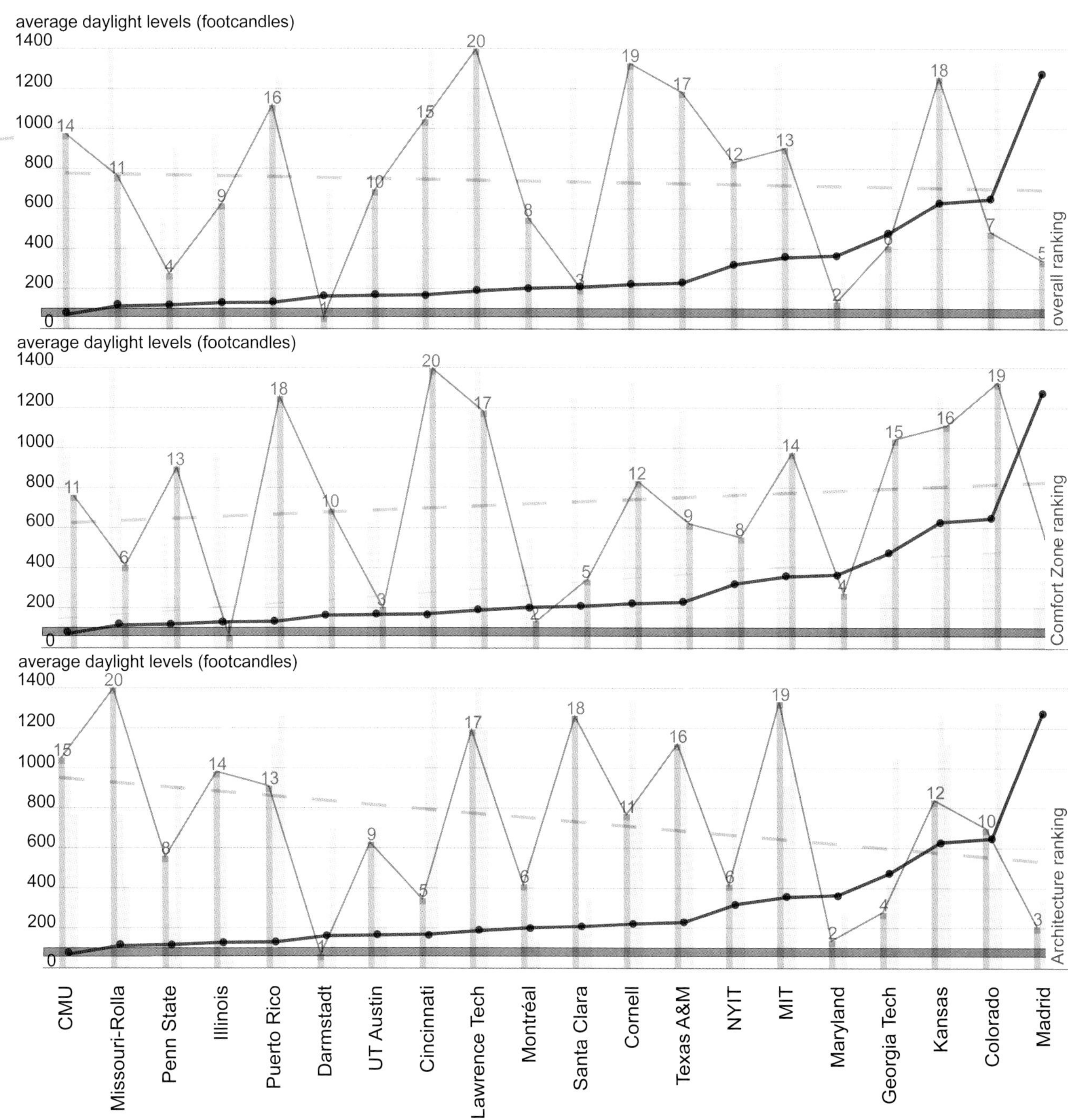
average daylight levels (footcandles)
1400
1200
1000
800
600
400
200
0
overall ranking
average daylight levels (footcandles)
Comfort Zone ranking
average daylight levels (footcandles)
Architecture ranking
CMU
Missouri-Rolla
Penn State
Illinois
Puerto Rico
Darmstadt
UT Austin
Cincinnati
Lawrence Tech
Montréal
Santa Clara
Cornell
Texas A&M
NYIT
MIT
Maryland
Georgia Tech
Kansas
Colorado
Madrid

Shading

All teams come from latitudes at which heat from the sun can overheat an interior space if the glazing on the south and west is not shaded during the overheated parts of the year. This investigation looks only at the effectiveness of shading on south-facing glazing in Washington, DC during the competition. The temperatures in Washington in October can vary significantly. In 2007, the days were relatively warm with a higher than average amount of sunlight hours. Those houses that didn't shade effectively suffered unwanted direct solar gain. The shading calculations are based on a profile angle of 42° at noon in early October. To effectively shade throughout the year, an adjustable solution is required and adjustability is noted where it occurred. In some houses, shading was shown in the construction drawings, but was not built. In those cases, the numbers address what was built.

Lawrence Tech

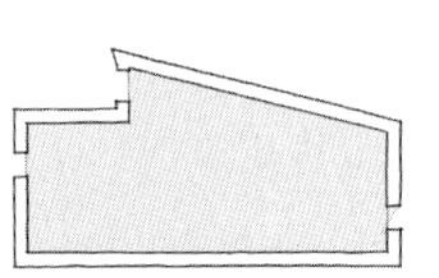

south glazing:	**87.6 sf**
shaded south glzg:	**26.1 sf**
percent unshaded:	**70.2 %**
shading type:	**recessed glzg**
adjustability:	**none**

Montréal

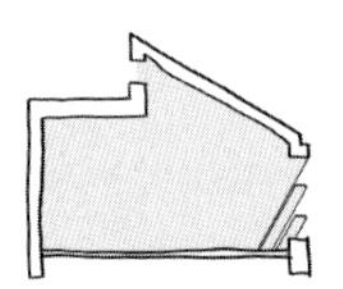

south glazing:	**121.9 sf**
shaded south glzg:	**18.3 sf**
percent unshaded:	**85.0 %**
shading type:	**recessed glzg**
adjustability:	**none**

Penn State

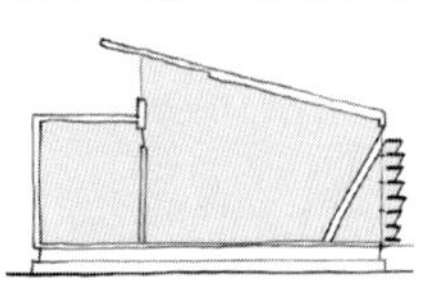

south glazing:	**128 sf**
shaded south glzg:	**111.3 sf**
percent unshaded:	**13.1 %**
shading type:	**louver**
adjustability:	**sliding, pivoting**

Georgia Tech

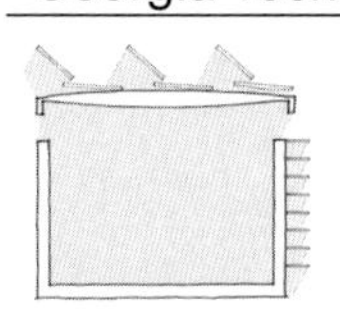

south glazing:	**236.2 sf**
shaded south glzg:	**190 sf**
percent unshaded:	**19.6 %**
shading type:	**louver**
adjustability:	**none**

Colorado

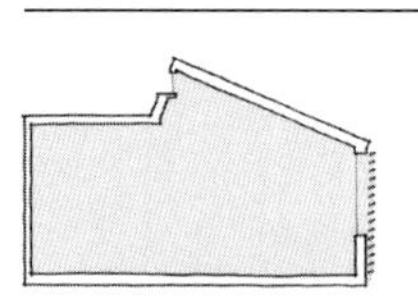

south glazing:	**91.8 sf**
shaded south glzg:	**10.4 sf**
percent unshaded:	**88.6 %**
shading type:	**louver**
adjustability:	**none**

NYIT

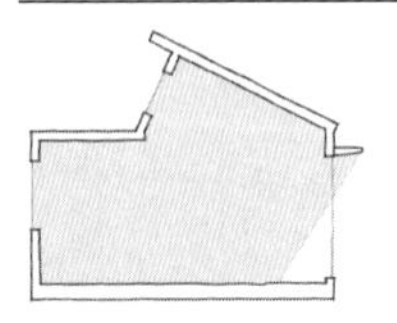

south glazing:	**259.6 sf**
shaded south glzg:	**51.6 sf**
percent unshaded:	**80.1 %**
shading type:	**louver**
adjustability:	**none**

Illinois

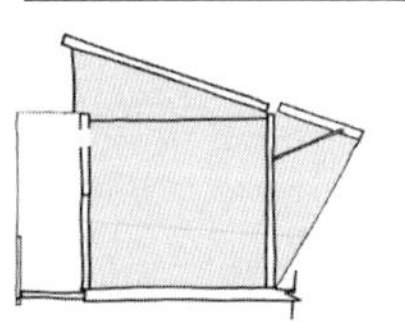

south glazing:	**128.8 sf**
shaded south glzg:	**40.6 sf**
percent unshaded:	**68.5 %**
shading type:	**overhang**
adjustability:	**none**

MIT

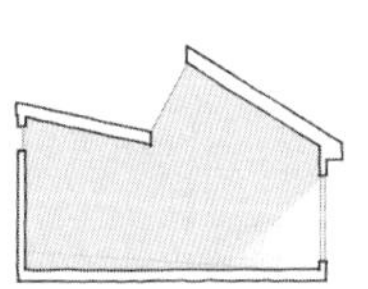

south glazing:	**73.3 sf**
shaded south glzg:	**0 sf**
percent unshaded:	**100 %**
shading type:	**none**
adjustability:	**none**

CMU

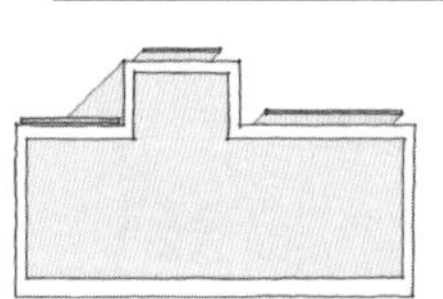

south glazing:	**151.9 sf**
shaded south glzg:	**0 sf**
percent unshaded:	**100 %**
shading type:	**none**
adjustability:	**none**

Darmstadt

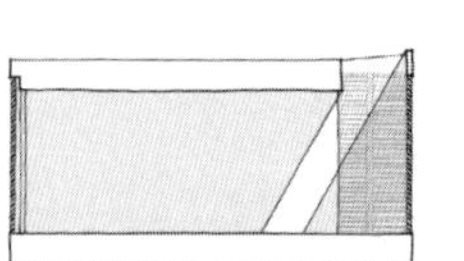

south glazing:	**234.8 sf**
shaded south glzg:	**234.8 sf**
percent unshaded:	**0 %**
shading type:	**shutters**
adjustability:	**pivoting doors**

LAYOUT: Location in Solar Village on National Mall

Missouri-Rolla

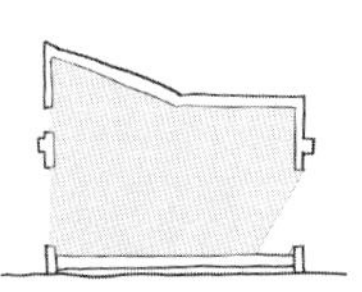

south glazing:	**124.6 sf**
shaded south glzg:	**0 sf**
percent unshaded:	**100 %**
shading type:	**none**
adjustability:	**none**

Maryland

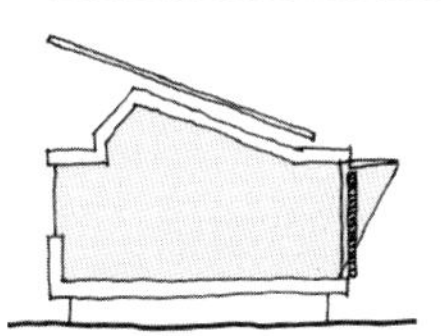

south glazing:	**163.5 sf**
shaded south glzg:	**150.5 sf**
percent unshaded:	**8 %**
shading type:	**shutters**
adjustability:	**pivot doors**

UT Austin

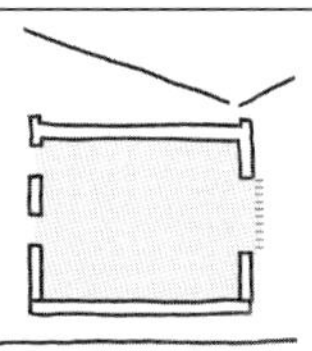

south glazing:	**54.0 sf**
shaded south glzg:	**45.2 sf**
percent unshaded:	**16.3 %**
shading type:	**louver**
adjustability:	**none**

Puerto Rico

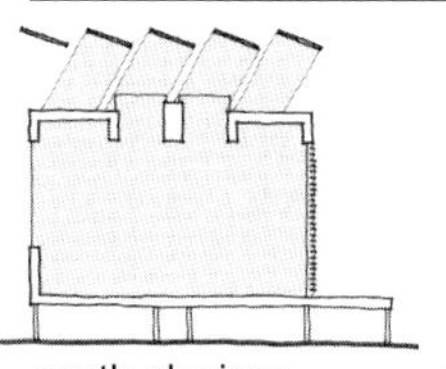

south glazing:	**132.1 sf**
shaded south glzg:	**132.1 sf**
percent unshaded:	**0 %**
shading type:	**louver**
adjustability:	**pivot**

Santa Clara

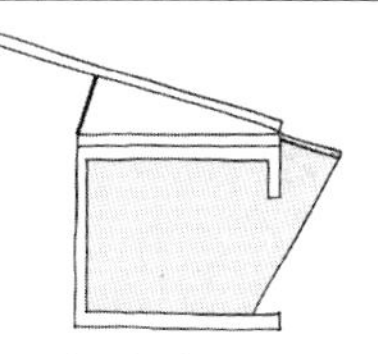

south glazing:	**114.9 sf**
shaded south glzg:	**0 sf**
percent unshaded:	**100 %**
shading type:	**louver (unbuilt)**
adjustability:	**none**

Kansas

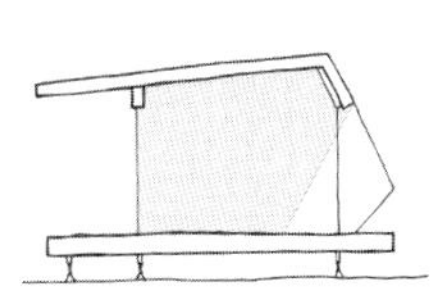

south glazing:	**52.1 sf**
shaded south glzg:	**0 sf**
percent unshaded:	**100 %**
shading type:	**none**
adjustability:	**none**

Cornell

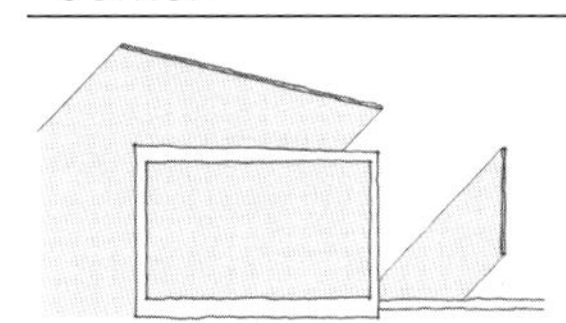

south glazing:	**125.6 sf**
shaded south glzg:	**0 sf**
percent unshaded:	**100 %**
shading type:	**none**
adjustability:	**none**

Cincinnati

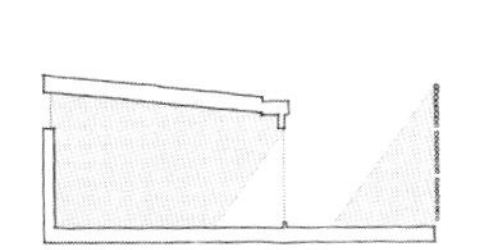

south glazing:	**149.9 sf**
shaded south glzg:	**0 sf**
percent unshaded:	**100 %**
shading:	**overhang (unbuilt)**
adjustability:	**none**

Texas A&M

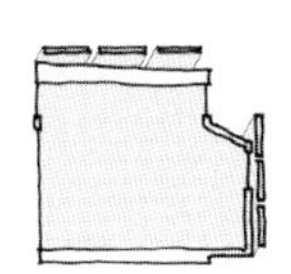

south glazing:	**145.0 sf**
shaded south glzg:	**17 sf**
percent unshaded:	**88.3 %**
shading:	**overhang (unbuilt)**
adjustability:	**none**

Madrid

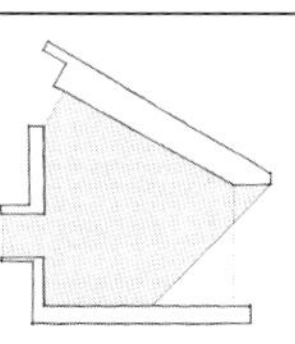

south glazing:	**213.0 sf**
shaded south glzg:	**205.0 sf**
percent unshaded:	**3.8 %**
shading type:	**shutters**
adjustability:	**sliding doors**

Shading: Unshaded Glazing Area

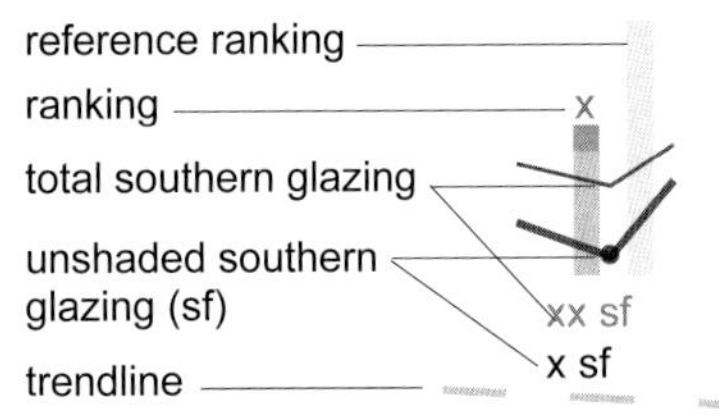

Simply shading a window to limit direct solar gain on the glazing can reduce heat gain by 75%. This was a consideration for those teams that shaded all or most of their southern glazing. Other teams had designed shading that was never built, or chose not to shade their southern glazing. This chart assesses the amount of unshaded glazing as compared to rankings in the Architecture and Comfort Zone Contests.

The hypothesis is that an increase in unshaded southern glazing would lead to a decrease in Comfort Zone rankings.

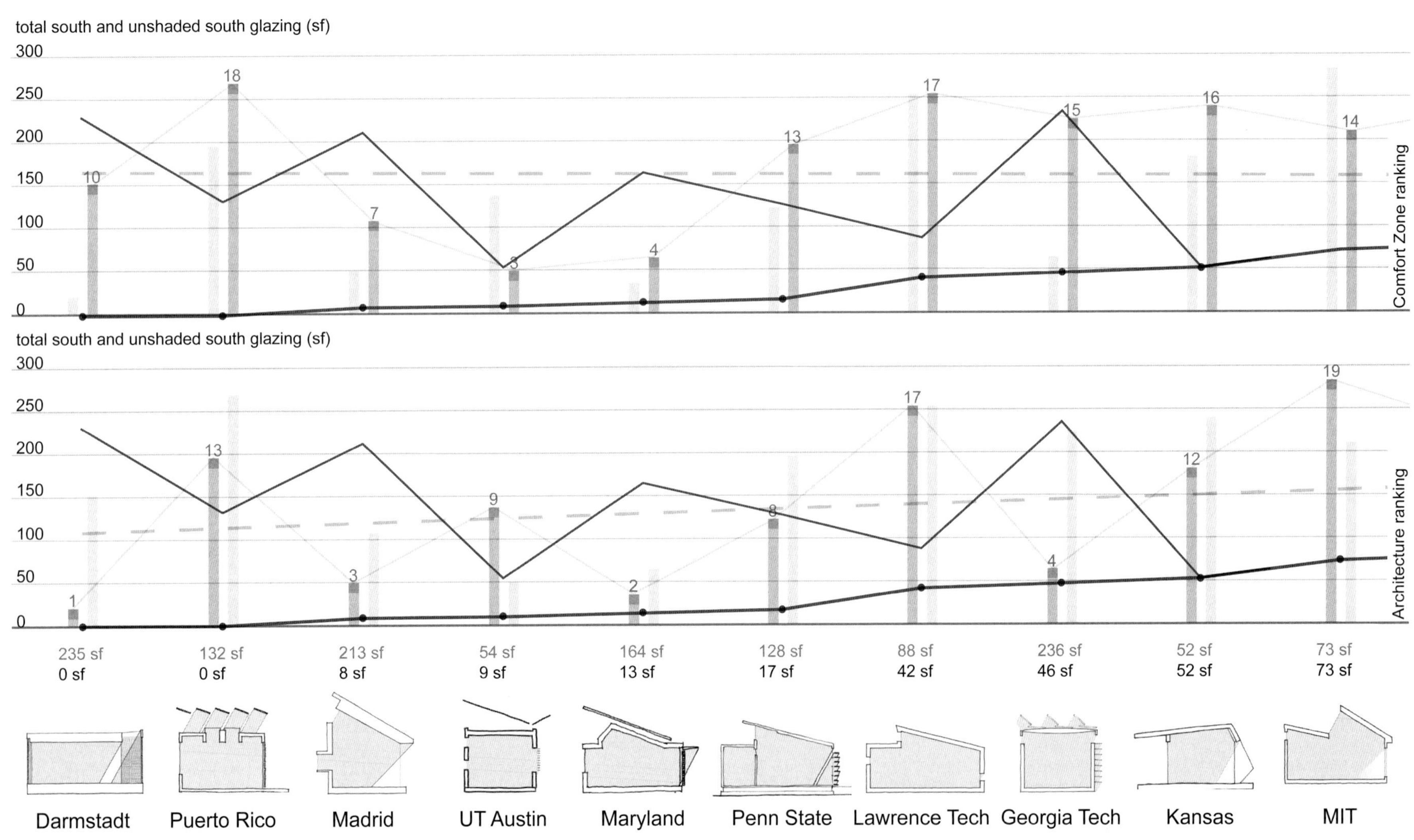

RANKING: Ascending Order of Unshaded Glazing (sf)

The trendline shows that there is basically no change in Comfort Zone rankings as a result of increased unshaded glazing. However, there is a noticable decrease in rankings in Architecture as a result of an increase in unshaded glazing. Shading had a greater impact on Architecture than Comfort Zone.

This chart also compares total southern glazing to unshaded southern glazing. There is a clear correlation between those teams which have less unshaded glazing and those which have more southern glazing overall.

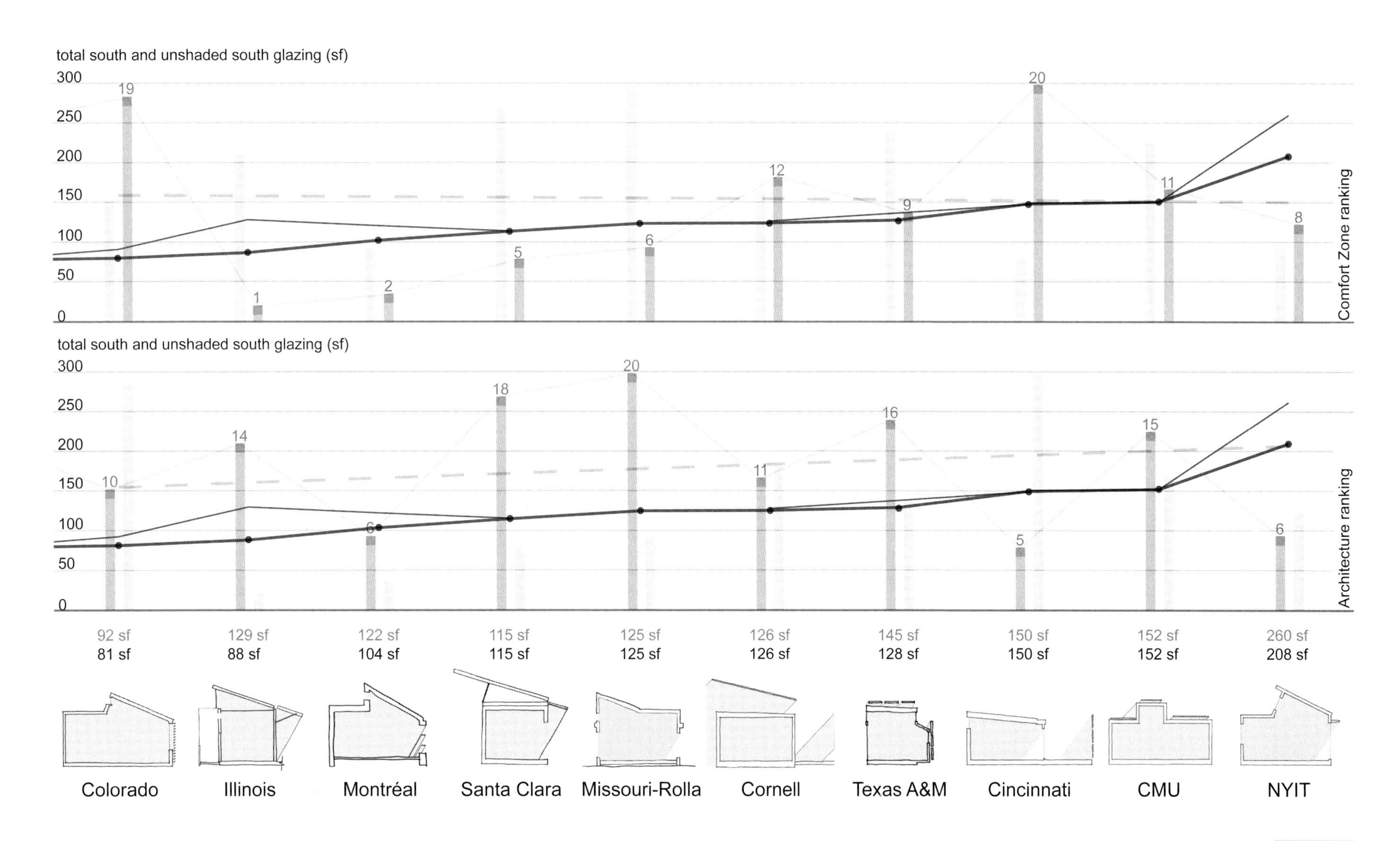

Ventilation

operable glzg (sf)
N: North
E: East
S: South
W: West
R: Roof
T: Total

Though there is no quantitative measurement of effectiveness of ventilation as a passive cooling strategy, several projects incorporated passive cooling through natural ventilation. The weather during the competition dates in Washington, DC was ideal for natural ventilation, with warm daily temperatures combined with relative humidity decreasing at night.

The arrows show possible air movement. The numbers show potential amounts of air movement through the wall in the form of operable windows, doors or skylights. We are assuming inhabitants would open these when the conditions are appropriate for natural ventilation.

Note: Clerestory and monitor glazing is included in the wall glazing data.

Lawrence Tech

Montréal

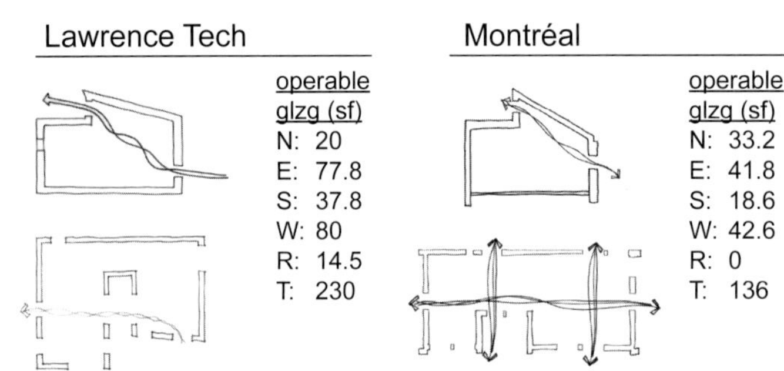

Lawrence Tech — operable glzg (sf)
N: 20
E: 77.8
S: 37.8
W: 80
R: 14.5
T: 230

Montréal — operable glzg (sf)
N: 33.2
E: 41.8
S: 18.6
W: 42.6
R: 0
T: 136

Penn State

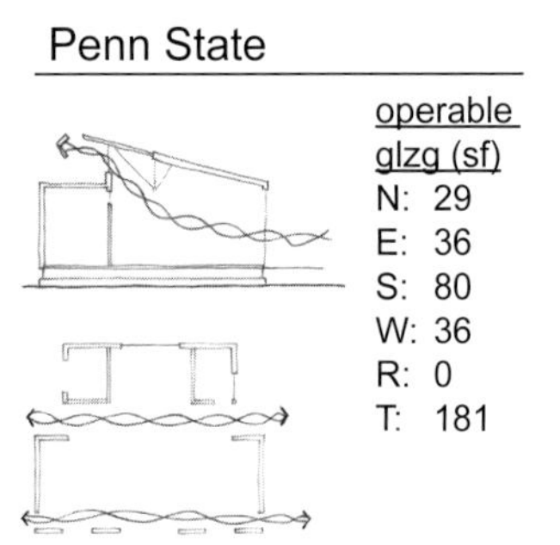

operable glzg (sf)
N: 29
E: 36
S: 80
W: 36
R: 0
T: 181

Georgia Tech

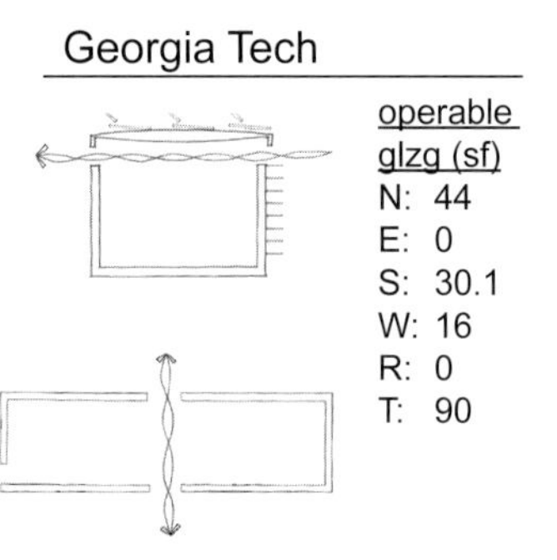

operable glzg (sf)
N: 44
E: 0
S: 30.1
W: 16
R: 0
T: 90

Colorado

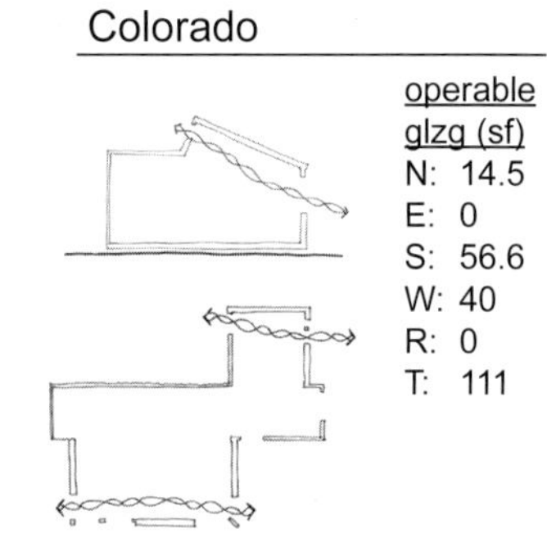

operable glzg (sf)
N: 14.5
E: 0
S: 56.6
W: 40
R: 0
T: 111

NYIT

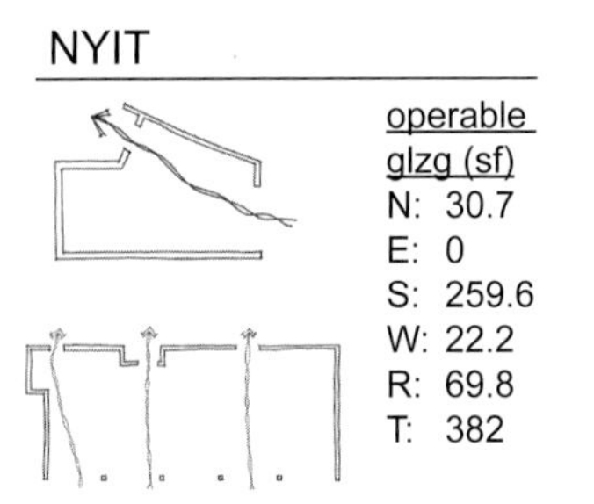

operable glzg (sf)
N: 30.7
E: 0
S: 259.6
W: 22.2
R: 69.8
T: 382

Illinois

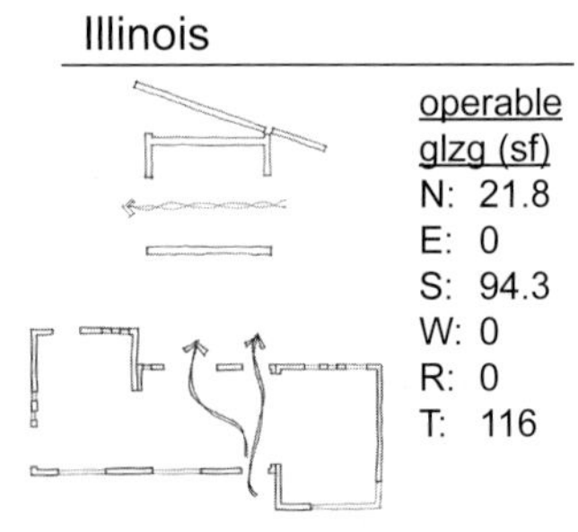

operable glzg (sf)
N: 21.8
E: 0
S: 94.3
W: 0
R: 0
T: 116

MIT

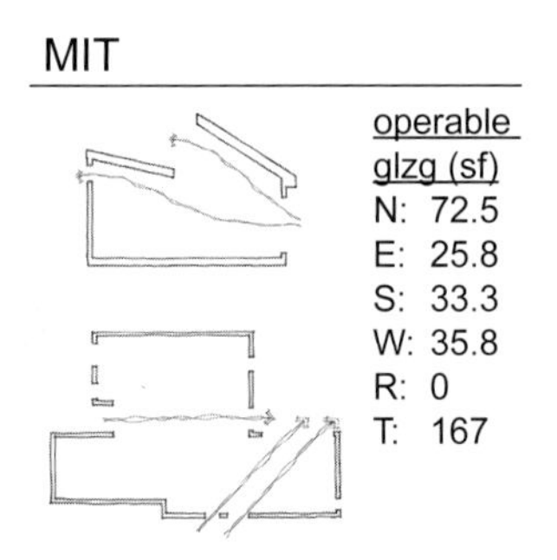

operable glzg (sf)
N: 72.5
E: 25.8
S: 33.3
W: 35.8
R: 0
T: 167

CMU

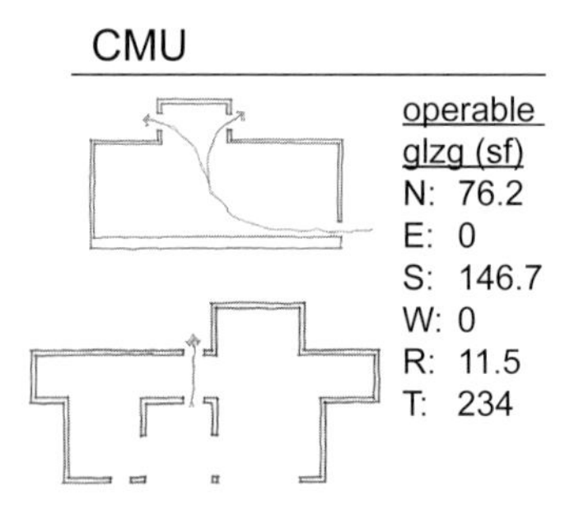

operable glzg (sf)
N: 76.2
E: 0
S: 146.7
W: 0
R: 11.5
T: 234

Darmstadt

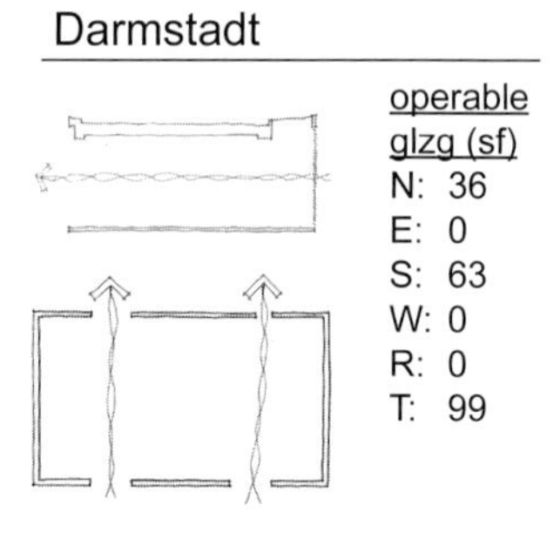

operable glzg (sf)
N: 36
E: 0
S: 63
W: 0
R: 0
T: 99

LAYOUT: Location in Solar Village on National Mall

Data from EERE weather file

Graph from Climate Consultant 4

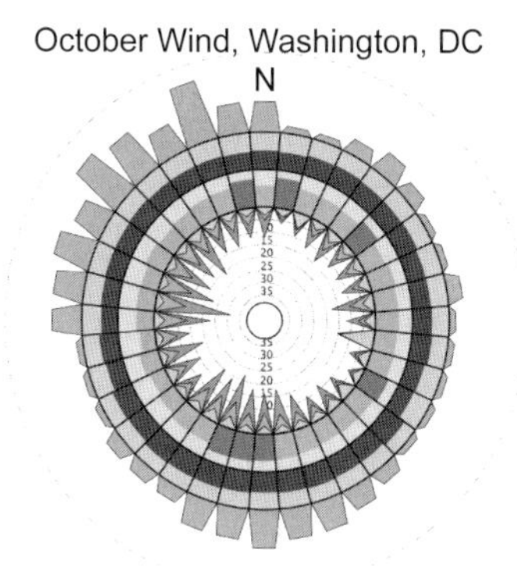

Wind measurements were taken throughout the competition. This data was collected by the DOE and is available from the Solar Decathlon 2007 website.

wind speed maximum: 6.63 mph
wind speed average: 1.85 mph
wind direction max: 323.6° East of North
wind direction avg: 191.9° East of North

Given the thin shape of the houses and their southern orientation, combined with the average and measured predominance of north–south wind direction, the following analysis focuses only on the north- and south-facing operable glazing.

Missouri-Rolla

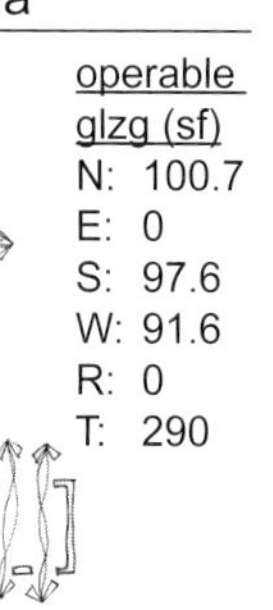

operable glzg (sf)
N: 100.7
E: 0
S: 97.6
W: 91.6
R: 0
T: 290

Maryland

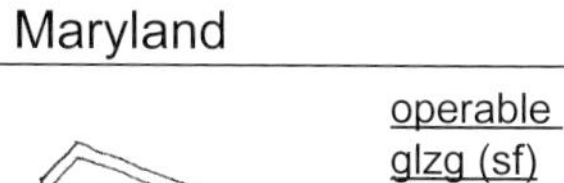

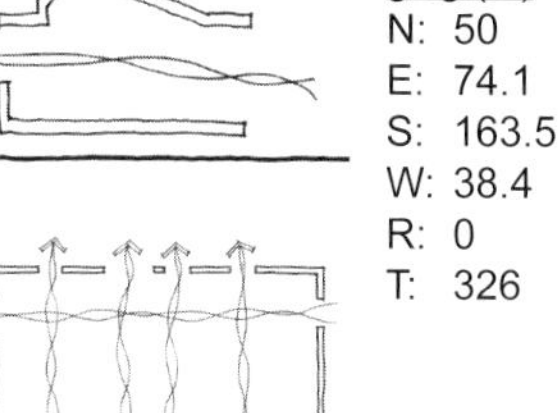

operable glzg (sf)
N: 50
E: 74.1
S: 163.5
W: 38.4
R: 0
T: 326

UT Austin

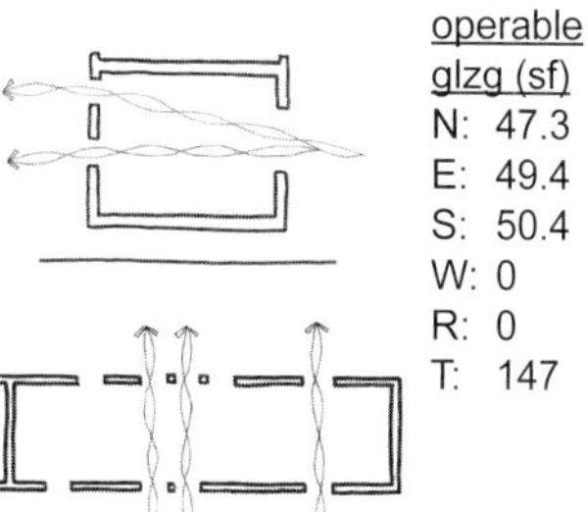

operable glzg (sf)
N: 47.3
E: 49.4
S: 50.4
W: 0
R: 0
T: 147

Puerto Rico

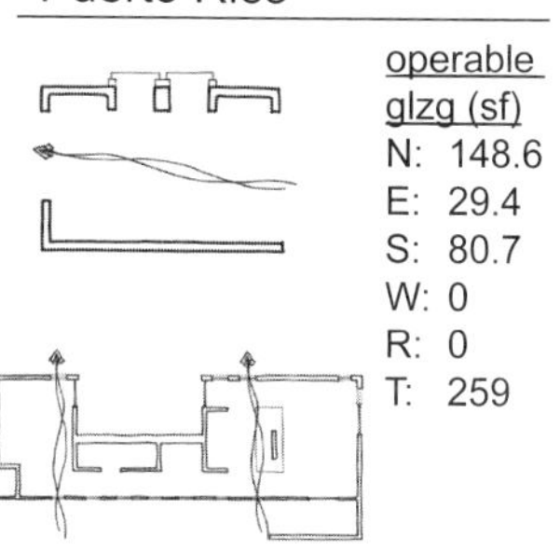

operable glzg (sf)
N: 148.6
E: 29.4
S: 80.7
W: 0
R: 0
T: 259

Santa Clara

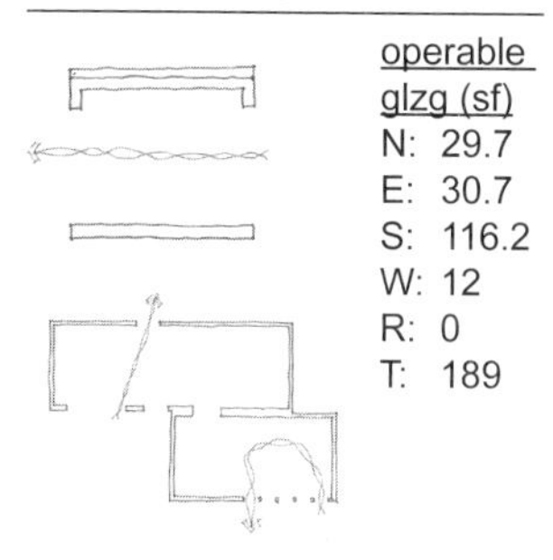

operable glzg (sf)
N: 29.7
E: 30.7
S: 116.2
W: 12
R: 0
T: 189

Kansas

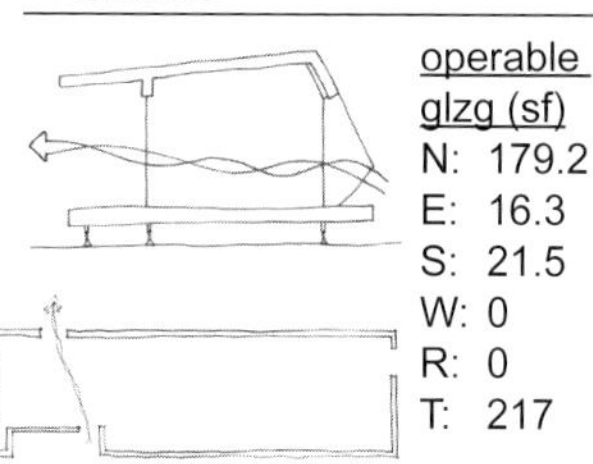

operable glzg (sf)
N: 179.2
E: 16.3
S: 21.5
W: 0
R: 0
T: 217

Cornell

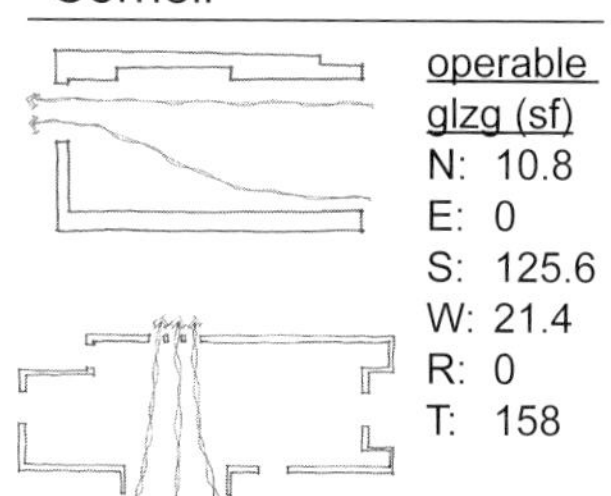

operable glzg (sf)
N: 10.8
E: 0
S: 125.6
W: 21.4
R: 0
T: 158

Cincinnati

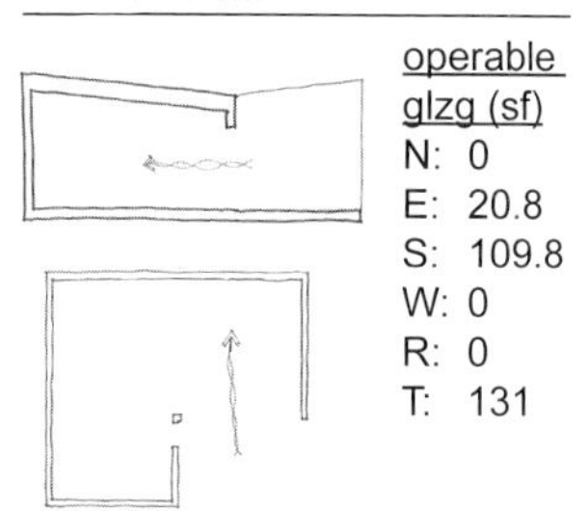

operable glzg (sf)
N: 0
E: 20.8
S: 109.8
W: 0
R: 0
T: 131

Texas A&M

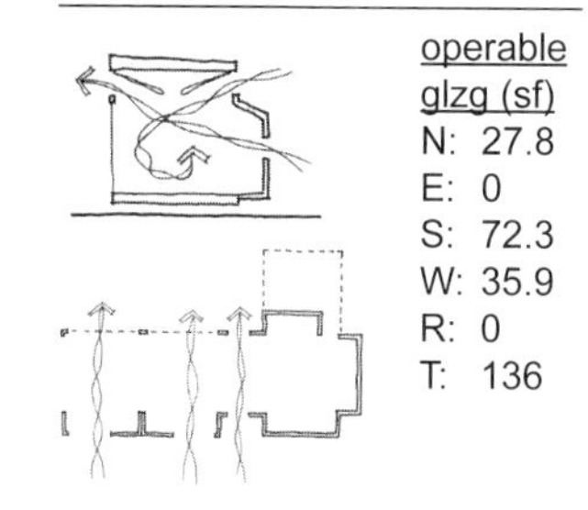

operable glzg (sf)
N: 27.8
E: 0
S: 72.3
W: 35.9
R: 0
T: 136

Madrid

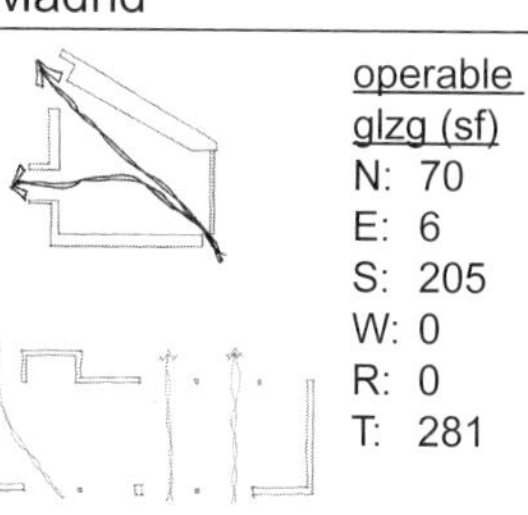

operable glzg (sf)
N: 70
E: 6
S: 205
W: 0
R: 0
T: 281

Ventilation: Operable Glazing Area

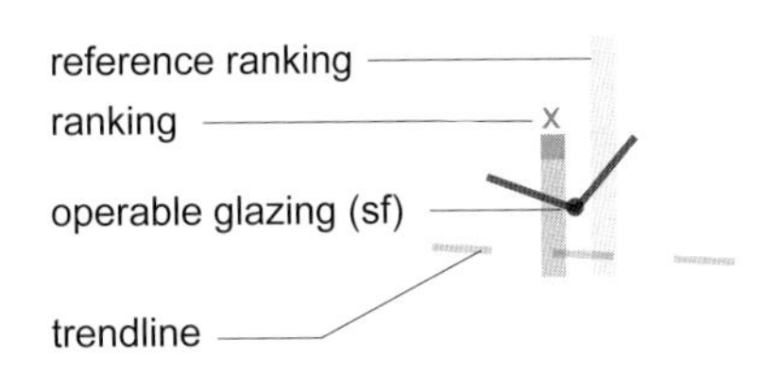

These charts compare area of potential glazing for ventilation with the rankings in the Architecture and Comfort Zone Contests. The hypothesis is that there would be an increase in Comfort Zone rankings as operable glazing area for ventilation increases.

When there are openings on both sides of the house for ventilation, the flow is limited by the smaller opening. Therefore, the numbers for each house represent the lesser square footage of operable glazing on the north or south side of the house as the limiting variable impacting the amount of ventilation moving through the space. Given the thin shape of the houses and their east–west orientation, combined with the average and measured predominance of north–south wind direction, the analysis includes only the north- and south-facade operable glazing.

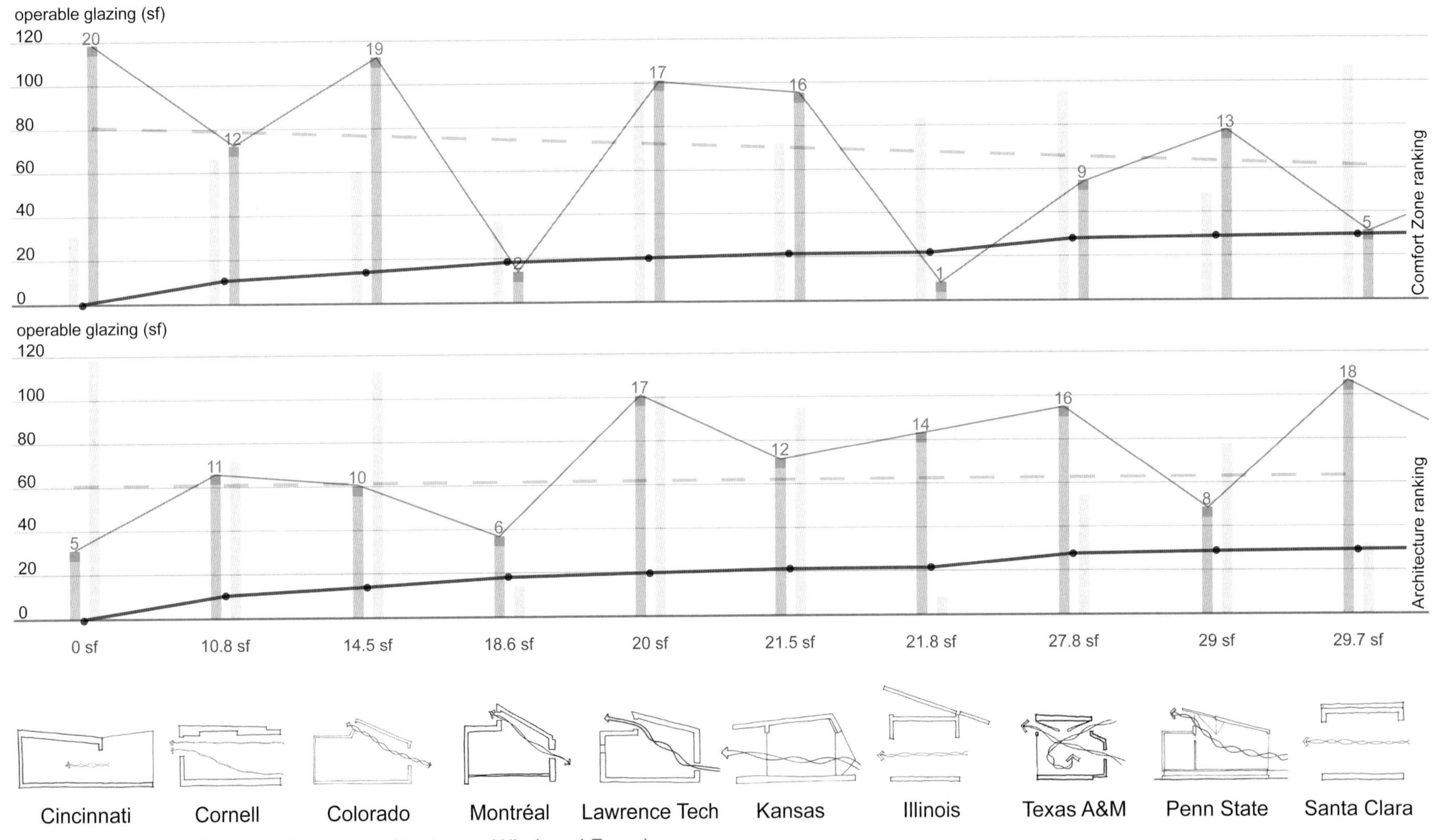

RANKING: Square Footage of Operable Glazing on Windward Facade

The weather during the 2007 competition in Washington, DC was ideal for natural ventilation. As a result, the teams that had increased operable glazing (and assumedly benefited from natural ventilation) ranked higher in the Comfort Zone Contest. In the Architecture Contest, there was very little correlation, though the trend was slight in the opposite direction. An increase in operable glazing correlated minimally with lower scores in the Architecture Contest.

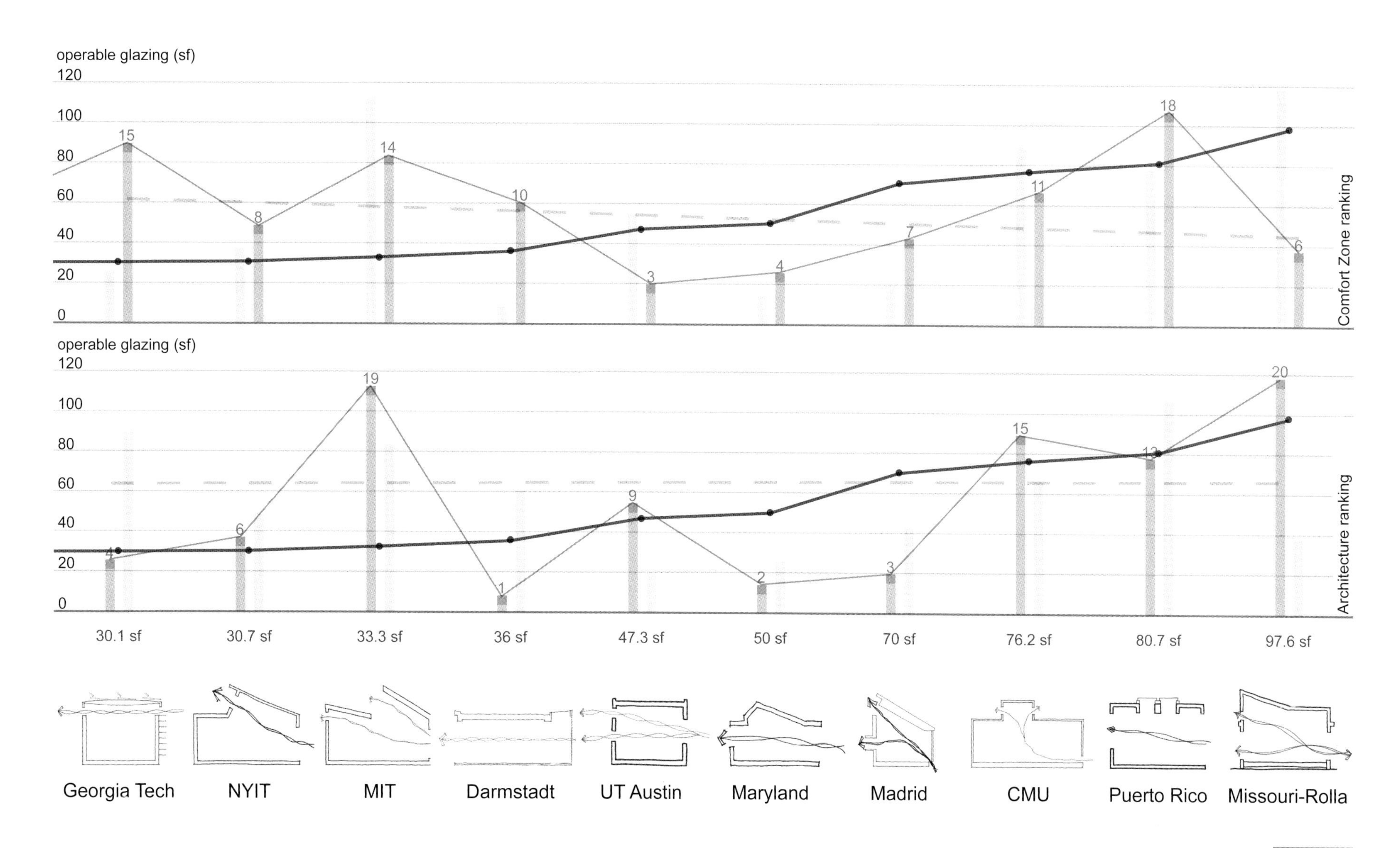

Photovoltaic Production

Every team was required to produce their own power for the competition events from photovoltaic arrays. Each team developed a different strategy to provide sufficient area and production for their PV array in an integrated architectural design. The choices, many of which were influenced by architectural design intentions, impacted the amount and efficiency of the PV arrays.

The following data comes from the DOE and the teams' documentation. These are ranked in terms of total electricity produced, following an adjustment for the angle of the photovoltaic array (solar angle). The solar angle of the array significantly impacted the efficiency of the PV array.

	Darmstadt	Madrid	MIT	Penn State	Colorado	Georgia Tech	Montréal	NYIT	Texas A&M	UT Austin
PV Area (sf)	1063	929	562	727	549	516	496	444	843	546
Electrical Production (kW)	12.3	9.6	9.03	9.4	8.8	8.2	8.2	7.5	7.5	7.6
Primary Solar Angle of PV Array (Secondary Angle in Parentheses)	3 (45)	30	31	15	22.6	42	30	26	25 (0)	18
Efficiency Ratio	0.78	0.98	0.98	0.89	0.94	1.00	0.98	0.96	0.96	0.91
Adjusted Electrical Production at Solar Angle (kW)	9.6	9.4	8.9	8.4	8.3	8.2	8.0	7.2	7.2	6.9
Conditioned House Area (sf)	638	512	655	732	659	672	581	782	465	632
Adjusted Production per House (watts/sf)	15.0	18.3	13.5	11.4	12.6	12.2	13.8	9.2	15.4	11.0

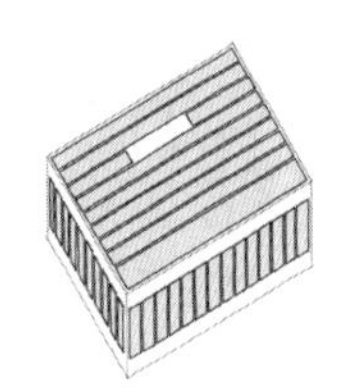
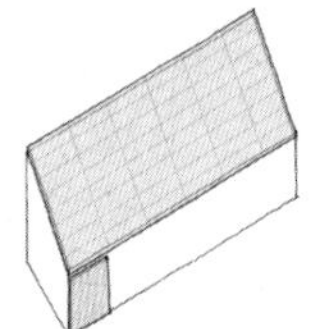
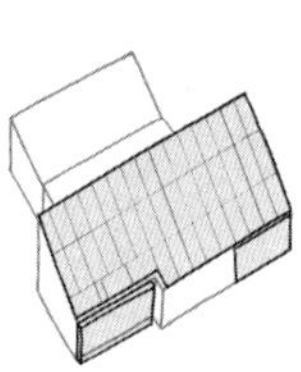
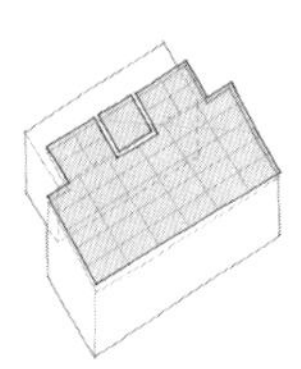
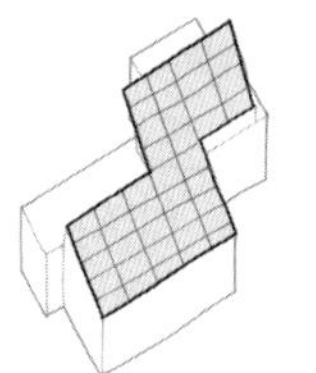
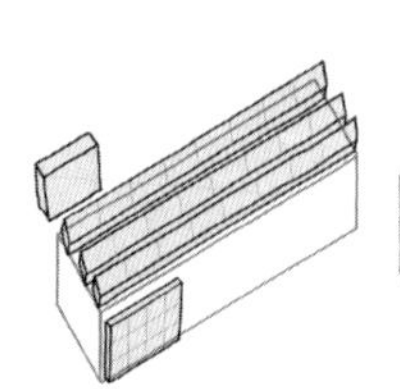
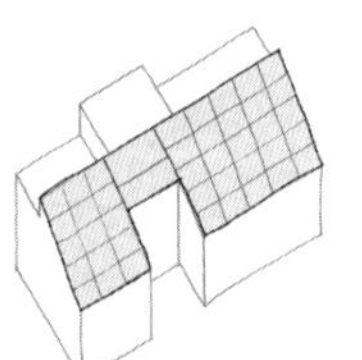
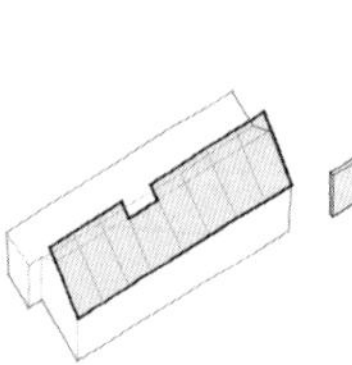
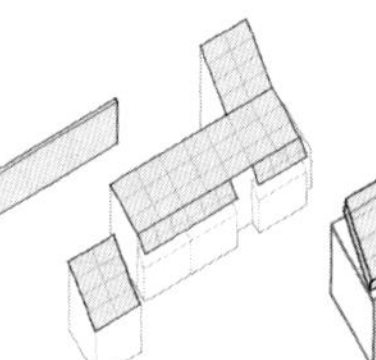
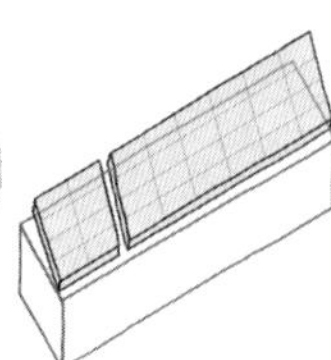

Darmstadt Madrid MIT Penn State Colorado Georgia Tech Montréal NYIT Texas A&M UT Austin

RANKING: Descending Order of Adjusted Electrical Production of PV Array at Solar Angle

PV area: total photovoltaic area providing power to the house
Electrical Production: electricity produced at ideal solar angle
Primary Solar Angle: predominant angle of photovoltaic arrays on house (some houses had multiple angles of PV)
Efficiency Ratio: adjustment of efficiency based on cosine of solar angle as compared to ideal solar angle for October in Washington (42°)
Adjusted kW Produced: product of total electrical production and efficiency ratio
Conditioned Area: mechanically conditioned area of house according to documents
Production per House: adjusted kW produced divided by the house square footage

	Puerto Rico	Cornell	Santa Clara	Maryland	Missouri-Rolla	Lawrence Tech	Kansas	Cincinnati	Illinois	CMU
PV area (sf)	394	796	665	417	591	770	428	514	544	428
Electrical Production (kW)	7.6	7.6	7.4	7.1	7	7.2	6.8	7.7	6.8	6.9
Primary Solar Angle of PV Array (Secondary Angle in Parentheses)	18	13	15 (30)	20	18.5	13.5	64 (46)	6	18	0
Efficiency Ratio	0.91	0.87	0.89	0.93	0.92	0.88	0.93	0.81	0.91	0.74
Adjusted Electrical Production at Solar Angle (kW)	6.9	6.6	6.6	6.6	6.4	6.3	6.3	6.3	6.2	5.1
Conditioned House Area (sf)	459	540	612	652	593	612	672	795	504	651
Adjusted Production per House (watts / sf)	15.1	12.3	10.8	10.1	10.8	10.3	9.4	7.9	12.3	7.9

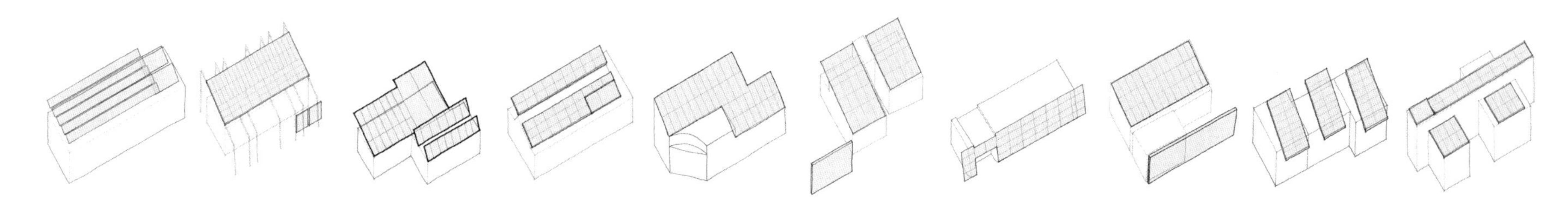

Puerto Rico Cornell Santa Clara Maryland Missouri-Rolla Lawrence Tech Kansas Cincinnati Illinois CMU

PV Production: Solar Angle

Architecture ranking — x
max electrical production of PV array
adjusted kW produced at solar angle
solar angle

The optimal angle necessary to maximize available solar energy for a given time and location is referred to here as the solar angle. The optimal solar angle in Washington, DC in October is approximately 42° from horizontal. Any variance of the angle from 42° impacts efficiency, though it only decreases significantly when the angle is approximately 15° more or less than the solar angle for the given location. This condition was the impetus for this set of charts.

The lower chart compares the relationship between the solar angle of the photovoltaics and the overall rankings in the competition. The upper graph compares the solar angle with the total electricity produced and the adjusted electricity produced at the given solar angle.

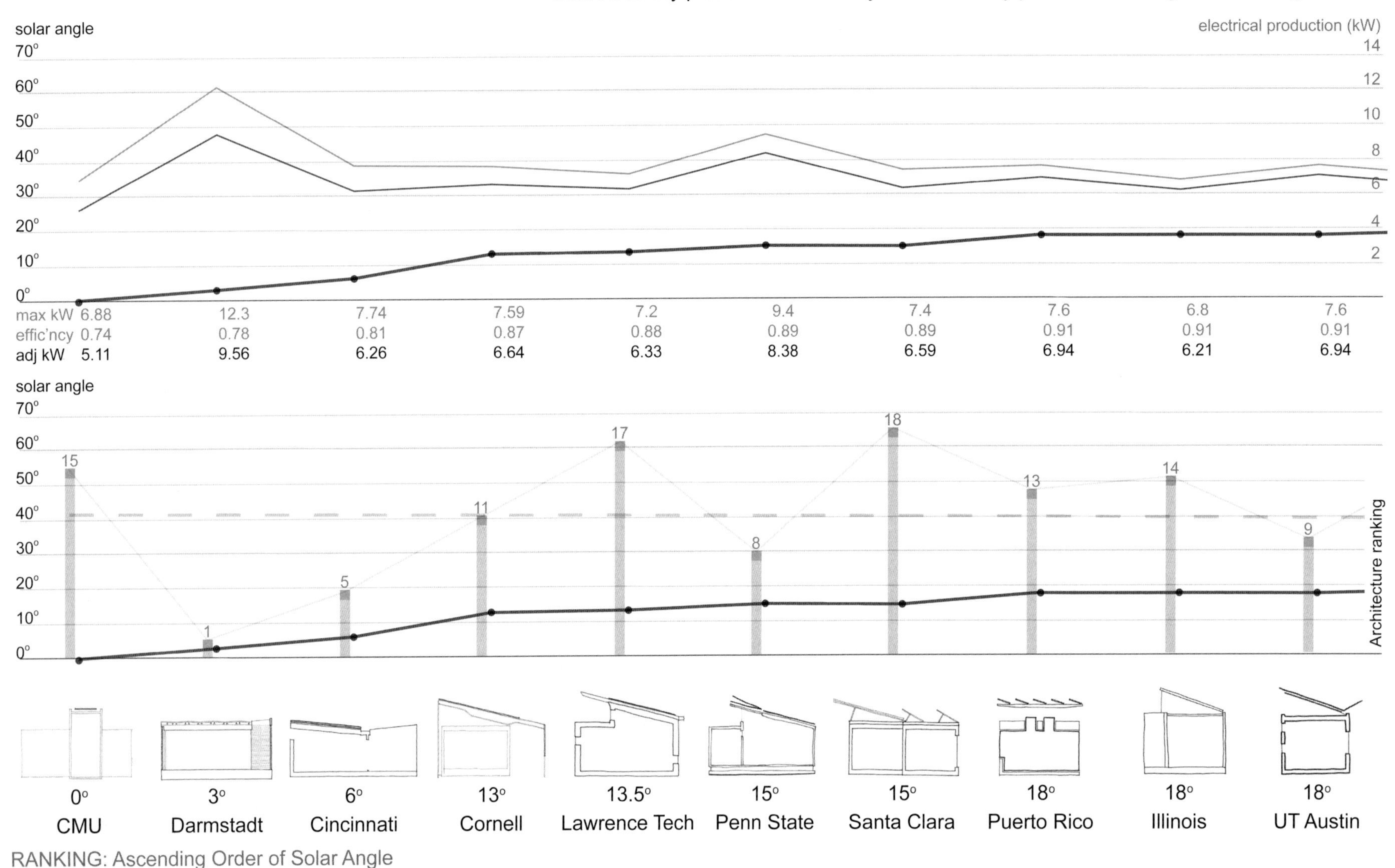

RANKING: Ascending Order of Solar Angle

According to the trendline of the lower chart, there is a slight increase in overall rankings as solar angle increases.

The upper chart shows production does improve slightly as solar angle nears 42° (though this doesn't account for total area of PV). It also shows the relationship between maximum kW production, efficiency and adjusted kW production at the solar angle of each house.

NOTE: Some teams had multiple angles for their solar array. This data accounted for these adjustments. Darmstadt in particular had many of its solar panels at a variable angle, however the majority were on the roof at 3°.

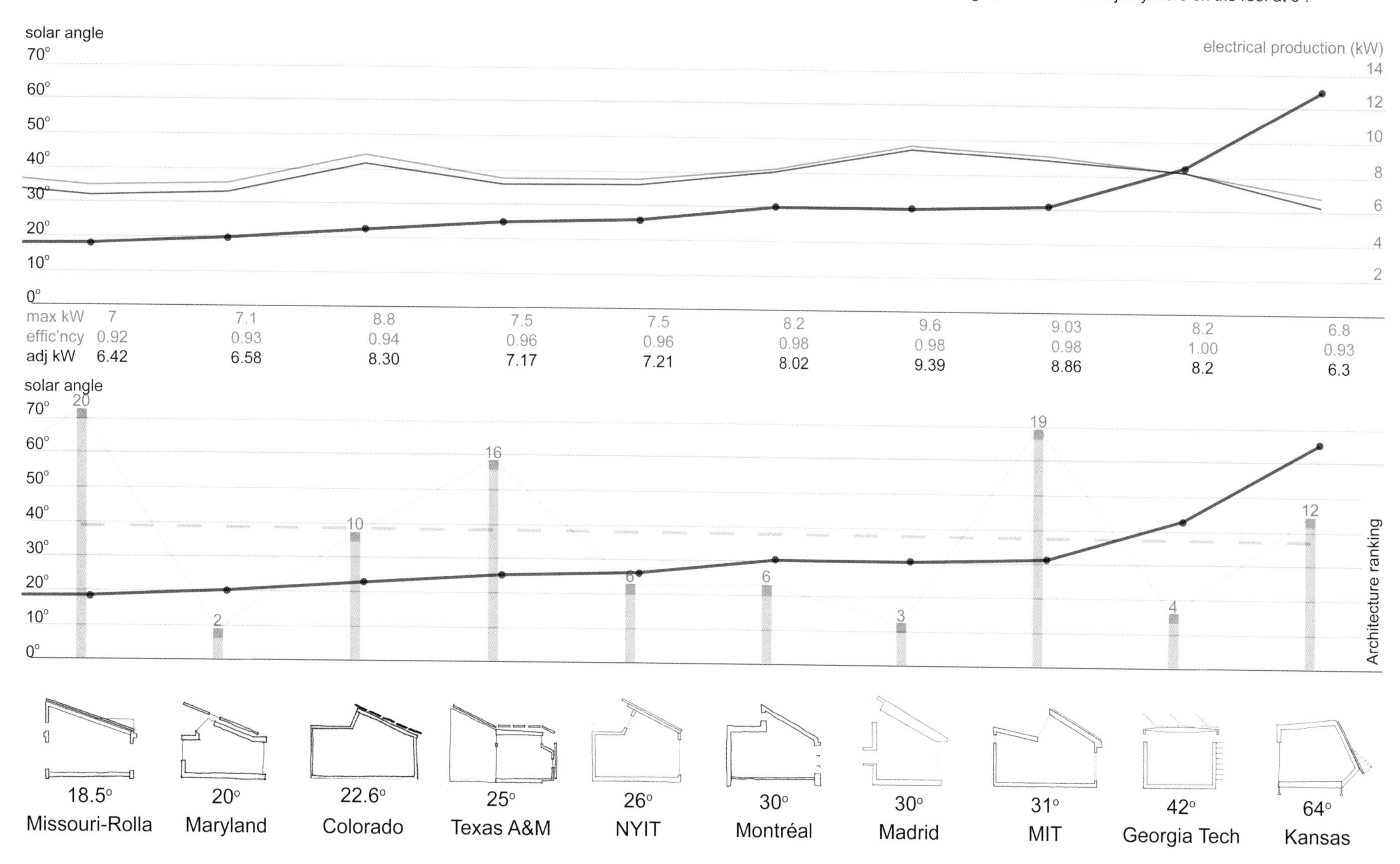

PV Area: PV Array and Rankings

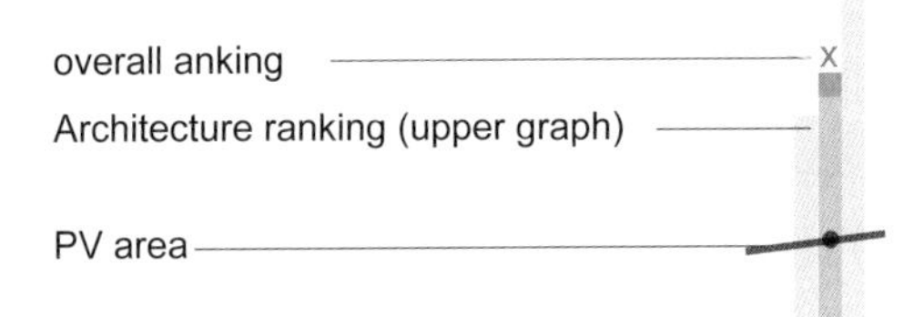

Each design team had choices about the angle and the amount of solar photovoltaic panels to integrate into their design. There was no limit on the square footage of photovoltaic panels that could be used on each house, although the large photovoltaic areas required to generate sufficient electrical power played a major role in the overall design and performance of the houses.

The lower chart compares the total area of solar photovoltaics (in square feet) with the overall competition rankings, while the upper chart compares the total area of PV with the Architecture Contest rankings.

NOTE: There are additional variations in electrical production based on the particular manufacturer producing the PV array. This electrical-production data comes from the team documentation, so this variation has been incorporated.

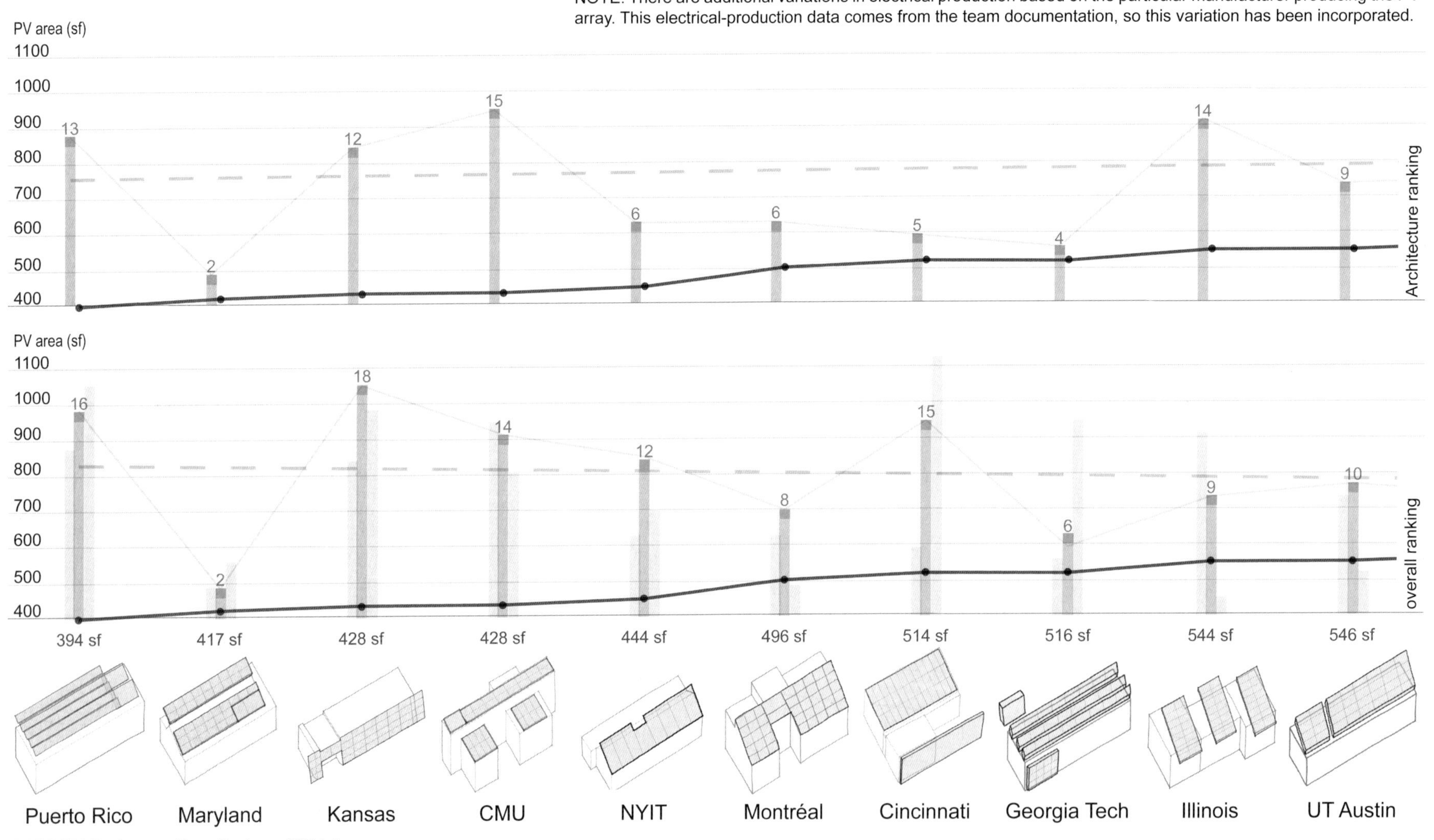

RANKING: Ascending Order of PV Area

The trendlines show that there is a minor decrease in Architecture Contest rankings, while there is a clear increase in overall rankings as a result of increased square footage of PV.

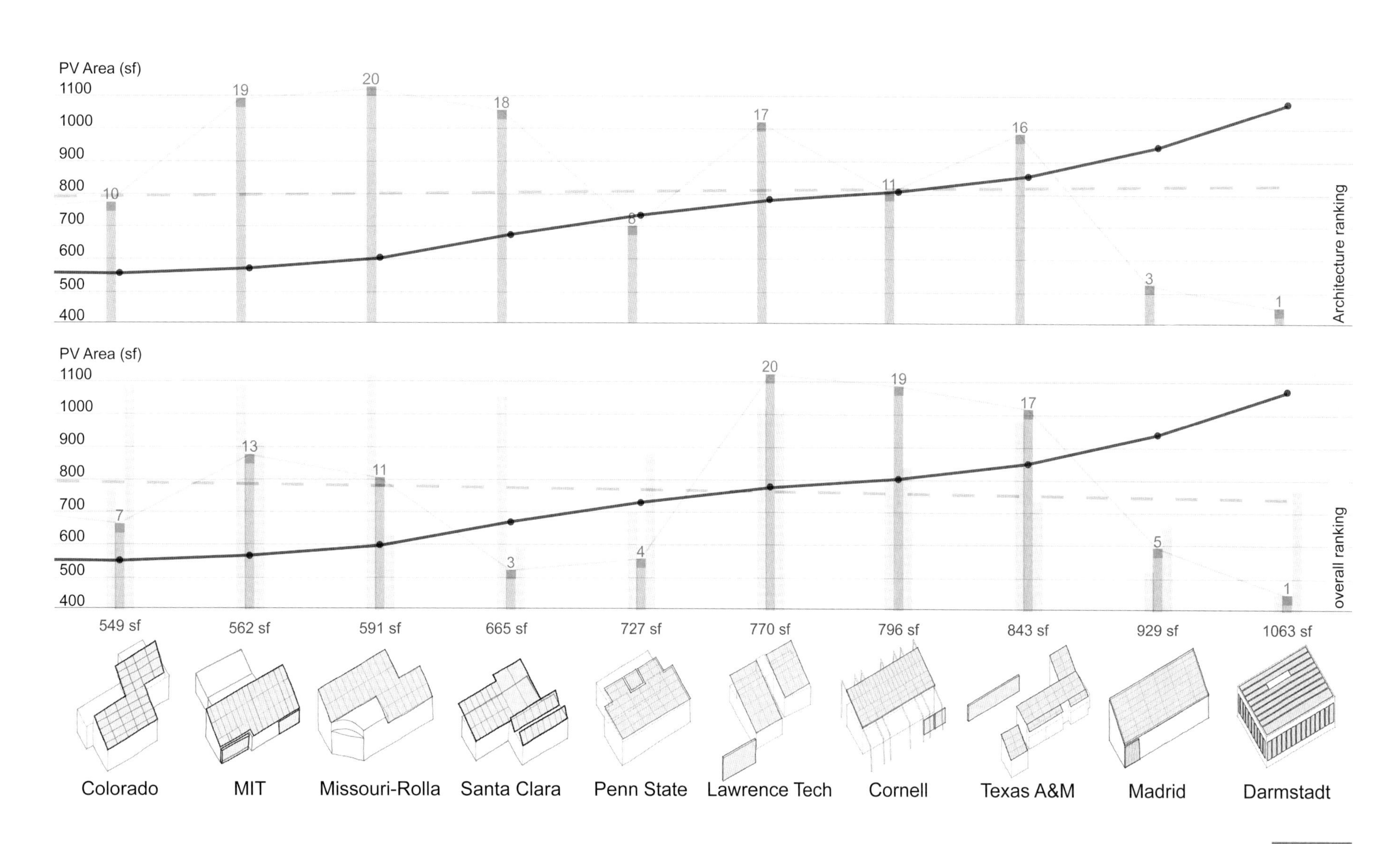

PV Area: PV Array and Areas

These charts compare the relationship between area of solar photovoltaic panels, conditioned floor plan area and electrical production (total and adjusted) per square foot. The top chart relates square footage of PV array to conditioned square footage of house. The hypothesis is that a larger house would incorporate more PV.

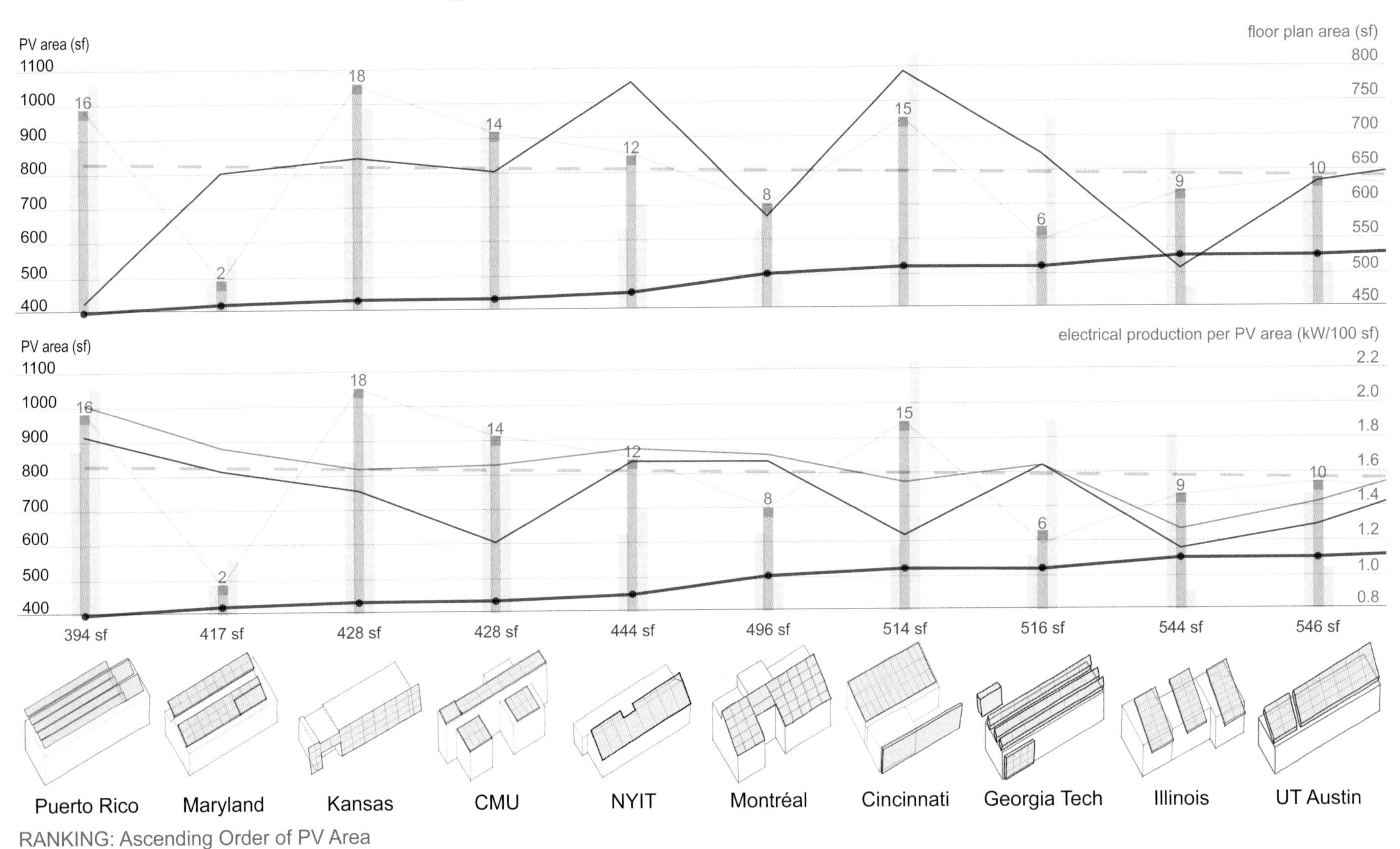

RANKING: Ascending Order of PV Area

Based on the upper chart, there does not appear to be any direct correlation between the conditioned floor area of a given house and its area of photovoltaics. Nor is there any recognizable correlation between the floor plan area and the overall rankings.

The lower chart illustrates that electrical production descreases as amount of PV per square foot increases. This indicates that efficiency decreases as the size of the PV array increases.

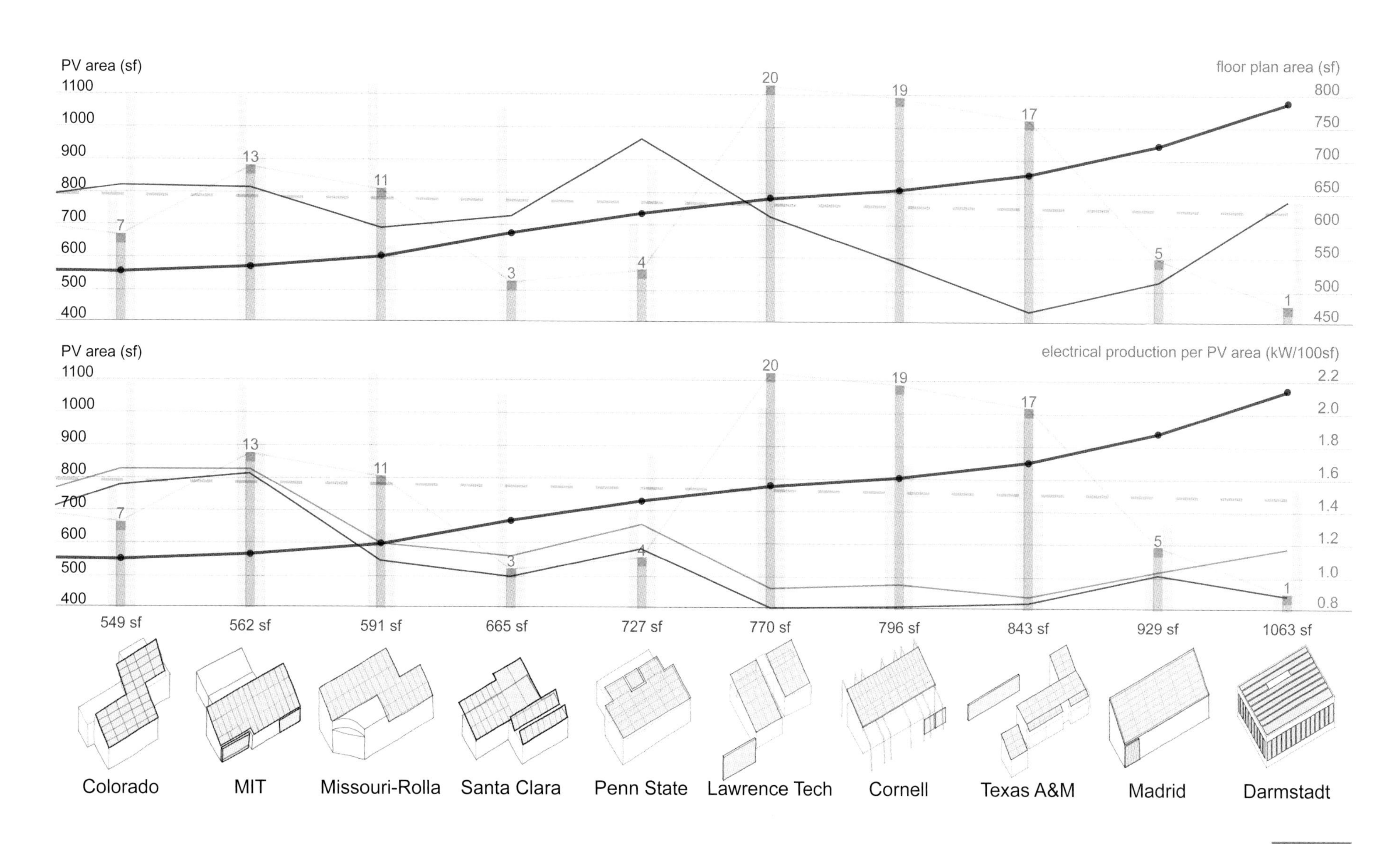

Precedents in Zero-Energy Design:
Conclusion

Conclusion

The presentation of the drawings, diagrams and analyses within this book was intended to provide insight and inspiration, though not necessarily to produce clear-cut conclusions. The intent was to provide a basis of comparison and analysis, so as to prompt additional questions and provoke further exploration and research into past and future competitions – as well as the disciplines of passive design and architecture. The Decathlon is a vehicle for considering the relationship between architecture and passive design through a comparable, comprehensible series of investigations.

The hypothesis that led to this research was that there would be a clear, measurable correlation between the teams that utilized passive design principles and improved rankings in the Comfort Zone Contest. There was also a hope that this would correlate with increased rankings in the Architecture Contest and overall competition rankings.

While the data does not conclusively state that the teams which incorporated passive design principles into their house designs excelled solely as a result of these strategies, there is evidence that these strategies had positive impacts and that several teams would have benefited from incorporating them to a greater degree.

There were some analyses that produced clear correlations:

- The teams that spent more money tended to fare better in the Architecture and Comfort Zone Contests, as well as in the overall competition.
- Teams located on the north side of Main Street in the Solar Village ranked higher in the overall competition.
- A simpler design *parti* led to improved rankings in the Architecture and Comfort Zone Contests.
- Larger houses (in area and volume) tended to fare better in the Architecture Contest and worse in the Comfort Zone Contest.
- Houses with low thermal conductance (i.e. that had better insulation) fared better in the Architecture and Comfort Zone Contests.
- Surprisingly, for the seven teams that utilized thermal storage there was a significant increase in rankings in the Architecture Contest and overall competition while there was a noticeable decrease in the Comfort Zone rankings. This result suggests a need for further research.
- Increased glazing for daylighting led to an increase in Architecture and Comfort Zone rankings.
- Increased interior *illuminance* readings correlated with increased Architecture and overall rankings, while the Comfort Zone rankings decreased.
- An increased use of shading strategies correlated with improved Architecture rankings.
- Increased areas of operable glazing led to significant improvements in Comfort Zone rankings.
- Increased area of photovoltaics (PVs) correlated with an increase in overall competition rankings.
- Increased areas of PV also correlated with a decrease in the efficiency of electrical production.

The results of the analyses are that passive design strategies and architectural sophistication correlated to increase rankings in the competition.

It is also clear that the amount of photovoltaics on each house may have been one of the more significant factors in the competition. The contests required energy, and that energy came from photovoltaic (PV) or solar thermal energy systems. Teams that had more energy available were at an advantage.

The amount of energy required to run the systems in each house was a result of many factors. The focus of this book has been the passive design decisions that the teams incorporated. If effectively designed and incorporated, passive design will inherently reduce the energy levels needed to keep a house within its thermal comfort zone. The teams that fared best in the competition incorporated shading, natural ventilation and thermal storage in order to mitigate their mechanical cooling loads. The Santa Clara team claimed they barely needed to turn on their mechanical system as a result of having a house with a low percentage of glazing (reducing heat gain), natural ventilation and thermal storage on the south and west walls. Though there were extensive mechanical systems in the house, ironically they proved unnecessary.

A Critique of Conclusions

There are clearly challenges associated with any attempt to assess the success of the subjective contests in the competition, such as Architecture or Market Viability. What became clear through this investigation, however, was that the objective contests were also laden with inconsistencies. There was a variety of unforeseen factors that impacted the rankings.

Significant impacting factors included:

- weather in Washington, DC in October 2007
- budgetary limitations of the specific institution
- institutional support for the resource-intensive project
- consistent student participation throughout the two-year design and construction process
- labor skills of the students and faculty involved
- disciplinary knowledge available to each team
- effectiveness of team structure
- inconsistencies in judging of contests
- distance from, and familiarity with, site of competition.

Weather

Of all the unforeseen factors, the weather in 2007 in Washington, DC probably had the greatest result on competition rankings. The 2007 temperature and humidity levels were higher than normal, and significantly higher than those during the previous competition in 2005. Had the weather been cooler, contest results would have been significantly different.

Budget

The overall project cost for the houses in the 2007 competition ranged from $270,000 to $1,378,000, with an average cost of $566,419. The team that won the Architecture Contest and the overall competition, Darmstadt, contributed the most expensive project. This did include approximately $250,000 for the cost of shipping the building from Germany, but the team did benefit from significant state and industry funding in comparison with other teams.

That said, the Maryland team, whose house cost $448,470 (well below the average cost), came in second in the overall rankings, second in the Architecture Contest and fourth in the Comfort Zone Contest.

Institutional Support

The commitment of time, funding and faculty varied immensely. Some schools immediately recognized the benefits of the competition, and offered funding and faculty time. Others faced far greater challenges when it came to gaining support from their home institutions. This disparity had a significant impact on the results.

The need for interdisciplinary collaboration on these projects was unparalleled. As most academics are aware, there are often bureaucratic challenges that arise when attempting to work collaboratively between departments. These difficulties are well documented. Several team members discussed the challenge of funding projects across multiple departments within a public university.

Faculty and Student Participation

The typical institutional curricular model involves an ever-changing course load for professors and students. To successfully design and construct a house for this competition required the consistent commitment of faculty and students. Inconsistent input leads to the need for retraining and increases chances for miscommunication.

Creating situations for professors to remain consistently involved was more feasible for those who could incorporate this project within their academic research agenda. Achieving scenarios in which students would be consistently available over multiple academic semesters or quarters was a greater challenge for many teams.

Darmstadt Technical University was able to create opportunities for a team of students to work on their house design and construction consistently for 18 months. In contrast, the University of Cincinnati had to coordinate with their departmental cooperative education program, in which students are regularly rotating from the university to internship positions around the world. This heavily impacted the consistency of student input on the project.

Available Knowledge and Skills

Every team included members from a wide array of disciplines, ranging from design and engineering to business and communications. However, the availability of knowledge, skills and resources at a particular institution or location inevitably had an impact on the success of each team.

One example of a knowledge deficit was the Santa Clara team, which did not have an architecture program to collaborate with at their institution. Their placement in the Architecture Contest was eighteenth. Their engineering program was fully involved in the project, and they scored fifth in the Comfort Zone Contest. They also scored well (sixth) in the Market Viability Contest, which is an assessment of the apparent appeal of the project within the marketplace. This disparity between

Architecture rankings and Market Viability rankings constituted a recognition that what appealed to designers was not always what appealed to the general public.

Team Structure

Every team had to coordinate faculty, students, administration, vendors and, in many cases, additional consultants or researchers. In addition, participants from the disciplines of design, engineering, business, marketing, graphic design and communication were all requisite members of these teams. This level of participation requires a tremendous level of organization and structure.

Having spoken with members of several teams, it was clear that the number of students, the variety of disciplines, the involvement of faculty and professionals and the team structure all had significant impacts on multiple aspects of the results. One faculty advisor suggested that the effectiveness of the organizational structure of this endeavor might have been the greatest single factor defining success or failure in the competition.

Inconsistency in Contest Judging

While the judges from the Department of Energy (DOE) made every attempt to assess the contests fairly, there were unintended inconsistencies that occurred throughout the competition. One example was the placement of the sensors for temperature and humidity. According to team members, in some houses these were located in spots that received direct solar gain on the sensor at some point during the day, thereby increasing the temperature-level readings.

Distance from Washington, DC

One conjecture that kept arising in our research was that inherent differences would arise in design, engineering and construction based on the location in which each building was designed. Every house is assessed in the same region, but other than for the brief period of the competition the buildings were designed, engineered and tested in their own region. These varied immensely, with Puerto Rico and Darmstadt being two notable examples.

It is likely that the University of Maryland, the institution closest to the location of the competition, had an inherent advantage. Provided it was effectively employed, their team's knowledge of the climate, the region and the culture would inevitably have had positive impacts on their project. They placed second in the overall competition.

The University of Virginia faculty advisor from 2002, John Quale, pointed out that past experience with the event may be a greater indicator of success than physical distance from Washington. Maryland had the advantage of not only being located close to the US capital, but also having participated twice previously (2002 and 2005).

Expecting the Unexpected

There were many direct correlations between the incorporation of passive design strategies and increased success in the Architecture, Comfort Zone or overall rankings. Yet, it is likely that unforeseen weather conditions, a surprising increase in the numbers of visitors and unexpected mechanical failures had as significant an impact as any specific strategies.

Nearly every team reported unforeseen surprises that impacted their design and construction. Many encountered mechanical systems that didn't work exactly as designed (or work at all). In some cases, system failures could be offset by natural energy flows through passive design, but in other cases these problems compounded each other.

The Cincinnati team made two unfortunate decisions during their design process that had an enormous impact on their capacity to remain within the thermal comfort zone during the competition. One decision was to exclude shading over their south porch, which would have shaded their 150 square feet (14 square meters) of south-facing glazing; and the other was to exclude operable windows from any location other than the south side of the house, thereby removing the possibility of using shading or natural ventilation as cooling strategies.

The omission of potential passive design features in the Cincinnati house would not have had as significant an impact if three things had been different. First, if the weather had been similar to the conditions of 2005 (cool and wet) the need for shading and natural ventilation would have been less important. Second, a misunderstanding about the rules led to the hot water storage tank from their 120-tube solar thermal heating system being placed inside the house, while they had been expecting to locate the tank externally. The tank held water at a temperature of 160° Fahrenheit, which added a tremendous internal heat load to the interior space. Third, there were complications with their air-handling unit that kept it from functioning properly. If their cooling system had worked properly they would have been able to adjust for the weather and internal heat gains more effectively.

Passive Design Conclusions

Though most teams employed multiple redundant mechanical systems, those teams that successfully incorporated passive strategies had a decreased need for their mechanical systems throughout the week. Thermal storage, ventilation, shading and daylighting clearly had an impact on the buildings that ranked well in 2007.

The winning entry for 2007, Darmstadt, produced an elegant solution that combined architectural clarity and engineering brilliance. The Darmstadt team benefited from dedicated technical research with a German photovoltaic engineering corporation, as well as having a group of students who had the opportunity to focus on the project for many months consecutively. The Darmstadt house scored well in nearly every respect. From its architectural conception to its mechanical integration, it was unique and effective. However, the project only scored tenth in the Comfort Zone Contest.

The Darmstadt team differed from others by conceiving of the building as a closed entity. There was basically no passive ventilation utilized for the interior of the building. They relied on significant shading, thermal storage, insulation and minimal internal heat gain. This strategy had the benefit of being effective in many different weather conditions. When the louvers were closed, there were minimal visual connections to the exterior.

In contrast, the second-place winner, the University of Maryland, turned the south wall into an operable aperture with pivoting louver doors that seemed to invite the outside in. They benefited from a tremendous amount of ventilation, as well as creating an explicit connection to the outdoors.

Both Darmstadt and Maryland utilized pivoting louvered doors, but for very different affects. For Darmstadt, each louver was covered with a photovoltaic panel that could be tuned to the angle of the sun. In the Maryland house, the louvers were designed to bounce in daylight, block direct solar gain and allow significant ventilation.

The research within this book is intended to be utilized for these types of specific comparisons. My hope is that others will use this resource as a starting point for their research in architecture, passive design or future Solar Decathlon competitions.

Architecture Conclusions

From my perspective, any attempt to assess architecture quantitatively is flawed. In this analysis, I am challenging the reliance on quantitative data by further quantifying unquestionably qualitative data (such as the level of simplicity of an architectural *parti*). What is beneficial in these analyses is less the conclusions than the simple process of developing the taxonomy. Others would certainly organize the order of simplest to most complex *parti* differently, but by providing one proposal, I am hoping to provoke others to consider these design strategies and develop other taxonomies.

What is beneficial is in-depth study and analysis of architectural decisions in the hope of understanding how to design a house that performs well, serves the needs of its inhabitants and inspires us with its beauty. The discipline of architecture relies on this type of case study in order to help us to further our design knowledge and abilities.

My original hypothesis was that an increased integration of passive design strategies would increase the quality of the architecture. Though I have no charts to prove it, I believe that this trajectory represents the future of integrated design.

Future Competitions

The Solar Decathlon competition has inspired tremendous knowledge, innovation and perspiration since its inception in 2000. This book was just one perspective on evaluating the houses of the Decathlon in order to further the opportunities for education and discovery beyond the personal experience of the participants or visitors. There is an immense body of research embedded within this competition, and for every diagram that was drawn there were hundreds of others that were not explored.

This competition is unquestionably worthy of greater study. The results could be more informative and potentially more conclusive if inconsistencies such as budget and location were reconsidered. Several critics have suggested introducing a cap on budget, and potentially running the competitions regionally. Obviously, there are advantages to having one competition setting where all visitors and participants can benefit from each other's knowledge and research, but local competitions would address specific local factors more directly.

Currently, for the average homeowner, the presence of photovoltaic panels is synonymous with *green design*. In reality, the potential positive impacts of passive strategies for design are much more attainable than the incorporation of expensive technological solutions.

If teams were required to address the regional specificities of climate, materials, resources and local market conditions, the results would likely have a greater impact within the local population than the current, centralized Decathlon. If the results included an assessment of passive design, the public would be offered tools and innovations that are readily available on every design and construction project – effectively incorporating the energy from the sun in all design decisions. These simple principles could lead to profound improvements within the design and construction industries as well as influencing the expectations of every homeowner.

Conclusions

It is clear that the teams that incorporated architectural design sophistication and passive design strategies tended to fare better in the competition. This will hopefully provide leverage for those future designers urging other team members, clients, students or professors that passive strategies are worthwhile.

There was a direct correlation between the amount of photovoltaic electricity produced and an increase in the overall rankings in 2007 as well as in 2002.[21] If the amount of photovoltaic surface area is one of the most critical factors in the competition, it illuminates a major flaw with the Decathlon as presently designed.

In the late twentieth century, our culture became aware that our environmental condition was a direct result of our post-Industrial Revolution-era attitude that any problem could be solved with "a bigger hammer." Nearly every environmental and ecological problem that we face can be traced backed to this belief.

If the key to winning the Solar Decathlon competition is to have the largest photovoltaic array, we are still working within the same paradigm – just with a new technology. Photovoltaics are an unquestionable improvement over non-renewable energy sources, but they are still extremely resource-intensive to produce and are only available to the wealthy.

We have the knowledge and skills to significantly reduce our energy needs by the integration of passive design strategies in every building constructed today. This is where the source of inspiration for the *good house* lies. Contemporary zero-energy designers must consider not only how to replace non-renewable energy sources, but also how to reduce our overall demand. This should be the first step in any integrated design process.

There is a critical role for technology in contemporary integrated design. The Solar Decathlon competition provides a venue for furthering the development of these technologies. However, the teams should be challenged and assessed on their successful integration of the active technological and passive design strategies. We need to incorporate the best of renewable technology with the opportunities that are readily available to every person designing or building on any budget.

Endnotes

1 US Green Building Council, "Green Building Research," http://www.usgbc.org/DisplayPage.aspx?CMSPageID=1718, accessed May 3, 2009.
Data Sources: Environmental Information Administration, 2008; EIA Annual Energy Outlook and Energy Information Administration, "Assumptions to the Annual Energy Outlook," 2008.

2 U.S. Department of Energy (DOE), "2007 Solar Decathlon Rules and Regulations," Overview, page 1 of 7, published February 16, 2007, http://www.solardecathlon.org/pdfs/2007_rules_regulations.pdf, accessed May 3, 2009.

3 U.S. Department of Energy (DOE), "2007 Solar Decathlon Rules and Regulations," Overview, page 2 of 7, published February 16, 2007, http://www.solardecathlon.org/pdfs/2007_rules_regulations.pdf, accessed May 3, 2009.

4 Bohm, Martha, "Solar Decathlon: Integrating Green Curriculum for Tomorrow's Professionals," Greenbuild Expo 2007; and U.S. Department of Energy (DOE), "Solar Decathlon 2009: Contests and Scoring," http://www.solardecathlon.org/contests_scoring.html, accessed May 4, 2009.

5 Peña, Robert, "Learning from the Solar Decathlon," ASES, 2008; and Stannard, Sandy, "Size Matters: Solar CalPoly and the 2005 Solar Decathlon," ACSA West Regional, 2006.

6 According to discussions with 2007-competition team members, one example of inconsistency in data collection was the placement of the temperature and humidity logging sensors in the house. These were placed by the jury, but in some cases they were purportedly positioned in spaces that were heavily impacted by the latent heat of the visiting crowds and in other cases in parts of the house that had very little foot traffic. These data loggers were, purportedly, receiving direct sunlight at certain times of the day in some houses, thereby significantly increasing the temperature-level readings.

7 U.S. Department of Energy (DOE), "Contests and Scoring - Architecture - Solar Decathlon 2007," http://www.solardecathlon.org/2007/contest_architecture.html, accessed November 11, 2008.

8 U.S. Department of Energy (DOE), "2007 Solar Decathlon Rules and Regulations," Overview, page 5 of 7, published Feb. 16, 2007, http://www.solardecathlon.org/pdfs/2007_rules_regulations.pdf, accessed May 3, 2009.

9 U.S. Department of Energy (DOE), "2007 Solar Decathlon - Architecture," http://www.solardecathlon.org/2007/contest_architecture.html, accessed May 4, 2009.

10 U.S. Department of Energy (DOE), "2007 Solar Decathlon Rules and Regulations," Contest 5: Comfort Zone, published February 16, 2007, http://www.solardecathlon.org/pdfs/2007_rules_regulations.pdf, accessed May 3, 2009.

11 Fraker, Harrison, "Formal Speculations on Thermal Diagrams," from Moore, Fuller, *Environmental Control Systems: Heating, Cooling, Lighting,* 1992, p. 165.

12 Fraker, Harrison, "Formal Speculations on Thermal Diagrams," from Moore, Fuller, *Environmental Control Systems: Heating, Cooling, Lighting,* 1992, p. 165.

13 Peña, Robert, "Learning from the Solar Decathlon," section 3.2 Passive Strategies.

14 Fraker, Harrison, "Formal Speculations on Thermal Diagrams," from Moore, Fuller, *Environmental Control Systems: Heating, Cooling, Lighting,* 1992, p. 170.

15 U.S. Department of Energy (DOE), "2007 Solar Decathlon Rules and Regulations," Competition Regulations, page 15 of 23, published February 16, 2007. http://www.solardecathlon.org/pdfs/2007_rules_regulations.pdf, accessed May 22, 2009.
16 ASHRAE Standard 55-66. ASHRAE – American Society of Heating, Refrigeration and Air conditioning Engineers. Standard 55, published 2004.
17 U.S. Department of Energy (DOE), "2007 Solar Decathlon Rules and Regulations," Contest 5: Comfort Zone, page 1 of 2, published February 16, 2007. http://www.solardecathlon.org/pdfs/2007_rules_regulations.pdf, accessed May 22 2009.
18 Brager, Gail, et al., "Operable Windows, Personal Control, and Occupant Comfort," ASHRAE Transactions 2004, vol. 110, pt. 2 http://repositories.cdlib.org/cgi/viewcontent.cgi?article=1004&context=cedr/cbe, accessed May 7, 2009.
19 Peña, Robert, "Learning from the Solar Decathlon," section 3.2 Passive Strategies.
20 U.S. Department of Energy (DOE), "Solar Decathlon - Contests and Scoring," Within the "Detailed Scores & Standings" download-able spreadsheet entitled "ScoringBackup.xls" is weather data for Washington, DC for the days of the event. http://www.solardecathlon.org/2005/final_results.html and http://www.solardecathlon.org/2007/final_results.html accessed November 11, 2008.
21 In 2005, the lack of sunlight during the competition in Washington, DC decreased the impact of PV production on overall rankings. Rankings for 2002 and 2005 were accessed November 24, 2008. http://www.solardecathlon.org/pdfs/techreport05/38264tp_equipment_summaries.pdf

Image Citations

img #	house	photographer
1	Darmstadt	Jim Tetro
2	Darmstadt	Michael Zaretsky
3	Maryland	Jim Tetro
4	Maryland	Michael Zaretsky
5	Santa Clara	Jim Tetro
6	Santa Clara	Michael Zaretsky
7	Penn State	Jim Tetro
8	Penn State	Michael Zaretsky
9	Madrid	Jim Tetro
10	Madrid	Michael Zaretsky
11	Georgia Tech	Jim Tetro
12	Georgia Tech	Michael Zaretsky
13	Colorado	Jim Tetro
14	Colorado	Michael Zaretsky
15	Montréal	Jim Tetro
16	Montréal	Michael Zaretsky
17	Illinois	Jim Tetro
18	Illinois	Michael Zaretsky
19	UT Austin	Jim Tetro
20	UT Austin	Michael Zaretsky
21	Missouri-Rolla	Jim Tetro
22	Missouri-Rolla	Michael Zaretsky
23	NYIT	Jim Tetro
24	NYIT	Michael Zaretsky
25	MIT	Jim Tetro
26	MIT	Michael Zaretsky
27	CMU	Jim Tetro
28	CMU	Michael Zaretsky

img #	house	photographer
29	Cincinnati	Jim Tetro
30	Cincinnati	Michael Zaretsky
31	Puerto Rico	Michael Zaretsky
32	Puerto Rico	Adam P. Fagen, Ph.D.
	http://www.flickr.com/photos/afagen/1573054279/	
33	Texas A&M	Jim Tetro
34	Texas A&M	Michael Zaretsky
35	Kansas	Jim Tetro
36	Kansas	Michael Zaretsky
37	Cornell	Jim Tetro
38	Cornell	Michael Zaretsky
39	LTU	Jim Tetro
40	LTU	Michael Zaretsky

NOTES: All photos by Jim Tetro are the property of the U.S. Department of Energy (DOE) and National Renewable Energy Labs (NREL).

As of June 2009, all photos by Jim Tetro can be found at: www.solardecathlon.org/2007/homes_gallery.html

Further Reading

Bibliography

Bohm, Martha, "Solar Decathlon: Integrating Green Curriculum for Tomorrow's Professionals," presentation from Greenbuild Expo, November 8, 2007, Chicago, http://2007.greenbuildexpo.org/ (Green Series), accessed November 24, 2008.

Brown, G. Z. and Mark DeKay, *Sun, Wind and Light, Architectural Design Strategies*, 2nd edition, Wiley, October 2000.

Clark, Roger and Michael Pause, *Precedents In Architecture: Analytic Diagrams, Formative Ideas, and Partis*, 3rd edition, Wiley, 2004.

Fraker, Harrison, "Formal Speculations on Thermal Diagrams," from Moore, Fuller, *Environmental Control Systems: Heating, Cooling, Lighting,* McGraw-Hill, 1992.

Kwok, Alison and Walter Grondzik, *The Green Studio Handbook: Environmental Strategies for Schematic Design*, Architectural Press, 2006.

LaVine, Lance, Mary Fagerson and Sharon Roe, *Five Degrees of Conservation: A Graphic Analysis of Energy Alternatives for a Northern Climate,* University of Minnesota Press, 1982.

Lechner, Norbert, *Heating, Cooling, Lighting: Sustainable Design Methods for Architects*, Wiley, 3rd edition, November 2008.

Moore, Fuller, *Environmental Control Systems: Heating, Cooling, Lighting,* McGraw-Hill,1992.

Peña, Robert, "Learning From The Solar Decathlon: High Performance Building Design, Operation and Evaluation," paper for American Solar Energy Society (ASES), 2008.

Quale, John, *Trojan Goat: A Self-sufficient House (Urgent Matters)*, University of Virginia Press, 2006.

Stannard, Sandy, "Size Matters: Solar CalPoly and the 2005 Solar Decathlon," ACSA 2006 West Regional Conference Paper.

Stein, Benjamin, John Reynolds, Walter Grondzik and Alison Kwok, *Mechanical and Electrical Equipment for Buildings*, 10th edition, Wiley, 2005.

Websites

U.S. Department of Energy, May 3, 2009:
http://www.energy.gov/
http://www.energy.gov/about/index.htm

U.S. Department of Energy Office of Energy Efficiency and Renewable Energy (EERE) Mission Statement, May 3, 2009:
http://www1.eere.energy.gov/
http://www1.eere.energy.gov/office_eere/mission.html

National Renewable Energy Lab (NREL), May 3, 2009:
http://www.nrel.gov/

Solar Decathlon 2007, May 3, 2009:
http://www.solardecathlon.org/highlights_2007.html

2007 Solar Decathlon Competition Teams

2007 Team Selection

From the U.S. Department of Energy (DOE), "2007 Solar Decathlon Rules and Regulations," Overview, page 4 of 7, February 16, 2007. http://www.solardecathlon.org/pdfs/2007_rules_regulations.pdf

The Solar Decathlon is an international competition open to all accredited colleges, universities, and other post-secondary educational institutions. Entrants are selected through a proposal process. All proposals are reviewed, scored, and ranked. Depending on the quantity and quality of submissions, a limited number of teams from all entries are selected.

For the 2007 Solar Decathlon, the U.S. Department of Energy's (DOE) National Renewable Energy Laboratory (NREL) issued a Request for Proposals in October 2005. Proposals were due in December 2005. After reviewing, scoring, and ranking the proposals, a team of reviewers from DOE and NREL selected the following teams to compete in 2007:

Carnegie Mellon University, Pittsburgh, Pennsylvania

Cornell University, Ithaca, New York

Georgia Institute of Technology, Atlanta, Georgia

Kansas State University, Manhattan, Kansas

Lawrence Technological University, Southfield, Michigan

Massachusetts Institute of Technology, Cambridge, Massachusetts

New York Institute of Technology, Old Westbury, New York

Santa Clara University, Santa Clara, California

Team Montreal (École de Technologie Supérieure, Université de Montréal, McGill University), Montreal, Canada

Technische Universität Darmstadt, Darmstadt, Germany

Texas A&M University, College Station, Texas

The Pennsylvania State University, University Park, Pennsylvania

Universidad de Puerto Rico, Río Piedras and Mayagüez, Puerto Rico

Universidad Politécnica de Madrid, Madrid, Spain

University of Cincinnati, Cincinnati, Ohio

University of Colorado, Boulder, Colorado

University of Illinois at Urbana-Champaign, Urbana, Illinois

University of Maryland, College Park, Maryland

University of Missouri-Rolla, Rolla, Missouri

University of Texas at Austin, Austin, Texas

2007 Team Websites: March 3, 2009

Carnegie Mellon University:
http://www.andrew.cmu.edu/org/SD2007

Technische Universität Darmstadt:
http://www.solardecathlon.de

Cornell University:
http://cusd.cornell.edu

Texas A&M University:
http://archone.tamu.edu/solardecathlon

Georgia Institute of Technology:
http://solar.gatech.edu

Universidad Politécnica de Madrid:
http://solardecathlon.ups.es

Kansas Project Solar House (Kansas State University and University of Kansas):
http://solarhouse.capd.ksu.edu

Universidad de Puerto Rico:
http://solar.uprm.edu

Lawrence Technological University:
http://solar.ltu.edu

University of Colorado at Boulder:
http://solar.colorado.edu

Massachusetts Institute of Technology:
http://mit.edu/solardecathlon

University of Cincinnati:
http://solar.uc.edu/solar2007

New York Institute of Technology:
http://iris.nyit.edu/solardecathlon

University of Illinois at Urbana-Champaign:
http://solardecathlon.uiuc.edu/2007

Penn State University:
http://solar.psu.edu/2007

University of Maryland:
http://solarteam.org

Santa Clara University:
http://scusolar.org

University of Missouri-Rolla:
http://solarhouse.mst.edu/2007_house.html#

Team Montréal (École de Technologie Supérieure, Université de Montréal, McGill University):
http://solarmontreal.ca

University of Texas at Austin:
http://soa.utexas.edu/solard

Solar Decathlon Competition Sponsors

Sponsors

The primary sponsor of the Solar Decathlon is the U.S. Department of Energy's (DOE's) Office of Energy Efficiency and Renewable Energy (EERE). DOE's National Renewable Energy Laboratory (NREL) sponsors and manages the event. DOE and NREL work in partnership with many additional sponsors at all levels to make this student solar housing competition and event a reality.

The U.S. Department of Energy (DOE) Mission Statement

The Department of Energy's overarching mission is to advance the national, economic, and energy security of the United States; to promote scientific and technological innovation in support of that mission; and to ensure the environmental cleanup of the national nuclear weapons complex. The Department's strategic goals to achieve the mission are designed to deliver results along five strategic themes:

1. Energy Security: Promoting America's energy security through reliable, clean, and affordable energy

2. Nuclear Security: Ensuring America's nuclear security

3. Scientific Discovery and Innovation: Strengthening U.S. scientific discovery, economic competitiveness, and improving quality of life through innovations in science and technology

4. Environmental Responsibility: Protecting the environment by providing a responsible resolution to the environmental legacy of nuclear weapons production

5. Management Excellence: Enabling the mission through sound management

Office of Energy Efficiency and Renewable Energy (EERE) Mission Statement

The Office of Energy Efficiency and Renewable Energy (EERE) works to strengthen the United States' energy security, environmental quality, and economic vitality in public–private partnerships. It supports this goal through:

1. Enhancing energy efficiency and productivity;

2. Bringing clean, reliable and affordable energy technologies to the marketplace; and

3. Making a difference in the everyday lives of Americans by enhancing their energy choices and their quality of life.

National Renewable Energy Lab (NREL) Mission Statement and Strategy

NREL's mission is to develop renewable energy and energy efficiency practices and to advance related science and engineering. The laboratory is charged with transferring that knowledge and subsequent innovations to the market to address America's energy and environmental goals.

NREL has forged a focused strategic direction to increase its impact on the U.S. Department of Energy's (DOE) and our nation's energy goals by accelerating the research path from scientific innovations to market-viable alternative energy solutions.

Glossary

Active solar – technologies that convert available solar energy into electricity or hot water. Often fans or pumps are incorporated into an active solar system. Energy may be stored or utilized as it is available.

Architecture Contest – the first of the ten Solar Decathlon contests from the 2007 competition. The competition rules state that this assessment is based on each team's capacity to design "*...high performance houses that integrate solar and energy efficiency technologies seamlessly into home design.*"

BIPV (Building Integrated Photovoltaic) – technologies that incorporate photovoltaic cells within the building skin, typically integrated into the skylight or window glazing, the roof or the exterior facade.

Building envelope / skin – the exterior walls, roof or floor of a building, where moisture and temperature are mediated between the interior and exterior conditions.

Comfort Zone Contest – the fifth of the ten Solar Decathlon contests from the 2007 competition. This contest measures the capacity of a team to design a house that remains within an extremely limited temperature and humidity threshold throughout the competition, as measured at 15-minute intervals during specific times when the house is closed to the public.

Daylight – interior illumination that is available during daytime hours within an interior space.

Diagram – "...a graphic design that explains rather than represents; *especially*: a drawing that shows arrangement and relations (as of parts)." (*Merriam-Webster*)

DOE (United States Department of Energy) – The DOE is a governmental department whose mission is to advance energy technology and promote related innovation in the United States. (http://www.energy.gov)

EERE (The Office of Energy Efficiency and Renewable Energy) – The EERE works to strengthen the United States' energy security, environmental quality and economic vitality in public–private partnerships. (http://www1.eere.energy.gov/office_eere/mission.html)

Energy efficiency – using reduced levels of energy to produce the same energy service (such as providing adequate light levels using natural daylight instead of electric light).

Environmental design – design decisions that are made in an effort to minimize the negative effects of a building on the environment.

Glazing (operable) – a transparent or translucent part of a building – window, skylight, etc. (which may be opened and closed by individual building occupants).

Green economy – a hypothetical green economy is one that is differentiated from a market-based capitalist economy in which natural resources have no inherent value other than as a commodity available in the service of profit-making. In a green economy, value is defined in consideration with the whole life cycle of a given product, including the energy expended to create it, resultant impacts on natural resources and costs to scrap or recycle the product.

HVAC (Heating, Ventilation and Air Conditioning) – mechanized environmental control systems that adjust thermal conditions within a given space.

Illuminance – a measure of the intensity of light striking a surface.

Insolation – available radiant energy from the sun.

Insulation – building materials that resist the transference of heat (typically through the building envelope).

NREL (National Renewable Energy Lab) – Facility of the U.S. Department of Energy (DOE) for renewable energy and energy efficiency research, development and deployment. (www.nrel.gov)

Non-renewable energy – energy that is created from limited resources, such as coal or natural gas.

Overall rankings – compiled rankings in the 2007 Solar Decathlon competition, resulting from the sum of each individual team's scores in the ten Decathlon contests.

Parti – the basic scheme of an architectural design, which could be sketched in a few lines.

Passive solar design – design strategies intended to mitigate the need for active mechanical systems in order to maintain thermal comfort.

(PV) Photovoltaic – the field of technology, research and product development related to the creation of products that turn solar energy into electricity.

Renewable energy – energy that is produced from renewable, natural energy sources (solar, wind, geothermal heat, hydro power, biomass).

Shading – devices that effectively block sunlight from reaching a particular destination (often glazing in a building envelope).

Solar angle – the ideal angle to receive maximum insolation available from the sun at a specific latitude and timespan. For Washington, DC in October, the ideal solar angle facing true south is approximately 42° from horizontal.

Solar Decathlon competition – A U.S. Department of Energy and National Renewable Energy Lab-sponsored biannual competition in which universities from around the world transport their "off-the-grid" (energy-independent) homes to the National Mall in Washington, DC, rebuild their homes and compete in ten contests over a period of three weeks in October.

Solar thermal – the field of technology, research and product development related to the creation of products that turn solar energy into thermal energy (heat).

Thermal comfort – *"that condition of mind which expresses satisfaction with the thermal environment" (ASHRAE 55-66, 2004).* Thermal comfort is a result of combinations of sensible and latent temperature and relative humidity, as well as psychological and behavioral factors.

Thermal conductance / Thermal transfer – the property of a material that indicates its capacity to transfer heat, either individually or as part of a combination of materials.

Thermal mass / thermal storage – refers to the utilization of materials that have the capacity to store heat (typically those of comparatively high density and volumetric heat capacity, such as stone and masonry).

Ventilation – the intentional movement of air into or through a building.

Zero-energy design – design that requires no net annual non-renewable energy resources in order to attain thermal comfort, maintain adequate light levels and provide water.

Index